ALSO BY MARK JONES LORENZO

Affront to Meritocracy: Stories of Overlooked Talents, Ignored Abilities, and Hidden Truths

Not Ok: A Requiem for GW-BASIC

Apophenia's Antidote

ȹ

Apophenia's Antidote

ΦΦ

A Probability and Statistics Primer

MARK JONES LORENZO

SE BOOKS
Philadelphia | Pittsburgh

Ψ

SE BOOKS
5307 West Tyson Street
Philadelphia, Pennsylvania 19107
www.sebooks.com

TI-83™ and TI-84™ Graphing Calculators are registered trademarks of Texas Instruments Incorporated.

Published in full-throated defiance of Yog's Law.

Cover design and art, as well as illustrations in the text, by Mark Jones Lorenzo (unless otherwise noted).

Manufactured in the United States of America.

10   9   8   7   6   5   4   3   2   1

Library cataloging information is as follows:

Lorenzo, Mark Jones
 Apophenia's antidote : a probability and statistics primer / Mark Jones Lorenzo.
p. ; cm.
 Includes bibliographical references.
 I. Title
1. Probabilities. 2. Mathematical statistics. 3. Philosophy of mathematics.
 QA273.H694 2016
 519.22'53—js22
20165357985
ISBN: 978-1-517-03664-5

For my students

Come now I will tell thee—and do thou hear my word and heed it—what are the only ways of enquiry that lead to knowledge. The one way, assuming that being is and that it is impossible for it not to be, is the trustworthy path, for truth attends it.

Parmenides, *On Nature* (trans. Arthur Fairbanks)

We still carry the historical baggage of a Platonic heritage that seeks sharp essences and definite boundaries…. This Platonic heritage, with its emphasis in clear distinctions and separated immutable entities, leads us to view statistical measures of central tendency wrongly, indeed opposite to the appropriate interpretation in our actual world of variation, shadings, and continua.

Stephen Jay Gould, "The Median Isn't the Message"

Statistics is a science in my opinion, and it is no more a branch of mathematics than are physics, chemistry and economics; for if its methods fail the test of experience—not the test of logic—they are discarded.

John W. Tukey, "The Growth of Experimental Design in a Research Laboratory"

Contents

ↄ⊃

Preface

꩜

Those suffering from *apophenia* tend to see patterns where there are none, like a looping Rorschach test with an infinite number of referents. A firm grasp on mathematics—specifically, statistical data analysis, along with its trusty handmaiden, probability theory—can help to separate the signal from the noise.

Robert Louis Stevenson said, "To be wholly devoted to some intellectual exercise is to have succeeded in life." In these pages we will meet a number of mathematicians, statisticians, and scientists who were wholly devoted to their intellectual craft. Biographical sketches, though numerous, are always brief and ancillary to the theories, methods, and practices that these great thinkers synthesized. Although there is some theory and proof proffered, many detailed, fully worked-out examples—written in a kind of Socratic dialogue—motivate the presentation of topics. As Ludwig Wittgenstein said, "Not only rules, but also examples are needed for establishing a practice."

But this is not a textbook.

In *Understanding Media: The Extensions of Man*, cultural critic Marshall McLuhan distinguishes between what he calls "hot" and "cool" media. "[H]ot media," he writes,

> do not leave so much to be filled in or completed by the audience. Hot media are, therefore, low in participation, and cool media are high in participation or completion by the audience. Naturally, therefore, a hot medium like radio has very different effects on the user from a cool medium like the telephone.

McLuhan adds that a "hot medium is one that extends one single sense in 'high definition.'" More effort is needed to decode the meanings of cool media, which can easily venture toward impenetrability.

McLuhan would probably classify a mathematics textbook as a cooler sort of medium: much effort and attention is required to decode the vocabulary, symbology, proof, and problem sets (worked-out or not).

Instead of a textbook, what you have in your hands is a *primer*. *Merriam-Webster's* defines a primer as "a small introductory book on a subject," which—although you may quibble with the "small" part—gets it about right. Although for a thorough understanding of the content much active participation and effort on your part will be required (aided considerably if you have a solid working knowledge of high school mathematics), *Apophenia's Antidote* is nowhere near as cool—or cold—as a textbook, but warmer, if not hot, and certainly less antiseptic.

Divided into nine key §'s, each describing a central mathematical topic or theme, the primer is organized as follows:

- The first §, "Foundations," dispenses with the ontological and epistemological foundations of statistics and probability primarily, and of mathematics in general. This § effectively is a long-form self-contained essay that the remainder of the primer unpacks.
- In §2, "Data and Its Discontents," the calculation of simple statistical summary measures, along with the construction of graphical displays, is presented.
- Next, §3, "What is Normal, Anyway?," picks apart the normal distribution, the most important distribution in statistics.
- The fourth §, "They Come in Pairs," focuses on quantitative measurements deriving from bivariate data sets.
- In §5, "Sometimes There's No Chance," probability theory is relayed.
- Then, the sixth §, "Sampling, Studies, &c.," tours the complex worlds of data gathering and experimental design.
- The seventh §, "From One, Many," explores sampling distributions.
- Building on these ideas, §8, "Not Lacking Confidence," delves into the notions of confidence intervals.
- Everything leads up to §9, "But is It Significant?," in which numerous forms of hypothesis testing are considered.

It is sincerely hoped that this primer promotes the firm probabilistic and statistical understanding necessary to dispatch many of the illusions of apophenia—and, in turn, to view the world with a clearer focus.

MARK JONES LORENZO

§1. *Foundations*

ɗ

What is Mathematics? Mathematical Reasoning, Invention, and Discovery.
What is Statistics? What is Probability? A Conceptual Framework.

§1a. *What is Mathematics?*

Pythagoras of Samos, best known for his eponymous theorem, preached the primacy of numbers and the superiority of geometry as a means to understanding the world. Euclid of Alexandria set down the foundations of Euclidian geometry in his *Elements*, the most widely printed and distributed book in human history besides the Bible, promulgating the axiomatic method and formal proof as the gold standards of mathematical thinking. Al-Khwārizmī, from whose name we get the word "algorithm," formalized arithmetic and algebra with the Hindu numerals. René Descartes—a mathematician, philosopher, and all-around-Renaissance man—bridged algebra and geometry utilizing, among other mathematical tools, a coordinate-plane system now named, appropriately, the Cartesian plane. Gottfried Wilhelm von Leibniz and Isaac Newton arrived independently at the calculus of infinitesimals. Georg Cantor systematized the concepts of infinity and set theory, while David Hilbert, with his list of twenty-three problems presented to the International Congress of Mathematicians in 1900, set the course of professional mathematicians for the next century. Bertrand Russell's rigorous fleshing-out of mathematical logic led to an *Elements* for the early twentieth century, called *Principia Mathematica*, which he coauthored with Alfred North Whitehead. And Kurt Gödel, the most important logician of all time, published his incompleteness theorems—which turned mathematics on its proverbial head.

But rather than turn to Pythagoras, or Euclid, or al-Khwārizmī, or Descartes, or Leibniz, or Newton, or Cantor, or Hilbert, or Russell, or Gödel, or some other mathematical eminence of antiquity, the Renaissance, or modern times—and there are innumerable we could call upon—perhaps the prolific contemporary mathematics textbook author Ron Larson best distills centuries' worth of answers to the time-honored epistemological question *What is mathematics?* To wit:

- *Mathematics = A Tool of Calculation.* Mathematics gives us techniques, procedures, and algorithms to calculate quantities.

- *Mathematics = A Field of Study*. Mathematics is a content-rich world of axioms, rules, and theorems, all of which are precisely defied and highly circumscribed.
- *Mathematics = A Language for Modeling Phenomena*. Mathematics links the highly formalized environment of calculations and symbols with the real-life events, actors, and actions.
- *Mathematics = A System of Logical Thinking*. Mathematics requires intensive symbolic manipulation, adhering to formalized rules, either mentally or on paper (or both).

Running as implicit throughout all of this bullet-pointed wisdom are even more foundational notions, such as *Is mathematics discovered or invented?* and *Why should mathematics be able to model any real-world phenomena at all?*[*] Other similar questions can take us even deeper down the rabbit hole.[†]

For example, take Gödel: his famous incompleteness theorems spoiled the dreams of grand-unifying-theory mathematicians like David Hilbert,[‡] exposing cracks in set theory's foundations; these set-theoretic complications are well illustrated by Bertrand Russell's *Barber paradox*: In a town, there's a single barber. He shaves only those men who don't shave themselves. (Men aren't permitted to have beards or other kinds of facial hair in the town.) So: who shaves the barber? Russell claims not to have conjured up this simple yet devastating mathematical critique of naïve set theory—namely, that there's an inherent contradiction in the statement *The set of all sets which are not members of themselves contains itself*—but, though it largely undermined his *Principia* project with Alfred North Whitehead, Russell didn't hesitate in communicating this "paradox" to the mathematical community at large. In Russell's words,

> [N]ormally a class is not a member of itself. Mankind, for example, is not a man. Form now the assemblage of all classes which are not members of themselves. This is a class: is it a member of itself or not? If it is, it is one of those classes that are not members of themselves, i.e., it is not a member of itself. If it is not, it is not one of those classes that are not members of themselves, i.e. it is a member of itself. Thus of the two hypotheses—that it is, and that it is not, a member of it-

---

[*] Note that the epistemological questions of physics overlap with those of mathematics, with additional ones such as *Why do the universal physical constants exist, and why are they "fine-tuned" for life? Or can the anthropic principle—namely, that us human beings being around to measure these constants necessitates their existence to begin with—neatly explain these constants?*

[†] An appropriate metaphor, since Lewis Carroll (real name: Charles Lutwidge Dodgson), author of *Alice's Adventures in Wonderland*, was by trade a mathematician.

[‡] Specifically, Hilbert wanted to know three things: Was any formal mathematical system *complete*? was it *consistent*? and was every statement provable? Gödel was able to answer the first two questions as no, which was not to Hilbert's liking. The third of Hilbert's questions is known as *decidability*, or the *Entscheidungsproblem*. The polymathic Alan Turing, who would later help crack the code of the German's Enigma machine at Bletchley Park, solved the third problem by sketching out a Turing machine, a theoretical precursor to the digital computer; the problem's answer hinged on finding noncomputable numbers—which did, in fact, exist, much to Hilbert's chagrin, delivering Hilbert's formalist program its coup de grâce.

self—each implies its contradictory. This is a contradiction.

Russell, along with logician Alfred Tarski, tried to work around such paradoxes with a *ramified theory of types*, which addresses such problematic impredicative definitions, to little avail.

In *Mathematics: The Loss of Certainty*, Morris Kline, the patron saint of popular mathematics writers (and a relentless critic of U.S. mathematics education),[*] relays a key truism: much of mathematics is built on quicksand, and it is nowhere near as complete and consistent as widely believed. In fact, Kline argues persuasively that the history of mathematics has been a series of cleanup jobs, each one papering over more and more mathematical results that were ontologically suspect to begin with. This "nothing to see here, keep moving along" approach has resulted in modern-day mathematics being a house of cards. For instance,

> mathematics developed illogically…. The creation of these new geometries and algebras caused mathematicians to experience a shock of another nature. The conviction that they were obtaining truths had entranced them so much that they had rushed impetuously to secure these seeming truths at the cost of sound reasoning. The realization that mathematics was not a body of truths shook their confidence in what they had created, and they undertook to reexamine their creations. They were dismayed to find that the logic of mathematics was in sad shape.

Through errors, "blunders," and a careless approach to logic, the towering edifice of mathematics was built, brick by shaky brick.

Admittedly, Kline's approach is somewhat cynical. How does Kline account for mathematics' efficacy at modeling real-world phenomena? First, the Age of Reason notion of mathematics as a perfect instrument with which to explain nature has since been discarded; in fact, the more ancient ideas of nature itself being built *using* mathematics was discarded. By Hilbert's presentation of his twenty-three problems, most mathematicians believed that the discipline had been tidied up; even if all conjectures hadn't yet been proven, at least there was a game plan to capture mathematical truth—until disaster struck in the form of Gödel's incompleteness theorems. (Disaster seemed to strike much earlier than that, with the negative numbers, the concept of zero, the irrational numbers, the so-called imaginary numbers, and the infinitesimals of the calculus. Famously, in the 1700s Bishop George Berkeley criticized the then-fuzzy calculus limit as infinitely small "ghosts of departed quantities" that must surely lean on religious doctrine rather than pure mathematical logic for their justification. But within about one hundred years, with the help of an ε (read: Epsilon), the notion of the limit was defined with sufficient mathematical rigor.) Mathematics has become fractured since, with mathematicians at odds over the nature of their discipline.

To Ron Larson's list of four answers to the question *What is mathematics?*, then, we can add two more, courtesy of Morris Kline:

- *Mathematics = A group of imperfect, mysterious tools that produce "miracles."* Some-

---

[*] Key points: "[M]athematics is expected either to be immediately attractive to students on its own merits or to be accepted by students solely on the basis of the teacher's assurance that it will be helpful in later life," coupled with a lack of real-world-problem presentation at all grade levels.

how, someway, mathematics is effective at modeling real-world events. Call this the black box theory of math: it all works smoothly, though without the innards exposed (since they're hidden inside an opaque box).

- *Mathematics = A discrete set of disciplines that are, at best, only loosely connected—and mostly irrelevant/inapplicable outside of their own domains.* Kline's most persuasive critiques of the nature of mathematics today revolve around his so-called loss of truth: "With a 'plague on all your houses' they [mathematicians] have retreated to specialties in areas of mathematics where the methods of proof seem to be safe."

But at least Kline ascribes meaning to the pursuit of mathematics, even if the mathematical foundations are suspect and the inner workings unclear. A more extreme view can be found in *Everything and More: A Compact History of Infinity*, by novelist-cum-thinker-extraordinaire David Foster Wallace. After briefly discussing abstraction, along with documenting the deep epistemological and metaphysical questions that mathematics raises, he homes in on the principle of mathematical induction,[*] only to arrive at this conclusion:

> But the conclusion, abstract as it is, seems inescapable: what justifies our confidence in the Principle of Induction is that it has always worked so well in the past, at least up to now. Meaning that our only real justification for the Principle of Induction is the Principle of Induction, which seems shaky and question-begging in the extreme.

Wallace's near-nihilistic take is, in many respects, even more disconcerting than Kline's: for if circularity is our strongest justification for mathematical methods, or for mathematics in general, or for any sort of abstraction, then—according to Wallace, at least—we should be petrified to even get out of our beds when the clock alarm rings in the morning, for fear the floor will collapse underneath our weight.

Courtesy of Wallace, let's tack on yet one more *What is mathematics?* answer to our list:

- *Mathematics = A set of abstractions about the world that seem to "work" (whatever that means, hence the scare quotes) but can only be justified by using circularity.* We can have confidence in the efficacy of mathematics as a tool because it's always come through for us in the past.

Especially dangerous is leveraging this circularity argument to justify induction—this is the "our only real justification for the Principle of Induction is the Principle of Induction" that Wallace laments—since the method of induction *itself* relies on inductive, or "bottom-up," reasoning.

Hundreds of years prior, philosopher David Hume noted the same flaw in this kind of reasoning: relying on past events to give us confidence about future events isn't particularly sound. This is formally called *the problem of induction.* Hume used the example of the black swan, unseen for generations by Europeans—who had only encountered the white variety of swan—until

---

[*] A common and important method of mathematical proof, often compared to the domino effect: tap one domino over, and not only does domino #2 fall, but #3, #4, … , #n and #(n + 1) fall over as well.

stumbling into black swans during sojourns in Australia. "No amount of observations of white swans can allow the inference that all swans are white," Hume noted, "but the observation of a single black swan is sufficient to refute that conclusion." Hume also argued that "the sun will not rise tomorrow is no less intelligible…and implies no more contradiction than the affirmation that it will rise," because despite the sun rising n times—where n is a very large number—this fact, *in and of itself*, offers us no guarantee about the $(n + 1)$th time. Twentieth century philosopher of science Karl Popper is even more general in his pronouncement: "A scientific idea can never be proven true, because no matter how many observations seem to agree with it, it may still be wrong"; instead, science progresses (in part) by testing if conjectures survive falsification. In other words, there's always a potential counterexample lurking around the corner ready to ruin your day, no matter how strong the evidence for the scientific *theory*—whether it be evolution or gravity.[*]

Even more recently, mathematician Nassim Nicholas Taleb bursts the notion that what's past is prologue with a tempest of counterexamples in his most famous book, *The Black Swan: The Impact of the Highly Improbable*. Taleb asks us to consider a farmer's turkey, well fed for nearly a thousand days. The turkey grows fond of its seemingly benevolent captor. But on the thousand-and-first day, the farmer chops the turkey's head off.

> Consider that the turkey's experience may have, rather than no value, a negative value. It learned from observation, as we are all advised to do (hey, after all, this is what is believed to be the scientific method). Its confidence increased as the number of friendly feedings grew, and it felt increasingly safe even though the slaughter was more and more imminent. Consider that the feeling of safety reached its maximum when the risk was at the highest![†]

We can quantify the turkey's "feeling of safety" using a formula arrived at by mathematician Pierre-Simon Laplace, called the *rule of succession*: the probability of an event occurring again if it has already occurred n times is given by $(n + 1)/(n + 2)$. The larger the n, the higher the probability.

Such are the dangers of induction, and also of empiricism—that is, of knowledge obtained mostly or entirely through sensory experience. Follow the dictates of past experience, and you just might get your head chopped off.

Also don't assume that a future worst-case scenario can be modeled off of extreme events from the past. In *The Black Swan* Taleb quotes the (first and only) captain of the *Titanic*, five

---

[*] Which also explains why when you take a modern medical diagnostic test the acronym NED (no evidence of disease) rather than END (evidence of no disease) is used: just because a malevolent black swan (e.g., a cancer cell) isn't found in your body doesn't necessarily preclude the *existence* of that black swan. Absence of evidence (a weak statement) is not evidence of absence (a strong statement). Think about it: for END to legitimately be declared, every cell in your bloodstream, every square inch of skin on your body, every molecule in your organs—from tip to tail—would have to be thoroughly examined and found to be disease free; that's simply unrealistic, despite continuing medical advancements.

[†] By the way, Wallace offers up a similar example in *Everything and More*, only with a "Mr. Chicken" rather than a turkey.

years before the disaster at sea: "But in all my experience, I have never been in any accident…of any sort worth speaking about. I have never seen but one vessel in distress in all my years at sea. I never saw a wreck and never have been wrecked nor was I ever in any predicament that threatened to end in disaster of any sort."[*] Morris Kline would have approved of treating the *Titanic* as a metaphor for mathematics—a discipline he argued was a sinking ship of logical contradictions, with the frantic mathematicians like rats scurrying onboard, fruitlessly rearranging the abstract deck chairs. Just because Gödel's proofs were the biggest calamity to befall the logical foundations of mathematics so far certainly doesn't necessarily mean that bigger calamities are yet to come.

§1b. *Mathematical Reasoning, Invention, and Discovery*

So where does that leave us? As Kline explains, mathematicians retreated to their respective territories of the discipline, only occasionally popping their heads out of the sand. The territories, as they evolved, generally can be mapped on two axes: a priori versus a posteriori and *math is discovered* versus *math is invented.*

Let's examine the a priori versus a posteriori distinction first.

- *a priori* = presupposed or derived without the benefit of experience.
- *a posteriori* = arrived at by experience.

Perhaps it's not immediately clear how an idea could be arrived at without leaning on experience—for instance, don't you have to understand a language, replete with its definitions and idiom, in order to conjure something up, regardless of having "experience" in your back pocket?

The answer is yes. Look at this infamous tweet by comedian Steve Martin:

> *REPORT FROM JURY DUTY: defendant looks like a murderer. GUILTY. Waiting for opening remarks.*

This is good example of reasoning a priori (though the tweet was intended solely for comic effect).[†] Martin made no attempt to gather empirical evidence; he simply made a snap judgment about the probability of guilt based on superficial characteristics. But in order to write such a tweet—and in order for his audience to understand it—a facility with the English language, the American legal system, and the cultural mores is required. (Those are just the prerequisites to comprehending the tweet; grasping why it's funny introduces another level of complexity into the mix.)

Suppose instead that Martin wrote the tweet after the defense had presented their evidence. It

---

[*] This is the sort of quote that is almost certainly apocryphal. But it's still worth quoting.

[†] And Steve Martin got into a bit of trouble for the public post—but it turned out that he was never tweeting from a courtroom anyway.

might have gone something like this:

> REPORT FROM JURY DUTY: *after hearing all the evidence, defendant seems like he did it. My vote: GUILTY.*

This new tweet turns the a priori assumption on its head (not to mention that it isn't at all funny). Evidence now plays a factor in Martin's judgments of the defendant's guilt. Thus the tweet is a posteriori—arrived at empirically, by experience.

Mathematics seems to be a natural home for a priori thinking, while the scientific method appears to be a good fit for a posteriori reasoning. A geometric statement like *A square has four equal sides and four right angles* doesn't rely on experience; in other words, we don't need to observe many squares out in the world to learn about their side lengths and angle measures, since a square is defined as a polygon with four equal sides and four right angles to begin with. Neither do the statements *All bachelors are unmarried* or *The three angles of a triangle are equal in measure to two right angles.*[*] But a statement like *All swans are white*, though, cannot be substantiated by a priori logic; it needs the benefit of experience to test it—and, as discussed above, to (potentially) falsify it.

But how far can such purely deductive a priori thinking take us? With reason alone, can we explain the world?[†] The Greeks thought so, making deductive, or "top-down," thinking the hallmark of their approach to both mathematics and philosophy. The seventeenth century philosopher Baruch Spinoza also thought so. "Logic alone, he argued," as Rebecca Goldstein explains in *Betraying Spinoza*, "is sufficient to reveal the very fabric of reality." She continues:

> [I]n the panoply of Western philosophers, Spinoza stands out as having made the strongest claims for the powers of pure reason, unassisted by empirical observation and induction. Anything which we can truly know is to be known through purely deductive thought, which begins with axioms and definitions.... Spinoza took as his model the system of Euclid's geometry....

Euclid's approach to geometry is indeed deductive, wrapping a logical system around the whole of mathematics known at the time. In the *Elements*, Euclid begins by listing five postulates, along with five so-called common notions, seemingly self-evident hypotheses that thus do not need further appeals to proof or experience.[‡] From this foundation of the (seemingly) self-evident, Euclid builds up a non-contradictory structure of mathematics—though, as would later

---

[*] In Euclidian geometry.

[†] In *The World Beyond Your Head: On Becoming an Individual in an Age of Distraction*, Matthew Crawford argues for a morality derived from an ecology of experience, in contradistinction to Immanuel Kant's: "Kant is after a *general* theory of morality, based on pure a priori reasoning—like arithmetic. That two plus two equals four is a fact that is impervious to experience; it will never have to be modified. In rejecting 'accidental circumstances' and 'the special constitution of human nature' as too parochial a basis for moral reasoning, Kant provides" a contrast to an *a posteriori* approach.

[‡] Euclid's first common notion: "Things which equal the same thing also equal one another."

be revealed by Gauss, Riemann, et al., the fifth postulate, known as the "parallel postulate," is not self-evident and therefore paves the way toward alternative, internally consistent, geometries besides the Euclidian (i.e., plane or flat-surface) variety. Much later, Descartes sharpened Euclid's, and the ancient Greek's, tools of thinking, disqualifying sensory experience (and inductive thinking) as a basis for capturing truth about the world—including metaphysical truths, such as the existence of a "Perfect Being," which Descartes arrived at deductively.

Contra Descartes and Euclid, enter the Renaissance thinker Francis Bacon. He took both inductive and deductive thinking to task; as Gavin Kennedy writes, "Bacon was as critical of experiments without thought as he was of thought without experiments." Bacon realized that although one black swan can indeed sink an inductive theory, such counterexamples can also refine the theory, bring it closer to some "true" law of nature.

John Stewart Mill agreed with Bacon—and fleshed out a much more formal theory of induction. Although detailed observation can capture correlations, Mill argued, it is only with controlled experiments that causation can be established. Although, Mill recognized, not every real-world situation—such as might occur in astronomy or social science—is amenable to experimentation, leaning on a priori reasoning, in such circumstances, is the only path forward.[*]

With that, let's return to Euclid. The *Elements* synthesized most of ancient Greek mathematics, but those mathematical results took centuries to become fully formed. The ancient Greeks initially believed in the commensurability of line segments: all line lengths (and, by extension, all numbers) could be written in the form a/b, where a and b are two natural numbers (the positive integers, sans zero). In modern terminology, any number that can be written as a ratio of two natural numbers is classified as *rational*.[†]

The Pythagoreans noticed that the ratio of the lengths of two line segments is not always rational but can instead be *irrational*, otherwise known as incommensurable.[‡] Famously, the irrationality of the square root of two was demonstrated as follows (but we'll use modern notation not available to the Greeks).

First, assume that $\sqrt{2}$ is in fact a rational number—that is, take $\sqrt{2}$ to be a ratio of two natu-

---

[*] Nonetheless, with the rise of social science and economics (in the person of *Homo economicus*) at the end of the nineteenth century, the pendulum swung back toward the deductive.

[†] A thought experiment about rational numbers goes something like this: imagine you have a rod of some length x and a cutting tool of near-infinite precision, and you wish to cut the rod into two pieces such that the ratio, expressed as a string of many digits, of the length of the first piece a to the second piece b map all known human knowledge. Note that a/b does not have to be expressed in base 10; using binary (base 2), for example, is also possible.

[‡] An irrational number, unlike a rational number, cannot be written as the quotient of two natural numbers. Thus, when an irrational number is written as a decimal expansion, its digits neither terminate nor house a repeating pattern. Georg Cantor proved, using what is called a "diagonal argument," that there are more irrational numbers than rational numbers, despite both sets having an infinite number of elements—hence, rather counterintuitively here, *infinity ≠ infinity*.

ral numbers, p and q. So we have the equation $\sqrt{2} = p/q$. Thus, we can assume that p and q have no common factors (if they did have any common factors, those factors would have been canceled out in the quotient p/q).

Next, square both sides of the equation: $\left(\sqrt{2}\right)^2 = \left(p/q\right)^2$, or $2 = p^2/q^2$. If we rearrange the simplified equation just a bit, we get $p^2 = 2q^2$. Why does that matter? Well, recall that all even numbers are divisible by 2; because the 2 appears on the right side of the equation, p^2 is divisible by 2 and therefore *has* to be an even number.

But this implies that p must also be an even number. For instance, take 36: it's an even number, and its square root, 6, is also even. Take 100: it's even, and so is its square root. So if p is even, then p^2, which is even, must be divisible not only by 2, but by $2^2 = 4$. So q and q^2 are even as well—but if both p and q are even, then our initial assumption was incorrect: p and q could not have had any common factors.

Since we have been led down the garden path to a contradiction, what we were trying to prove—namely, that $\sqrt{2}$ is rational—cannot be correct. Thus, the square root of two is an irrational number.

This *proof by contradiction*—demonstrate a proposition true by showing that it being false would force a contradiction[*]—not only proved the irrationality of the $\sqrt{2}$ but also gave the Greeks quite a few nightmares. The ancients didn't know what they didn't know.[†]

Abstract geometry, performed exclusively by compass and straight edge, was king amongst the greatest mathematical thinkers of ancient Greece, with commensurable quantities their manna from heaven. Yet when $\sqrt{2}$ was shown to be an incommensurable measurement—geometrically, the diagonal of a unit square could never be constructed from a ratio of two commensurable quantities—key assumptions of these ancient mathematicians were burst. In fact, the proof ma have even led to the execution of its discoverer, Hippasus of Metapontum, although, like so much other recorded history from the faraway past, the story may be apocryphal; after all, the Library of Alexandria, that physical repository of so much ancient knowledge, was completely destroyed long ago by the order of Caliph Omar—since, as he reputedly said, "If those books are in agreement with the Quran, we have no need of them; and if these are opposed to the Quran,

---

[*] There are a number of ways to mathematically prove a proposition: by contradiction (or *reductio ad absurdum*, Latin for "reduction to absurdity"), proof by induction (discussed earlier), direct proof, and proof by counterexample, among others. The Greeks helped synthesize deductive proof, courtesy of the logical arguments of their *syllogisms*—the most well-known of which is surely "All men are mortal. Socrates is a man. Therefore, Socrates is mortal," an airtight flow of deductive propositional logic of the *modus ponens* variety. But proof by intimidation—where the jargon, results, and/or authority is supposed to speak for it/themselves—is fallacious, as is proof by example, which inappropriately generalizes a single instance. (You may have been on the receiving end of a proof by example by your supervisor at work: even if you've performed one particular task flawlessly save one observed instance, you are chastised *as if* you *always* make that same mistake.)

[†] An epistemology with a long pedigree, well expressed recently by (unsurprisingly) Taleb as the "unknown unknowns," but more famously, at around the same time, by Donald Rumsfeld.

destroy them"—so there's so little we know.

It is here, with the controversy over $\sqrt{2}$, that we explore our second axis: *math is discovered* versus *math is invented*, a debate that is still far from settled.

Were irrational numbers like $\sqrt{2}$ —or, for that matter, transcendentals (special irrationals, like π, which don't solve whole-number-coefficient equations), rationals, integers, or any numbers at all—discovered or invented? Are numbers and axioms and lemmas and proofs and abstractions just floating in the ether, waiting for those intrepid explorers to snatch them down and transcribe them to paper, or are they woven entirely from whole cloth by the most creative among us?

Perhaps it all comes back to the brain, that three-pound mass of trillions of interwoven neurons. Oliver Sacks, the neurologist and memoirist, offers his take on the mystery of math's creation in *On the Move: A Life*:

> We think of science as discovery, art as invention, but is there a "third world" of mathematics, which is somehow, mysteriously, both? Do the numbers—primes, for example—exist in some eternal Platonic realm? Or were they invented, as Aristotle thought? What is one to make of irrational numbers, like π? Or imaginary numbers, like the square root of –2? Such questions exercised me, fruitlessly, from time to time....

The human brain seems to have an innate concept of number—perhaps not of multiplying four-digit by four-digit numbers, but certainly of counting. This *number sense*, which even infants possess, permeated into spoken language early on. Take Hebrew: it is one of many ancient languages which linguistically differentiate three kinds of quantities: one, two, and many. (Although humans have difficulty estimating large numbers of objects—which is why guessing the number of jelly beans in a jar is very difficult—smaller groups of objects can usually be compared in quantity without conscious thought.)

Eventually, number systems, in various bases, were developed. In 1202, Leonardo Pisano, better known as Fibonacci, wrote a book entitled *Liber Abaci*, or *Book of the Abacus*, that disseminated to Western readers the great utility of the Hindu-Arabic numbering system, helping to jettison Roman numerals.[*] Base ten, which made possible precise navigation and financial accounting methods, has stood the test of time (humans having ten fingers and ten toes, certainly a call to everyday experience), although other bases are still commonly used today (such as base two, with computer systems; and base sixty, a Babylonian construct, with time and angles).

Along with bases, zero came to be fully integrated into systematic number systems. Initially just a placeholder, by the ninth century the concept of zero—that of a quantity meaning nothing,[†] as well its use in place-value notation—came into being, likely first in India, although there is evidence of modern-style use in other countries around that time. Negative numbers didn't show up until much later.

The ancient Babylonians and Egyptians arrived at the basics of arithmetic and measurement.

---

[*] It is in *Liber Abaci* that the famous Fibonacci series appears for the first time, as a solution to a problem about rabbit breeding.

[†] "No one goes out to buy zero fish," so zero is an especially abstract number, as Alfred North Whitehead noted.

For the ancient Greeks, though, the road to understanding the nature of the world was through the practice of geometry. In fact, engraved above the door of Plato's Academy was the following statement: "Let no one ignorant of geometry enter." Plato himself practiced mathematics a priori, and held that the universe was assembled mathematically; think of the Platonic solids, thought to constitute the classical elements of matter, as a salient example. Notice the deductive flow: from idealized shapes and images to (potentially) real-world application—unlike the Egyptians, who may have built the pyramids but didn't employ a "generalized pyramid theory" to construct their Wonder of the World. Also recall Plato's Theory of Forms: a perfect world of concepts and things lurks behind our everyday experiences, ready to be discovered by the philosopher-kings. For another take on this, again consider *Betraying Spinoza*: "Logic alone, [Spinoza] argues, is sufficient to reveal the very fabric of reality. In fact, logic alone *is* the very fabric of reality." This, his *radical objectivity*, is an unmitigated math-as-discovery viewpoint.

Plato's "discovery" approach has been termed *Platonism*, and some mathematicians note that the experience of doing mathematics *feels* more like discovery than invention.[*] Barry Mazur, a Harvard University mathematician who had a hand in proving Fermat's Last Theorem, has said that when proving theorems he sometimes feels like "a hunter and gatherer of mathematical concepts."

Platonism has held sway, but, because of the many crises of mathematics, by the twentieth century math had roughly split into three camps (although there are exceptions, and subdivisions, and minutiae that we don't need to get into here), with the discovered-versus-invented debate perpetually lurking in the background: *Logicism*, *Formalism*, and *Intuitionism*.

Briefly: Logicism ties mathematics to logic—mathematics is simply a subset of logic, and thus can be arrived at a priori, or deductively. Bertrand Russell subscribed to this view, but Gödel's incompleteness theorems called it into question. Last century, Zermelo–Fraenkel set theory with the axiom of choice helped to resolve paradoxes and contradictions within mathematics.

David Hilbert was a proponent of Formalism, which, as the *Stanford Encyclopedia of Philosophy* explains, argues that mathematics "is not a body of propositions representing an abstract sector of reality but is much more akin to a game, bringing with it no more commitment to an ontology of objects or properties than ludo or chess." Science author Robert Lamb suggests that For-

---

[*] And necessity may not be the mother of invention, as the old adage goes; *laziness* may be. For instance, Admiral Grace Murray Hopper invented the compiler—software that translates the English of computer code into machine language—because she was "lazy" about doing the translation herself. We could even argue that the development of mathematical notation itself was, in part, due to laziness. Look at how Fermat had to write his famed "Last Theorem" in the margin of Diophantus' *Arithmetica*, sans symbols: "It is impossible to separate a cube into two cubes, or a fourth power into two fourth powers, or in general, any power higher than the second, into two like powers. I have discovered a truly marvelous proof of this, which this margin is too narrow to contain." Fermat wrote this at the tail-end of the *rhetorical period* in mathematics, where symbols by and large weren't used; by the Renaissance's full flowering, the *symbolic period* would come to dominate, through fits and starts and what Marshall McLuhan (courtesy of Kenneth Boulding) terms a "break boundary," a system passing a point of no return. By the way, Fermat wrote of his "marvelous proof" (which, even if he wasn't bluffing, was probably logically flawed) in 1637; the theorem (technically, a conjecture) was finally proved, by Andrew Wiles of Princeton University, in the mid-1990s. In the interregnum, symbolic notation took hold, making life easier for mathematicians everywhere—especially for Wiles, who, even with the benefit of modern compact notation, needed hundreds of pages to complete his proof.

malism "propose[s] that math is a kind of analogy that draws a line between concepts and real events."

Finally, mathematical Intuitionism presents the strongest case against mathematics existing in some apart-from-humans objective reality. Intuitionists believe that mathematics exists solely in the mind of mathematicians; think of it as anti-Platonism.[*] There are consequences to this approach. For instance, Aristotle's *law of excluded middle*—"it will not be possible to be and not to be the same thing," i.e., either *A* or not *A* is true—does not hold, which has a cascading effect on the logical foundations of mathematics.

Most mathematicians, like Harvard's Barry Mazur, take an agnostic view and in practice don't necessarily conform rigidly to these philosophical schools of thought any more than U.S. Republican or Democratic voters' opinions hew perfectly to their respective voting party's lines. Instead, these mathematicians, standing on the shoulders of giants, toil away at their proofs, heads in the sand, blithely hoping that the edifice of mathematics doesn't one day completely collapse on them. To further demythologize: mathematicians are like the rest of us, then, Swords of Damocles hanging over all of our heads as we press on each day despite the finitude of life and the certitude of death.

§1c. *What is Statistics?*

According to Morris Kline, "Statistics is the mathematical theory of ignorance." Not an auspicious *aperçu*. And it gets worse: the use of statistics, especially in the popular press, oftentimes gets labeled as "lying with numbers" or with some other nefarious moniker.[†] Before we can unpack these criticisms, let's explore what, exactly, statistics is—and what it isn't.

Statistics is inextricably linked with probability, which we will deal with shortly. As David Bergamini, in his masterful slim volume entitled simply *Mathematics*, explains,

> Probability and its helpmate statistics are, in a sense, like two people approaching the same house from opposite ends of the street. In probability the contributing factors are known, but a likely result can only be predicted. In statistics the end product is known but the causes are in doubt.

This sounds an awful lot like the a priori-a posteriori discussion we were having earlier. Bergamini goes on to proffer three examples of statistics in action: when calculating life insurance ("…the actuary knows only that ultimately the policy holder will die…"), when studying genetics ("From a statistical sampling…the scientist can make accurate forecasts of the probable char-

---

[*] For *true* anti-Platonism, though, look no further than mathematical Fictionalism, which denies the existence of abstract mathematical objects altogether.

[†] Or the practice of statistics is made light of in an ironic manner, with quips like: "97.3% of all statistics are made up on the spot."

acteristics of unborn offspring…"), and when engaging in random sampling ("…the characteristics of a few individual items, chosen entirely by chance, are tested in order that probability then can be used to forecast with accuracy the characteristics of large numbers of similar items…").

Probability and statistics are intimately connected. But is statistics, as a discipline, completely a subset of mathematics? Or is it something else entirely, a hybrid of different disciplines? Psychologist Richard Nisbett notes that "[s]ome theorists believe statistics isn't a branch of mathematics at all but rather a set of empirical generalizations about the world." Consider the two Venn diagrams below. Which diagram better illustrates the relationship that statistics has to mathematics?

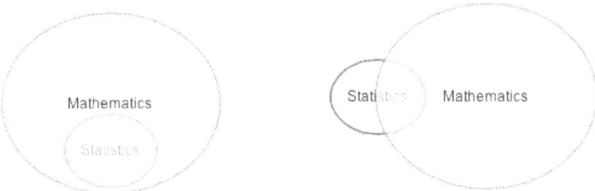

John W. Tukey, one of the most prolific statisticians of the last century (and someone whom we'll be reading more about in these pages), serves up an interesting answer.

> Statistics is a science in my opinion, and it is no more a branch of mathematics than are physics, chemistry and economics; for if its methods fail the test of experience—not the test of logic—they are discarded.

Tukey argues for an empirical, falsifiable, a posteriori approach. He argues for the Venn diagram on the right. He argues for an attention to that which is grittily *real*, in line with Parmenides' poem *On Nature*: "Come now I will tell thee—and do thou hear my word and heed it—what are the only ways of enquiry that lead to knowledge. The one way, assuming that being is and that it is impossible for it not to be, is the trustworthy path, for truth attends it."

Don't be lured into complacency, though; statistics has plenty of mathematical proofs—a few of which we'll work through later—but our method should not be one of Platonism, of proof-for-proof's sake, of reveling in the elegance of abstractions. This seems logical on its face: statistics borrows tools from other disciplines, and leans heavily on the scientific method,[*] to help draw proper conclusions from real-world data. (As Gavin Kennedy, author of *Invitation to Statistics*, notes, "The rule under which [statistics] gathers and operates on…data cannot be isolated from the theories, and indeed the practice, of science.") Furthermore, since statistics is at the crossroads of numerous "applied" academic disciplines, those who practice statistics are permitted to "play in everyone else's backyard," according to Tukey.

---

[*] In the book *Superstition: Belief in the Age of Science*, Robert L. Park maps out the beginnings of the scientific method, which he traces to a solar eclipse in 585 BCE over the ancient Greek city of Miletus: "What distinguished the eclipse…from every previous eclipse of the Sun by the Moon was that it had been predicted [by Thales of Miletus].… Thales understood what had happened and made use of the event to state the law of cause and effect, perhaps the most brilliant insight of all time…." So, perhaps for the first time in recorded history, superstition took a back seat to the scientific method.

But what actually *is* statistics? First off, it is important to note that the practice of statistics is not simply the compiling of data points, or the brute-force tabulation of summary measures (like the mean, median, or mode), or the cooking of mathematical recipes followed thoughtlessly, or the utilization of techniques to manipulate or deceive an gullible audience.

Statistics, rather, is the mathematical discipline concerned with the collection, presentation, and analysis of real-world data.[*] The collection of the data is especially paramount: no matter how sophisticated your statistical tools, data collected in a non-representative (usually taken to mean non-random) way will not lend themselves[†] toward revealing empirical truths. And in our world of analytics, big data, and data mining (and other buzzwords *du jour*),[‡] data analysis drives decision-making, so data collection had better be methodologically sound—since these decisions must be made under varying degrees of uncertainty.

Some examples will help to better flesh out the practice of statistics.

Example 1c.-1

A conservative radio talk-show host asks his listeners if *Roe v. Wade*, the U.S. Supreme Court decision legalizing abortion nationwide, should be repealed. Of those who called in, 79% supported the repeal. Do you believe that 79% of *all* Americans believe the decision should be repealed?

Solution. Because the data were gathered in a biased manner—namely, those who not only were listening to the radio program but those who called in likely were of a conservative bent and thus probably leaned toward repeal—the results of the survey are not representative of the American public at large. In fact, the 79% figure may not be representative of *conservatives* in the U.S., either, because it was obtained by surveying individuals who chose to be in the sample. This is

---

[*] But statistics didn't always have as expansive of a scope as it does today. Although historical anecdotes about statistics will be weaved into this text later to offer additional insight, suffice it to say that in ancient history (e.g., Roman) there was little distinction between enumeration (counting) and statistics; several hundred years ago talk of doing "statistics" meant merely the tabulation of census data by nation states, perhaps with an eye toward decision-making; whereas by the nineteenth century statistics had linked up with probability theory in games of chance and scientific research. The twentieth century development of computers facilitated the still-ongoing explosion of statistics' utility in many walks of life.

[†] A word about singulars and plurals here: *data* is a plural noun, with *datum* its singular. In *The Sense of Style: The Thinking Person's Guide to Writing in the 21st Century*, linguist Steven Pinker argues for keeping data plural—and not using the word this way: "The data is clear…." Just as persuasively, a recent discussion on idiom on *NPR* compared sand's usage to data's: grains of sand, pieces of data; here, data functions more like a collective noun. Throughout this text, despite occasional sentence awkwardness, I will make use of the traditional (plural) form of data.

[‡] Although, to be fair, Marshall McLuhan noticed big data coming way back in the early 1960s: "We have reached a similar point of data gathering when each stick of chewing gum we reach for is accurately noted by some computer that translates our least gesture into a new probability curve or some parameter of social science."

termed a voluntary response, or self-selected, sample.

So much of statistics involves critical thinking and reasoning, and being comfortable with numbers and simple probabilities. In fact, statistics seems to be one of the most challenging subjects to learn largely because an intuitive sense of number doesn't come naturally to most. In his short but insightful book *Innumeracy: Mathematical Illiteracy and Its Consequences*, Temple University mathematics professor John Allen Paulos laments this "inability to deal comfortably with the fundamental notions of number and chance," using the term *innumeracy* as a catchall descriptor of this "plague" (as he, with a touch of exaggeration, calls it). Paulos describes this widespread phenomenon as afflicting "otherwise knowledgeable citizens"—in other words, educated adults—adding that he is "distressed by a society which depends so completely on mathematics and science and yet seems so indifferent to the innumeracy and scientific illiteracy of so many of its citizens."

Example 1c.-2

Town A's murder rate doubled from last year to this year. Town B's murder rate decreased by 10% from last year to this year. In which town would you rather live?

Solution. At first, you may be tempted to simply go with Town B—after all, its murder rate *decreased*, while Town A's went up significantly. But the problem here is one of context. Let's say that both Towns A and B had 1000 people living in them last year. And last year in Town A, one person was murdered. How many people would have then been murdered this year in Town A? Two, which satisfies the doubling condition. (Note that after the one person was murdered last year, a new person either moved into Town A or was born in Town A.) Contrariwise, in Town B last year, 500 people were murdered—that's half of the Town wiped out! This year, after 500 people move back into Town B to keep Town B's population at a steady 1000, "only" 450 people are murdered. Four hundred and fifty people is 90% of 500 people.[*] The point of this example isn't to convince you to live in Town A or B per se; rather, it is for you to see that without having numbers that satisfy the conditions, you have no context for decision-making.

It's not only rates and percentages[†] that, without context, can be deceptive. Averages of data sets often don't tell the whole story. "Statisticians joke about the man with his feet in the oven and his head in the refrigerator," Peter L. Bernstein observes. "On the average he feels pretty good." As Darrell Huff, in his classic text *How to Lie with Statistics*, puts it, "That's why when you read an announcement by a corporation executive or a business proprietor that the average

---

[*] Actually, a 10% reduction here is a bit more subtle than it appears at first blush. Although it's true that 450 is 90% of 500, also notice that $(500 - 450) / 450 = 11.111\%$. We will have to be very careful with rates of change going forward.

[†] Going forward, we will be precise on whether we want frequencies (counts), proportions, and percentages, as well as clarifying what "rates" means. For this introduction, though, nothing's yet been formally defined.

pay of the people who work in his establishment is so much, the figure may mean something and it may not." Case in point:

Example 1c.-3

The average salary at Company A is $50,000 per year. The average salary at Company B is $75,000 per year. Both companies currently have 5 employees. Company A and Company B both offer you a job—which offer do you accept?

Solution. There's simply not enough information to make an informed decision. For instance, perhaps Company A's leader is a radical egalitarian and pays everyone in his (or her) employ—including himself—the same salary, namely, $50,000 per year; thus, the mean of these five salaries is the requisite $50,000. But let's say Company B is headed by an unreconstructed capitalist: $247,000 for him (or her), and $32,000 for the remaining four employees. Since you'd likely be on the bottom of the food chain too, expect a paltry salary at Company B.

As Huff take pains to mention, the term "average" doesn't necessarily refer to the mean—it could function as a placeholder for the median or the mode as well—though it commonly represents the mean. However, the mean, as we'll explore in detail later, carries with it its own set of issues; "[i]f the average is a median, you can learn something significant from it: Half the employees make more than that; half make less," Huff writes. The median is a more useful measure, Huff explains, especially when there are outliers—such those usually contained in a data set of company salaries.

But we also have to be careful to recognize a Lake Wobegon effect. The lake, a fictional creation of novelist Garrison Keillor, is a place where "all the women are strong, all the men are good looking, and all the children are above average." When random samples of college students are asked about their driving skills, usually between 80% to 90% of them reply that they are "more skillful and safer" drivers than others in their class. Apparently most of these students took a bath in Lake Wobegon.

So it's important to have an intuitive sense of number, a key part of being numerate that Paulos preaches in *Innumeracy*, when examining a statistic. Does the number make sense? If so, why? If not, why not? Estimating real-life population sizes and event frequencies are part and parcel of being adept at the analysis of data.[*]

Example 1c.-4

A congressman claimed, during official congressional testimony, that 24 homeless people die every second in the United States. What's wrong with this statement?

---

[*] So-called Fermi problems, like *How many piano tuners are there in Chicago?*, make an art form out of back-of-the-envelope calculations.

Solution. If 24 homeless people die every second, then consider how many die in a year:

$$24 \times 60 \times 60 \times 24 \times 365 = 756,864,000$$

Considering that there are only around 320,000,000 people living in the U.S. as of this writing, that's an awful lot of homeless people shuffling off this mortal coil. (Also note that there are roughly two to three million homeless in the United States.) It would have made more sense for the congressman to claim 24 homeless people die *every day*, although that might be undercutting things.

In addition, that an authority figure—in this case a congressman—relayed the statistic imbues it with extra weight. Recall the Red Scare. In 1950, demagogic U.S. Senator Joseph McCarthy proclaimed, "I have here in my hand a list of 205 state department employees that were known to the Secretary of State as being members of the Communist Party and who are nevertheless still working and shaping the policy of the State Department." Needless to say, the number of Communists changed from week to week, and McCarthy's accusations became slipperier and slipperier. As Aristotle said, a man's habits become his character—and McCarthy's character, along with his habits of accusation, was ultimately found lacking. It's similar to the great respect you accord a friend or family member who claims to know something that goes against the grain: we oftentimes don't separate the message from the messenger, giving more undue credence to claims from those whom we perceive to be of higher stature.

Sometimes, though, a number (or numbers) is calculated correctly but interpreted incorrectly. Here is a well-known, albeit comedic, example:

Example 1c.-5

Your boss is furious! Apparently, 40% of all sick days you've taken at the job have fallen on a Monday or a Friday. She accuses you of trying to extend your weekends, but that's simply untrue. How do you explain yourself?

Solution. Mondays and Fridays make up 40% of a five-day workweek, so, all things being equal—in other words, you being as likely to get sick on a Monday as any other day of the week[*]—40% of your sick days *should* have been taken on a Monday or a Friday. This silly example has a famous pedigree: it was, with a couple alterations, a featured joke in a *Dilbert* comic strip many years ago. What's not silly, however, is the idea of *expectation*: the differential between what we should obtain (in theory) and what we actually do (when collecting data) is an important idea to keep in mind when drawing inferences from data.

Grappling with cause and effect is an especially challenging enterprise. Sometimes, it's fairly easy to see that one variable is only *correlated* with the other.

---

[*] For simplicity's sake, let's assume that this is true; in the real world, however, it probably isn't. Who knows what sort of trouble you got yourself into on the weekend which might necessitate an extended holiday....

Example 1c.-6

In the fourth quarter of game six of the 2013 NBA Finals, while he was going for a dunk, LeBron James' headband was knocked off by another player and lost amid the chaos of the moment. In those final, post-headband minutes, LeBron took over the game—and scored an incredible number of points. Did LeBron take over the game *because* he lost his headband?

Solution. No. Most superstitions can be explained by looking for faulty conflations of correlation and causation,[*] but correlation is *never* causation until proven otherwise (absence of correlation doesn't mean absence of causation, either). The burden of proof, as we shall see, is very high.

Human reasoning succumbs to faulty superstition all the time. In his book *Alfred Hitchcock: The Man Who Knew Too Much*, author Michael Wood explains the MacGuffin, "the mechanical element that usually crops up in any story" that Hitchcock frequently made use of, with a story from his own childhood:

> When I was a child, the local Midland form of the MacGuffin story went like this: Two men are sitting on a railway train passing through England (or sometimes Scotland or Wales). One of them is crumpling up pages of newspaper and throwing them out the window. The other man finally cannot contain his curiosity and asks the first man what he is doing. He says, "Oh, this keeps the elephants away." The second man says, "But there are no elephants in England (or Scotland or Wales)." The first man says, "See."

Even animals can develop such superstitions. In the 1940s, the psychologist B. F. Skinner fed a group of captive hungry pigeons several times per day using a mechanical device. After a while, the pigeons started acting strangely—such as turning once, twice, or three times, in a counterclockwise manner—believing that their behaviors would *cause* the device to spit food pellets out. The birds were conditioned to superstition.

Example 1c.-7

The data show that the vast majority of traffic accidents occur when vehicles are travelling at moderate speeds; very few accidents occur at high rates of speeds. Does this mean that travelling at a high rate of speed is safer?

Solution. This fallacy comes to us courtesy of mathematics puzzle maker/author Martin Gardner and is rooted in faulty cause and effect. Here, again, expectation is the key: since few of us drive at very high rates of speed, there are few accidents at high speeds.

Gardner presents us with two other similar gems. "It is often said that most car accidents occur near the home," he writes. "Does this mean that travel on highways, many miles from home, is

---

[*] For much more on this, see *Superstition: Belief in the Age of Science* by Robert L. Park.

safer than driving around town?" Since you're more likely to drive around your home—after all, you have to both leave and return home no matter where you go—there's a greater chance of getting into accidents near your home.

And children who have bigger feet tend to spell better, so does this mean that the bigger feet *causes* the increased spelling aptitude? Of course not. Here, a third variable, *age*, explains the relationship between the other two variables: children who are older tend to have bigger feet and also tend to have larger vocabularies and a greater facility with words.

Lots of potential *hidden* or *lurking* or *confounding variables* can influence a statistical relationship through time. Walter Mischel conducted a unique set of experiments in the 1960s with preschoolers at Stanford University. He would set one and two marshmallows (or whatever food the toddler seemed to be fond of) on separate plates in front the preschoolers, telling them that if they waited a certain length of time, they could have the set of two marshmallows—but if they didn't manage to exercise self-control, they were entitled *only* to the smaller reward. As the years passed, Mischel realized that these toddlers' performance on the so-called Marshmallow Test correlated neatly to different life outcomes, and the study became longitudinal. For instance, kids who were able to restrain themselves for the bigger reward tended to have lower BMIs, better SAT results, and higher self-reported life satisfaction scores later on than those who more impulsively went for the single marshmallow. But even strong correlations don't necessarily apply to individual kids. As Mischel explains in his aptly titled book, *The Marshmallow Test: Mastering Self-Control*,

> "Myra," a friend who was a senior researcher at the institute and had heard my talk, contacted me [because she was very worried about her son].… At age four, her son had consistently refused to wait for more cookies (his favorites), no matter how hard she tried to get him to do so.…

> When Myra calmed down, she of course realized how incorrectly she had interpreted the results: correlations that are meaningful, consistent, and significant statistically can allow broad generalizations for a population—but not necessarily confident predictions for an individual. Look at tobacco use, for example. Many people who smoke die early from tobacco-induced diseases. But some—indeed many—don't.

At its heart, such use of correlations are really about using the past to predict the future, though no perfect tools to do this exist.[*] The siren song of gathering perfect information with which to predict the future goes at least as far back as *Laplace's demon*: if there were an entity that could somehow know the positions and motions of all of the atoms in the universe, then "for such an intellect nothing would be uncertain and the future just like the past would be present before its eyes," according to the seventeenth century mathematician Pierre-Simon marquis de Laplace (from *A Philosophical Essay on Probabilities*). Laplace believed that every event followed some identifiable cause: "Present events are connected with preceding ones by a tie based upon the

---

[*] Consider the reverse problem, expressed well by this aphorism: "If your only tool is a hammer, all of your problems will be nails." Just because we have these glorious mathematical tools doesn't mean we should shoehorn reality to conform to them.

evident principle that a thing cannot occur without a cause that produces it…," and his demon was in prime position to untangle this mighty web of cause and effect.[*] It was all simply a matter of measurement.[†]

But sometimes such measurements were used in nefarious ways, such as with craniometry, a pseudoscience that made use of the measurement of human skulls to make predictions about intelligence and character traits; see Stephen Jay Gould's *The Mismeasure of Man* for a brilliant explication of this "biological determinism."

Furthermore, conclusions about causation are made erroneously by inadvertently systematically excluding a portion of the population. This is usually termed *selection bias*. Nassim Nicholas Taleb gives it a better name: "silent evidence," the tendency to only consider "surviving" elements when examining some process.[‡] Think of the old saw, "History is written by the victors."

Example 1c.-8

Suppose the performance of a mutual fund company over several decades shows, on average, 12% growth every year. Why should you be suspicious of the company's performance?

Solution. Funds that have underperformed, poorly performed, or collapsed were likely either excluded from the long-term analysis or merged into other funds, painting a rosier picture of the company than is warranted.

Taleb cautions us to read stories of "success," like from the book *Rich Dad Poor Dad*, with a grain of salt:

> Consider the thousands of writers now completely vanished from consciousness: their record did not enter analyses. We do not see the tons of rejected manuscripts because these have never been published, or the profile of actors who never won an audition—therefore cannot analyze their attributes. To understand successes, the study of traits in failure need to be present. For instance, some traits that seem to explain millionaires, like appetite for risk, only appear because one does not study bankruptcies. If one includes bankrupt people in the sample, then risk-taking would not appear to be a valid factor explaining success.

---

[*] The great French mathematician Jules-Henri Poincaré agreed with Laplace: "A mind infinitely powerful, infinitely well-informed about the laws of nature, could [foresee events] from the beginning of the centuries." But then Poincaré subtly extended Laplace's argument to one of gambling, noting that "[i]f such a mind existed, we could not play with it at any game of chance, for we would lose."

[†] Although, with respect to cause and effect, mathematician and economist John Maynard Keynes argued that although "[t]here is a relation between the evidence and the event considered,…it is not necessarily measurable."

[‡] I am reminded of a riddle first presented to me in elementary school. A plane crashes on the U.S.-Mexican border. Where should the survivors be buried?

Example 1c.-9

You require heart surgery. Obviously, you want the best cardiac surgeon for the job, so you obtain some mortality statistics for two nearby surgeons; let's call them A and B. Approximately 90% of Surgeon A's patients survive one year after the operation, while only 65% of Surgeon B's patients do. Which surgeon should you select?

Solution. Although the question at first glance appears rhetorical, these basic summary measures do not paint a full picture of each surgeon's efficacy in the operating room. Perhaps Surgeon A only takes on the easiest cases, simply to artificially inflate his mortality statistics. More information is needed to make an informed decision. After all, as Aristotle noted, medicine is a "stochastic art."

Such summary numbers clearly don't always tell the whole story, and not only in the medical profession; business, government, sports, and education all use numbers that either suffer from the flaw of silent evidence or have non-obvious assumptions built into them.

We can refashion the previous example using teachers instead of doctors—since teachers, more and more, are subject to the vagaries of sometimes-questionable data collection and analysis—thusly: high school Teacher A and Teacher B both teach a single section of grade-level Algebra 1 in the same public high school; both sections have twenty students in them, equally split between males and females. At the end of the school year, a standardized test is administered to both classes. Approximately 75% in Teacher A's class achieve a passing score, while only 50% in Teacher B's class manage to do so. Even if the pattern more or less repeats, year after year, with no other data—*quantitative* (numerical) and qualitative—can we definitively label Teacher A the more effective teacher?

Only provisionally; there are a multitude of other factors that make up those summary scores, many of which are simply out of any teacher's control. Even things within a teacher's control don't ensure that the comparison will be accurate. For example, Teacher A may consciously be teaching to the test, causing problems best explained by Campbell's Law, named for social psychologist Donald Campbell: "The more any quantitative social indicator is used for social decision-making, the more subject it will be to corruption pressures and the more apt it will be to distort and corrupt the social processes it is intended to monitor." When analyzing data, we need to carefully consider not only how the data were collected but also exactly *what* data were collected. The data *must* present a comprehensive picture of whatever it is we're trying to analyze.[*]

When examining a set of data, we can easily be misled. Statistics—and statisticians—have had to live down a famous quote supposedly by the British statesman Benjamin Disraeli, as relayed by Mark Twain: "There are three kinds of lies: lies, damn lies, and statistics." And this: "Figures don't lie, but liars figure." In short, statistics has not enjoyed a good reputation because, as Gavin Kennedy notes, the discipline has had to juggle the certainties of mathematical proof with the

---

[*] And collecting more and more data in and of itself won't necessarily help to fix some systemic problem. As Dana Goldstein, in *The Teacher Wars*, her thorough history of public education, notes pithily, "The hope that collecting more test scores will raise student achievement is like the hope that buying a scale will result in losing weight."

uncertainties of real-world data. Paul F. Velleman of Cornell University says that "[o]ne can wield the tools of statistics to mislead. But even those who repeat [this] quotation don't believe the purpose of the discipline of statistics is to mislead, or that there is something fundamentally dishonest about statisticians." But there is also no question that "[t]he secret language of statistics, so appealing in a fact-minded culture, is employed to sensationalize, inflate, confuse, and oversimplify," as Huff warns. He continues: "Statistical methods and statistical terms are necessary in reporting the mass data of economic trends, business conditions, 'opinion' polls, the census. But without writers who use the words with honesty and understanding and readers who know what they mean, the result can only be semantic nonsense."

But no matter what the data show, we need to be careful when drawing inferences. When analyzing data, we tend to gravitate toward explanations that are most readily apparent or familiar. In psychology, this cognitive bias is called the *availability heuristic*. Watch the news enough, and you'll come to believe that being bit by a shark, getting struck by lightning, or being killed in a terrorist attack is many times more likely than meeting a more prosaic fate, like being injured in a traffic accident or dying of heart disease. The novel or the spectacular is more easily recalled, leading most people to have a very shaky grasp on judging risk (yet another theme in Paulos' *Innumeracy*). As the textbook *Introductory Statistics with Randomization and Simulation* notes, "In February 2010, some media pundits cited one large snow storm as evidence against global warming. As comedian Jon Stewart pointed out, 'It's one storm, in one region, of one country.'"

Even medical doctors aren't immune; an old aphorism instructs medical interns, "When you hear hoofbeats, think of horses, not zebras"—the idea being that zebras, like exotic illnesses, are rarely encountered and thus should not be the go-to patient diagnoses. But a doctor must also be judicious when arriving at an explanation for symptoms or signs, recognizing that rare doesn't mean impossible as well as not ignoring disconfirming evidence. In his remarkable book *Complications: A Surgeon's Notes on an Imperfect Science*, Atul Gawande documents the frightening case of a woman's leg being attacked aggressively by necrotizing fasciitis, a flesh-eating bacteria. Although extremely rare—a true diagnostic zebra—necrotizing fasciitis is usually fatal without treatment.

Sometimes, though, the spectacular is not only common—it is willfully ignored. Case in point: the infamous Ford Pinto. The book *Justice*, by Harvard professor of philosophy Michael J. Sandel, details the dilemma that faced the Ford Motor Company:

> [I]n the 1970s, the Ford Pinto was one of the best-selling cars in the United States. There were 12.5 million of the vehicles sold. Unfortunately, though, the Pinto's fuel tank likely exploded if another car collided with it from the rear. More than 500 people died when their Pintos burst into flames, and many more suffered burn injuries. When Ford was sued for the faulty design, it emerged that Ford engineers had been aware of the exploding gas tank danger and determined that the benefits of fixing it (in lives saved and injuries prevented) were not worth the $11 per car it would have cost to equip each Pinto with a device that would have made the gas tank safer.

Roughly a decade earlier, the National Highway and Traffic Safety Administration (NHTSA) ran a mortality calculation: "Counting future productivity losses, medical costs, funeral costs, and the victim's pain and suffering, the agency arrived at $200,000 per dead person (a related calculation found a cost of $67,000 per injured person)," according to Sandel.

Example 1c.-10

If no changes were made to the Pinto, Ford estimated 180 deaths and 180 burn injuries. Conduct a cost-benefit analysis below to show why Ford chose *not* to fix each Pinto's gas tank.

Solution. Here is the cost to Ford if they had fixed every gas tank:

$$11 \times \$12,500,000 = \$137,500,000$$

Alternatively, this is what Ford estimated it would cost them if no gas tanks were repaired:

$$180 \times \$200,000 + 180 \times \$67,000 = \$48,060,000$$

Because the first total is much greater than the second, Ford decided not to fix the tanks.

Of course, Ford's calculation was flawed. First, more than 500 people, not "just" 180, ended up dying in their Pintos; and second, the public relations nightmare—resulting in lawsuits and other assorted bad publicity—was incalculable.[*]

Speaking of bad publicity, in the popular book *Freakonomics* by economist Steven D. Levitt, a very controversial hypothesis, based on analysis of data, was posited: that national legalized abortion, courtesy of *Roe v. Wade*, lead to a significant decrease in crime around twenty years later—because those young men who *would have* committed the crimes *never had a chance to be born*. (Those future criminals would have supposedly been "unwanted" by their parents.)[†]

Levitt leans on the idea of a *natural experiment* to support his claim. A natural experiment is what it sounds like: an experiment that occurs "naturally," subject to forces and conditions that are not in the control of the researchers, but that nonetheless create conditions for legitimate scientific study—such as randomization—and perhaps allow for the drawing of important inferences. (Think of it as the earth conspiring to help move science forward.) The 1854 Broad Street pump cholera outbreak in London resulted in one such natural experiment, which we will examine next §. Another natural experiment occurred in the Netherlands in 1944, during the famine known as the *Hongerwinter*, which was caused by both a German embargo and a devastating

---

[*] Contemporary valuations for human life, by the way, vary considerably, from the FDA's $7.9 million to the EPA's $9.1 million, to accounting for "quality adjusted life." See Richard Nisbett's book *Mindware* for the details.

[†] To be more precise, *Freakonomics* brought these ideas to the public, but the theory was initially proffered by both Steven Levitt and John Donohue. The duo wrote an article entitled "The Impact of Legalized Abortion on Crime." As they explain it, "The drop in the proportion of unwanted births during the 1970s and early 1980s appears to be the result of the increasing availability and resort to abortion. The evidence we present is consistent with legalized abortion reducing crime rates with a twenty-year lag. Our results suggest that an increase of 100 abortions per 1000 live births reduces a cohort's crime by roughly 10 percent."

winter. During the famine, children who had celiac disease mysteriously experienced symptom relief. When famine relief arrived and the bread came back, those same celiac-stricken children got sick again. From this, for the first time, the ingestion of wheat was linked to celiac disease.[*]

Other instances of natural experiments come from charter schools, where students in many cases are selected to attend based on some sort of lottery.[†] Standardized test scores from before and after the lottery, coming from students who were selected to attend charters and from those who ended up stuck at their old, traditional schools, can be compared. Although not quite the "gold standard" of scientific research—namely, the completely randomized experiment—in a real world full of messy data, it's not too far afield.

But does Levitt's claim lend itself to a natural experiment as well? Stunning in its audacity, the abortion/crime-decline hypothesis is perhaps a case where the inferences are taken too far, since there are so many confounding factors when traveling from point A (legalized abortion) to B (crime reduced), and the hypothesis cannot be tested or the conclusions verified in a laboratory setting or some other "sterilized" environment free of the myriad vagaries of real-world influences.

Not only Levitt's studies have been controversial. Other researchers have suggested that HIV does not cause AIDS, a bit of radiation exposure may in fact be beneficial to human beings, and more guns means less crime.[‡]

Which leads us to an obvious question: if we have summary data over time, can we draw *any* valid conclusions at all? When we explore studies and experimental design later, we'll go into more detail but, for now, let's take a look at an example of a more modest analysis.

Example 1c.-11

The following table displays a summary of "favorite sports." Describe at least three characteristics or trends shown.

	1947	1960	1974	1983	1999	2013
Football	17%	21%	36%	38%	42%	51%
Baseball	39%	34%	21%	19%	15%	12%
Basketball	10%	9%	8%	9%	13%	10%
Other	34%	36%	35%	34%	30%	27%

Solution. Baseball's popularity looks to be declining (baseball is termed "America's pastime,"

---

[*] Less dramatic instances of natural experiments involving diet, culture, and health, such as studies of Japanese transplants to Hawaii and Los Angeles, can be found in *The Man Who Touched His Own Heart: True Tales of Science, Surgery, and Mystery* by Rob Dunn.

[†] See the documentary *Waiting for "Superman"*, for example.

[‡] Found in the book *Nine Crazy Ideas in Science: A Few Might Even Be True* by Robert Ehrlich.

after all). Football, on the other hand, is on the upswing. Note the *majority* (> 50%) figure in 2013. And basketball experienced a spike from the mid-1980s onward, perhaps due to the Michael Jordan effect.

Let's not confuse the term "majority" with "plurality." In an election with more than two candidates, if one of those candidates has received more votes than anyone else but has still not crossed the threshold of 50% of the vote total, he or she has obtained a *plurality*, also sometimes called a *relative majority*.[*]

Understanding how pluralities and majorities work is the *sine qua non* of appreciating the presidential voting system in the United States, referred to simply as the Electoral College. In most of the fifty U.S. states, the presidential candidate earning the most popular votes receives the entirety of the state's electors; this type of zero-sum game is deemed a winner-take-all system: to the victor go the spoils.[†]

Usually, the candidate with the most votes nationwide captures the most electors, but not always. Controversies usually ensue when the popular vote and the electoral totals produce opposing winners, such as in 1876 (Hayes versus Tilden) and 2000 (Bush versus Gore). More statistically "extreme" cases than the voting totals from those two elections could, theoretically, be seen. A candidate in a winner-take-all system could win well under 50% of the popular vote but still capture the presidency, even in a two-candidate field, because he or she could lose every vote in some big states (like Texas or California) but make up for it, elector-wise, by slim one-vote majorities in many other states, large and small.

By analogy, think of it this way: two teams, A and B, compete in the NBA basketball finals, a best-of-four seven-game championship series. Suppose the following scorecard results (shaded cells represent the winning team):

	Game 1	Game 2	Game 3	Game 4	Game 5	Game 6	Game 7	*sum*
A	102	131	110	55	53	59	59	**569**
B	1	1	0	56	56	60	60	*234*

Putting aside whether the scores are realistic (they're not), notice how although team A scored more than twice as many points as team B through the series, team B won the requisite four games—and the championship. That's sort of what the Electoral College is like: each state represents one "game" in a fifty-game championship series.[‡]

All of the worked-out examples and germane discussions in §1c necessitate virtually no prior

---

[*] What we call a "majority" here is sometimes also referred to as an "absolute majority."

[†] Other methods of voting run into Arrow's impossibility theorem, among other problems.

[‡] Okay, not exactly; the analogy isn't perfect for a whole host of reasons, including (1) different states have different relative "weights" (i.e., the number of a state's electors is dependent on that particular state's population) whereas in a sports best-of-seven-game series each game is equally weighted in value and (2) more than two candidates can compete on the national stage, among other issues.

knowledge about statistics as a discipline of study. But statistics can fruitfully be thought of as a handmaiden of probability.* So before going any further with statistics, we need to take a careful look at statistics' master: probability.

§1d. *What is Probability?*

Statistics deals with the collection, presentation, and analysis of data. Probability, the measurement of the likelihood of the occurrence of events (where any particular event's probability must be between 0 and 1), underpins much of the theory behind statistics and risk management. In fact, "The revolutionary idea that defines the boundary between modern times and the past is the mastery of risk: the notion that the future is more than a whim of the gods and that men and women are not passive before nature," according to Peter L. Bernstein in *Against the Gods: The Remarkable Story of Risk*. Without the development of probability theory and risk management, the modern world would not exist, since "engineers would never have designed the great bridges that span our widest rivers, homes would still be heated by fireplaces or parlor stoves, electric power utilities would not exist, polio would still be maiming children, no airplanes would fly, and space travel would be just a dream."

Although typical discussions of the history of probability—and the eventual mastery of risk— begin with Gerolamo Cardano in the sixteenth century and the epistolary exchanges of Pierre de Fermat and Blaise Pascal in the seventeenth, let's instead go back to the ancient Greeks again— this time, though, focusing on Aristotle.

Aristotle's sense of probability does not quite correspond with the modern sense of the term, but he did put forth several interpretations. "For what is improbable," he says, "does happen, and therefore it is probable that improbable things *will* happen. Granted this, one might argue 'what is improbable is probable.'" *What is improbable is probable*: here, Aristotle paints himself into a corner, because his notion of probability has not been systematically defined. In the article "Aristotle's Treatment of Probability and Signs," Edward H. Madden elaborates. "Aristotle believes that one can avoid this imposture by distinguishing between 'general' and 'specific' probability, and apparently intends by the former the statistical sense of frequent occurrence; but he does not establish what he might mean by the probability of an individual event and so leaves the notion of 'specific probability' unclear."

Things are defined a bit more clearly, however, as a *enthymematic argument*—an argument where the conclusion or premise is implied—in Aristotle's *The Prior Analytics*. "A probability is a generally approved proposition: what men know to happen or not to happen, to be or not to be, for the most part thus and thus." Again Madden explains:

---

* You may noticed that I earlier (in the "Preface") labeled probability a handmaiden of statistics. Now, I reversed the designation. Let me explain: Carl Gauss observed that mathematics is the handmaiden of science. And, according to Professor Velleman, "John Tukey taught that Statistics is more a science than it is a branch of Mathematics…. Statistics is held to the additional standard imposed by science. A model for data, no matter how elegant or correctly derived, must be discarded or revised if it doesn't fit the data or when new or better data are found and it fails to fit them." So science, mathematics, and statistics all are served by and serve one another in a kind of symbiosis.

An enthymematic argument from probability, [Aristotle] says, is one in which the major premise is almost but not quite universal, and so the conclusion is almost but not quite universal, and so the conclusion only probable. Aristotle's example: "Most men who envy hate; this man envies, so he probably hates." [The argument can be refuted] not by proving the original probability argument is not *bound* to be true, but by showing that it is not really *likely*—i.e., one must rather state what is more usually true than the statement attacked.

So, as Aristotle writes, "It [the refutation] will be most convincing if it does so in both respects"—i.e., in terms of frequency or "exactness"—"for if the thing in question *both* happens *oftener* as we represent it *and* happens more *as* we represent it, the probability is particularly great."[*] Essentially, this is called a "frequency approach," since we have to account for the number of times some event occurs. But it's important to refrain from interpreting Aristotle in the modern sense: such as attaching a likelihood outcome, like high or low or in between, to some event, or even, for that matter, calling events "events" when describing his framework. Thus, Aristotle's notions of probability are not clearly elucidated enough for us to use fruitfully in practice.

Beyond Aristotle, there is little in the way of the notions of probability threatening to overrun the Greek's deductive approach to mathematics or chance events. Plato's Socrates said, "[A] mathematician who argues from probability in geometry is not worth an ace." As Peter L. Bernstein notes, "[T]*hinking* about games and *playing* them remained separate activities," suggesting that their "clumsy numbering system based on their alphabet" explains at least part of the reason for their dismissal of probability as a valid discipline of study. In games and in life, the Greeks left their fates to the mercy of the gods.

Flashing forward more than one thousand years brings us to Cardano. A physician and mathematician ("mathematician" as a standalone profession is a much later creation) in the sixteenth century, Cardano is best known for his work in algebra[†] and anticipated many later results in probability. He is often overlooked as a leading light in the creation of probability theory, however, for two key reasons: (1) His probabilistic results, which were published posthumously in *Liber de Ludo Aleae* (*Book on Games of Chance*), weren't completely correct, and (2) He predated the symbolic language necessary for mathematical codification, language that Pascal and Fermat were able to take advantage of a century later.

Cardano's key methodological error is that he conflates unscientific luck and the supernatural with chance.[‡] Here is Cardano:

---

[*] All of Aristotle's quotes in this § have been from Madden's article.

[†] Cardano engaged in a drawn-out battle with Niccolò Fontana Tartaglia over the solution to a cubic equation.

[‡] Centuries later Mark Twain satirized games of chance (i.e., those that rely on blind luck, such as had been played by the ancient Egyptians and Romans) and games of science (i.e., those that make use of skill) in his short story "Science vs. Luck" (1870). "At that time, in Kentucky…the law was very strict against what it termed 'games of chance,'" the piece begins. Several boys in the town, who were caught playing "old sledge," are accused of breaking the law by taking part in (ostensibly) a game of chance. The plot hinges on the boys' lawyer's successful strategy

In these matters, luck seems to play a very great role, so that some meet with unexpected success while others fail in what they expect.... If anyone should throw with an outcome tending more in one direction than it should and less in another, or else it is always just equal to what it should be, then, in the case of a fair game there will be a reason and a basis for it, and it is not the play of chance; but if there are diverse results at every placing of the wagers, then some other factor is present to a greater or less extent; there is no rational knowledge of luck to be found in this, though it is necessarily luck.[*]

As biostatistician Prakash Gorroochurn, author of the reputation-rehabilitating article "Some Laws and Problems of Classical Probability and How Cardano Anticipated Them," explains, "[Cardano] fails to recognize such fluctuations are germane to chance and not because of the workings of supernatural forces."

On the positive side of the ledger, however, Cardano offered up a definition of *classical probability* (also called theoretical or mathematical or a priori probability), ahead of nearly everyone else. Cardano again:

So there is one general rule, namely, that we should consider the whole circuit, and the number of those casts which represents in how many ways the favorable result can occur, and compare that number to the rest of the circuit, and according to that proportion should the mutual wagers be laid so that one may contend on equal terms.[†]

Gavin Kennedy notes that Cardano realized each of the sides of a die had an equal chance of landing: "That nobody, as far as I know, worked this out beforehand *and wrote it* down, highlights the significance of Cardano doing so." Cardano, an inveterate gambler, used his games of chance as means of experimenting with probability—and won (at least some) eternal recognition for his efforts.

About two hundred years later—in the interregnum Galileo published a bit about dice games, extending Cardano's work—Pierre-Simon Laplace formally defined classical probability in his *Théorie analytique des probabilités*. "The probability of an event is the ratio of the number of cases favorable to it," he wrote, "to the number of all cases possible when nothing leads us to except that any one of these cases should occur more than any other, which renders them, for us, equally possible." In other words, if all simple events are equally likely to occur, the probability of an event is equal to the number of "successful" simple events divided by the number of total events in the *sample space*, or the list of all possibilities (what Cardano termed the "whole circuit"). Laplace makes use of Jacob Bernoulli's Principle of Indifference (alternatively, Principle

---

securing their acquittal, by altering the judge's and jury's perceptions of old sledge as a game not solely predicated on blind luck but instead requiring foresight and a faculty with numbers. (And lest you think this chance-versus-science distinction doesn't concern enlightened society anymore, note that the Unlawful Internet Gambling Enforcement Act of 2006 outlawed betting on online poker in the U.S. because poker is a "game subject to chance," even though that judgment is suspect at best.)

[*] Quoted in Prakash Gorroochurn's article about Cardano.

[†] Ibid.

of Insufficient Reason) in his definition. The Principle states that, having no better information, all outcomes in a sample space should be assigned equal probabilities, as long as the outcomes are mutually exclusive (i.e., cannot occur together). In the twentieth century, economist John Maynard Keynes, in *A Treatise on Probability*, reexamines the Principle (he in fact named it the Principle of Indifference), and criticizes it:

> Let us suppose as before that there is no positive evidence relating to the subjects of the propositions under examination which would lead us to discriminate in any way between certain alternative predicates. If, to take an example, we have no information whatever as to the area or population of the countries of the world, a man is as likely to be an inhabitant of Great Britain as of France, there being no reason to prefer one alternative to the other. He is also as likely to be an inhabitant of Ireland as of France. And on the same principle he is as likely to be an inhabitant of the British Isles as of France. And yet these conclusions are plainly inconsistent. For our first two propositions together yield the conclusion that he is twice as likely to be an inhabitant of the British Isles as of France....
>
> ... It is not plausible to maintain, when we are considering the relative populations of different areas, that the number of *names* of subdivisions which are within our knowledge, is, in the absence of any evidence as to their size, a piece of relevant evidence.

Keynes is arguing that by changing your definitions of the events, using the Principle of Indifference may lead you to contradictory probability calculations.[*]

So Cardano brought us closer, but a full treatment of the discipline would have to wait until a correspondence between Pascal and Fermat regarding the so-called Problem of Points. As Gorroochurn justifiably notes, "The Problem of Points (POP) is not only the first major problem of probability but it is also the one responsible for its foundation."[†] The POP, originated by the Franciscan monk Fra Luca Bartolomeo de Pacioli in 1494, and partially answered by Tartaglia, was first presented to Pascal by Antoine Gombaud (a/k/a the Chevalier de Méré); later, Fermat was brought into the discussion by Pascal, with the two exchanging letters about the problem.

The Problem of Points is simple enough. There are two players playing (let's say) a fair game of dice. (Variants of dice, such as *astragali* made of bones, were used in gambling games for

[*] The Principle of Indifference may go back even further—to ideas of probability expressed in the Talmud, a Jewish source of liturgy. Suppose there are nine shops in town that supply kosher meat, and one that does not; if a man finds some meat on the street, what are the chances that the meat is kosher? "All that is stationary is considered as half and half.... If nine shops sell ritually slaughtered [kosher] meat and one sells meat that is not ritually slaughtered and he bought in one of them and does not know which one, it is prohibited because of the doubt; but if meat was found, one goes after the majority." In other words, "[I]f meat is found in the town at large, the chances that it comes from any one of the ten shops are equal and therefore the probability that it is *kasher* [kosher] is 9/10," according to Nachum L. Rabinovitch (from the article "Studies in the History of Probability and Statistics XXII: Probability in the Talmud").

[†] From the article "Thirteen Correct Solutions to the 'Problem of Points' and Their Histories."

centuries prior.) Rounds of this dice game are played, with each player contributing equally to a pot. The winner of x rounds wins the entire pot. Suppose, though, that the game is interrupted before x rounds is reached—how should the pot be divided? Should it all go to the player with more rounds won thus far, or should the pot be apportioned by each player's rounds won?

The solution involves finding the sample space of the possible game outcomes if the game *had in fact not been interrupted* and, from there, calculating the probability of each player winning the entire pot. (The pot should not be split, however, based solely on the rounds won by each player before the interruption of the game.) Here is, in part, how Pascal explains a specific case of the problem to Fermat, in which the winner of at least three rounds of a game takes the pot (taken from an August 24, 1645, correspondence):

> Therefore, to see in how many ways four games are combined between two players, it is necessary to imagine that they play with a die with two faces (since there are only two players) as in heads and tails, and that they cast four of these dice (because they play to four games); and now one it is necessary to see in how many ways these dice have different states. This is easy to calculate; they are able to have sixteen which is the second degree of four, that is the square. For we figure that one of the faces is marked a, favorable to the first player, and the other b, favorable to the second; thus these four dice are able to be turned up on one of the following sixteen states:[*]

a a a a (P1)	a b a a (P1)	b a a a (P1)	b b a a (P1)
a a a b (P1)	a b a b (P1)	b a a b (P1)	b b a b (P2)
a a b a (P1)	a b b a (P1)	b a b a (P1)	b b b a (P2)
a a b b (P1)	a b b b (P2)	b a b b (P2)	b b b b (P2)

> and since the first player lacks two games, all the faces which have two a make him win; therefore there are 11 of them for him; and since the second lacks three games, all the faces where there are three b are able to make him win; therefore there are 5 of them. Therefore it is necessary that they divide the sum as 11 to 5.[†]

Notice how, since each of the sixteen permutations is equally likely, the POP solution is an exercise in classical probability à la Laplace. Fermat and Pascal extended these ideas to more general versions of the game, including Pascal's mapping of the problem to his eponymous triangle (also called the Arithmetic triangle); and key mathematicians, some contemporaries of Fermat and Pascal and others born after Fermat's and Pascal's death, weighed in on the POP as well.[‡]

---

[*] In the table, each possibility in the sample space is labeled parenthetically with P1 or P2, indicating which player won that particular permutation.

[†] From "Correspondence on the Problem of Points," available online at http://www.cs.xu.edu/math/Sources/Pascal/Sources/pasfer.pdf.

[‡] And a key mathematician before their birth: Cardano. The POP didn't originate with Pascal and Fermat; rather, again, the POP first appeared in *Summa de Arithmetica, Geometrica, Proportioni, et Proportionalita*, written in 1494 by Fra Luca Pacioli—a book which also usefully contained multiplication tables and methods of double-entry bookkeeping, and helped to teach Leonardo da Vinci mathematics.

Example 1d.-1

In a family with eight children, what is more likely: an equal number of boys and girls, or seven kids of the same gender?

Solution. Listing the possibilities of equal numbers of boys and girls, we have only one: four boys and four girls. But if seven children are of the game gender, then we could have seven boys or seven girls—two possibilities. Thus, seven kids of one gender is more likely. (Note that the order of the births doesn't matter here.)

With the imprimatur of Fermat and Pascal, probability came into its own as a legitimate subject of mathematical inquiry, untainted by its associations with seedy games of chance. Luminaries such as Christiaan Huygens, Jacob Bernoulli, Carl Friedrich Gauss, Thomas Bayes, and John Venn all got into the act, tightening the axioms and helping to formalize the discipline. By the twentieth century, Andrey Kolmogorov codified and systematized probability, bequeathing us the abstract structures mathematicians still in use to this day.

In his perceptive essay "A Philosopher's Guide to Probability," philosopher Alan Hájek splits the study of probability into two parts: its axiomatization (in other words, its formal mathematical rules) and its ontological foundations (namely, what, precisely, it all means). Let's spill some ink now on the ontology of probability, since most of the rest of the text will take these foundations for granted.

There are two overarching interpretations of probability. The first, called *objectivism in probability*,[*] takes the approach that probabilities exist empirically. More specifically, the *frequentist interpretation* examines the ratio of the frequency (count) of an event to the number of times an experiment is repeated—this quotient is equal to the probability of the event. As the number of trials approaches infinity, the relative frequency of an event converges to the actual, or true, probability of the event.

Example 1d.-2

You are handed a gold coin. What could you do to test if the coin is unfair (i.e., weighted)?

Solution. Taking out a scale isn't going to help. But repeatedly flipping the coin and recording the results of the flips should, according to the frequentist approach, give you a good idea if the coin is fair or unfair—the farther from 50-50 heads-tails, the more likely the coin is weighted, assuming many, many trials are conducted.

In his book *Risk Savvy: How to Make Good Decisions*, author and psychologist Gerd Gigeren-

---

[*] Don't worry: this has nothing to do with Ayn Rand.

zer poses a simple question: If the weather forecast calls for a 30% chance of rain tomorrow, what exactly does this mean? He relays several possibilities: (1) That it will rain approximately 30% during the course of the day; (2) That it will rain in roughly 30% of the region in which the forecast was made; (3) That 30% of meteorologists believe it will rain; or (4) That, given a set of days of similar conditions, 30% of them can be expected to have rain at some point during the day. Mathematician Jordan Ellenberg, in *How Not to be Wrong: The Power of Mathematical Thinking*, explains how to "shoehorn the weather into frequentist model: maybe we mean that among some large population of days with conditions similar to this one, the following day was rainy [30%] of the time. But then you're stuck when asked, 'What's the probability that the human race will go extinct in the next thousand years?'" In which case we have a new problem.

Instead of many trials of a single experiment, suppose that, for whatever reason—like in the humans-going-extinct question—we only have an opportunity for a single trial. What can we deduce from this single trial? If a coin, unknown to be fair or unfair, is flipped once, and it lands on heads, what does this tell us about the potential weightedness of the coin? Does it tell us anything? In fact, consider any particular unrepeatable event, such as the Red Sox's blistering comeback against the New York Yankees in the 2004 Major League Baseball American League Championship Series—what, exactly, were the chances of that comeback? The frequentist approach breaks down here, since there is no way to conduct many trials of what is, by its very nature, a one-off event.

Another issue: the *reference class problem*—understood as far back as 1662 by Antoine Arnauld and Pierre Nicole in their probability treatise *Logic, Or the Art of Thinking*[*]—perhaps an even greater challenge to frequentism than the problem of a single case. Reconsider the Red Sox example. Even if we were to attempt to calculate the probability of a comeback, what would our *reference class*, the group to which we're assigning potential trials, be here? Would it be all baseball championship games, or just American League Championship games? Perhaps it's just games the Red Sox played in the playoffs, or games the Red Sox played that season (regular and/or playoffs). Or maybe it should be just games in which the Red Sox faced the Yankees that season, or over the course of the last several seasons. Should we restrict our attention instead to merely "close" games in which teams, or exclusively the Red Sox, played? And, if we do that, how far back in time should we consider? Thus, the reference class problem really is a problem of classification.[†] The probability of the proposition "Chances that the Red Sox Make a Comeback" changes depending on which reference class is examined. The problem comes from mathematician John Venn. "It is obvious that every individual thing or event has an indefinite number of properties or attributes observable in it," he wrote, "and might therefore be considered as belonging to an indefinite number of different classes of things...."

Another *objectivism in probability* interpretation is called the *propensity interpretation*, which is similar to frequentism in that probability is kept within the realm of real-world experience but also has a number of notable differences. As the *Stanford Encyclopedia of Philosophy* explains,

---

[*] Probability is explicitly measured for the first time in writing in this text.

[†] Gerd Gigerenzer urges his readers to "[a]lways ask for the reference class: Percent of what?"

Like the frequency interpretations, propensity interpretations locate probability 'in the world' rather than in our heads or in logical abstractions. Probability is thought of as a physical propensity, or disposition, or tendency of a given type of physical situation to yield an outcome of a certain kind, or to yield a long run relative frequency of such an outcome....

For him [Karl Popper, widely credited with fully articulating the propensity interpretation, though Charles S. Peirce originated it], a probability p of an outcome of a certain type is a propensity of a repeatable experiment to produce outcomes of that type with limiting relative frequency p. For instance, when we say that a coin has probability 1/2 of landing heads when tossed, we mean that we have a repeatable experimental set-up—the tossing set-up—that has a propensity to produce a sequence of outcomes in which the limiting relative frequency of heads is 1/2.

So, in attempting to head off some of the criticisms of the frequentist approach, propensity theory takes a step back; for instance, rather than claiming that a coin is fair because after many flips its heads-to-tails ratio converges to an approximately stable 1-to-1, propensity theory says that the fair coin's ratio arrives at stability *because* of its propensity, even though such a definition risks falling into the trap of circularity. Popper doesn't clarify exactly how a coin has a "disposition" or "tendency" to a "limiting relative frequency" of heads of some value, nor is it clear how bearing down on some long-run frequency using a propensity interpretation approach offers advantages as compared to the frequentist approach.

As opposed to objectivism, *subjectivism in probability*, as Hájek explains, "see probabilities as *degrees of belief*, or *credences* of appropriate agents. These agents cannot really be actual people, since, as psychologists have repeatedly shown, people typically violate probability theory in various ways,[*] often spectacularly so.... Instead, we imagine agents to be ideally rational."

The person most responsible for subjectivism is Thomas Bayes, in a paper entitled "An Essay towards solving a Problem in the Doctrine of Chances," although his mathematical results were published posthumously. Richard Price,[†] who published Bayes' essay, explains the motivation behind Bayes' thinking:

[H]is [Bayes'] design at first in thinking on the subject of it was, to find out a method by which we might judge concerning the probability that an event has to happen, in given circumstances, upon supposition that we know nothing concerning it but that, under the same circumstances, it has happened a certain number of times, and failed a certain other number of times. He adds, that he soon perceived that it would not be very difficult to do this, provided some rule could be found, according to which we ought to estimate the chance that the probability for the happening of an event perfectly unknown, should lie between any two named degrees of probability, antecedently to any experiments made about it; and that it appeared to him that the rule must be to

---

[*] We will see examples of this later on, specifically with the work of Kahneman and Tversky's Prospect theory.

[†] Price published plenty of his own mathematical results; he is considered by many to be the father of actuarial science.

suppose the chance the same that it should lie between any two equidifferent degrees; which, if it were allowed, all the rest might be easily calculated in the common method of proceeding in the doctrine of chances.

Whereas with the frequentist approach, we can test some process without assigning it a probability a priori, using a Bayesian framework we go forward with a probability in hand, "antecedently to any experiments made about it"; that probability may be a "personal belief," called the subjectivist view, or rooted in rationality, called the objectivist view. No matter which view taken, during a probability experiment Bayes tells us to update our prior probabilities, or *priors*, with each new data point. Our revised probabilities are called *posteriors*.[*]

Bayes' Theorem, as his framework is usually referred to (sometimes it is called Bayes' Rule instead), is summarized by Bayes as

> *Given* that the number of times in which an unknown event has happen and failed: *Required* the chance that the probability of its happening in a single trial lies somewhere between any two degrees of probability that can be named. [Bayes' italics]

Bayes' Theorem is especially useful when examining medical tests for the presence of disease. Conditional probabilities—calculations of probabilities that rely on prior events—are crucial to understand when working with Bayes' Theorem. We'll explore some examples later on.

Subjective probability, at its core, deals with personal belief. Mathematician Bruno de Finetti, who helped codify the subdiscipline, wrote that subjective probability is "the degree of belief in the occurrence of an event attributed by a given person at a given instant and with a given set of information," and, furthermore, noted that *all* approaches to probability have a subjective component. To wit: "when one pretends to eliminate the subjective factors one succeeds only in hiding them."

John Allen Paulos relays an amusing subjective probability anecdote in *Innumeracy*.

> A man who travels a lot was concerned about the possibility of a bomb on board his plane. He determined the probability of this, found it to be low but not low enough for him, so now he always travels with a bomb in his suitcase. He reasons that the probability of two bombs being on board would be infinitesimal.

Here, Paulos' silly example illustrates flat-out mistaken personal belief in the calculation of a probability. Other times, the arrival of a subjective probability harbors logical errors.

Example 1d.-3

Recall the fairy tale of Sleeping Beauty. Suppose we subject her to an experiment, which she fully agrees to, eyes wide open. We put her to sleep this Sunday evening, and then we flip a fair coin. If the coin lands on heads, we wake her on Monday—but only on Monday. If the coin lands

---

[*] Notice how Bayes combines the two seemingly mutually exclusive approaches to mathematics discussed earlier—a priori and a posteriori—into a single probability calculation method.

on tails, though, she is woken up on both Monday *and* Tuesday. But if she is woken twice, on Monday she is given an amnesia pill before going back to sleep so that she completely forgets she was awakened. Whether the coin lands on heads or tails, every time that she wakes she is asked the following: "What side, Sleeping Beauty, do you think the coin landed on: heads or tails?"

Solution. In order for Sleeping Beauty to make a good guess, she needs to calculate the probability of the coin landing on heads, given that she has woken up. But it's here that subjective probability, and Bayes' ideas, seem to break down. *The Sleeping Beauty Problem* presents a paradox to the would-be Bayesian: Should Sleeping Beauty believe the probability of heads is 1/2, since when she wakes up she doesn't know whether or not it's Monday (recall the amnesia pill), or should she believe the probability of heads is 1/3, since she would be woken up twice with tails but only once (just on Monday) with heads? People who believe the answer is 1/2 are called "Halfers" and people who instead feel the answer is 1/3 are called "Thirders."

You may already realize the epistemic issue here. Remember, Bayes says that we should update our probabilities whenever given additional information—but *is* Sleeping Beauty given additional information just by waking? In fact, as Berry Groisman notes in the essay "The End of Sleeping Beauty's Nightmare," Sleeping Beauty doesn't have to be put back to sleep only once (on Monday evening) if tails shows: she could be put to sleep night after night after night for a year, given an amnesia pill each evening, reducing the probability of tails from 1/3 to 1/366. Although Groisman seems to offer up a nice solution by shifting the question, the problem is unsettled in mathematical literature.

Of course, like all of mathematics, probability is a living discipline and continues to evolve, even as we account for its relative lateness on the scene.[*] Some of the changes are in reaction to ambiguities in definitions or logic or from paradoxes that have developed, like Sleeping Beauty's. For instance, John Maynard Keynes took an oppositional stance to subjective probabilities in his *A Treatise on Probability*, writing that "in the sense important to logic, probability is not subjective. It is not, that is to say, subject to human caprice. A proposition is not probable because we think it so"; the brilliant (and short-lived) mathematician Frank Ramsey disagreed, saying probability "is a measurement of belief *qua* basis of action." That debate took place more than eighty years ago.

Around that same time, Andrei Kolmogorov mathematically formalized probability. As Slava Gerovitch, author of "The Man Who Invented Modern Probability," tells it, "Kolmogorov drew analogies between probability and measure, resulting in five axioms, now usually formulated in six statements, that made probability a respectable part of mathematical analysis." Most of the mathematical framework that's used today, such as elementary (or simple) events, sample spaces (or universal sets), the additional rule, conditional probabilities, and independence, takes its cue

---

[*] There is even a Marxist explanation for the late birth of probability: that, as the commerce of capitalism emerged, a theory of probability arose with it. More persuasive than its appearance solely as a tool of the bourgeoisie is the notion that probability theory patched up, deductively, the philosophical flaws with empirical reasoning through a *systematic* study of chance (and hence at least partially unpredictable) events.

from Kolmogorov's theory of probability. Each random event is mapped to a sample space's measurable set, and a probability is calculated off of that set.

Kolmogorov's original path of study was not mathematics but history. As a teenager, he wrote and presented a statistics paper detailing how taxes in medieval Russia were calculated not based on individual households, which were usually expressed as fractions of taxes, but on villages in their entirety, which were expressed in whole numbers of taxes—and then simply divided up. (Hence the fractional components of the households' money due.) Gerovitch sums up the reaction to his paper thusly: "'You have found only one proof,' was his professor's acid observation. 'That is not enough for a historian. You need at least five proofs.' At that moment, Kolmogorov decided to change his concentration to mathematics, where one proof would suffice." A little while later, Kolmogorov was inspired by another professor, this time a mathematician by the name of Nikolai Luzin,[*] eventually leading to the mathematical renaissance that was his rebranding of probability theory on firm(er) foundations.

§1e. *A Conceptual Framework*

When building a bridge to connect probability and statistics, we need to carefully examine the idea of *variability*.

Definition 1e.-1

Variability. Even if we repeatedly gather data using similar processes and conditions, our results from the data will not necessarily be identical (or perfectly predictable).

Perhaps Galileo was the first to systematically document the variability inherent in real-world observations. The differences in the measurements of a single astronomical body (such as a star), he said, were the result of unavoidable errors in the observations of the measuring instrument.[†]

We don't need a telescope, though, to better understand the consequences of variability. Suppose we have a fair die. We would expect the probabilities of the simple events—rolling a 1, 2, 3, 4, 5, and 6—to be equally likely, as shown in the following *probability distribution*:

Simple Event	1	2	3	4	5	6
Probability	1/6	1/6	1/6	1/6	1/6	1/6

If we rolled a die 60 times, however, it's extremely unlikely that this would turn up:

---

[*] Luzin was later declared an enemy of the Soviet state, with Kolmogorov ultimately turning on his old professor.

[†] Galileo also noticed that measurement errors tended to be symmetrical about the (true) mean, leading to distributions of errors that have normal shapes. Much more on the normal distribution later.

Simple Event	1	2	3	4	5	6
Frequency	10	10	10	10	10	10

despite the probabilities of each simple event being the same. Instead, we'd probably see a distribution more like this, with mostly unequal frequencies in each cell:

Simple Event	1	2	3	4	5	6
Frequency	8	12	6	11	10	13

Even though we gathered results in a similar way—by repeatedly rolling a single fair die—the results aren't the same: that's variability.

But variability doesn't satisfy as an explanation for the variation in results of *any* data set. What if the probability distribution of the fair die had instead looked like this:

Simple Event	1	2	3	4	5	6
Frequency	5	4	37	5	6	3

It's tough to hold the line and say: This fair die, after 60 rolls, produced this distribution, and any differences from the expected frequencies are due to chance variation. A much more likely explanation: the die is not fair, weighted to land much more often on the number 3. Which brings us to our next definition.

Definition 1e.-2

The Rare Event Rule for Inferential Statistics. If, when assuming hypothesis A to be true, the probability of some event B occurring is very small, then in all likelihood hypothesis A is in fact not true.

Could the probability distribution have been that seemingly heavily biased to the number 3 and yet *still* have been from flips of a truly fair die? Sure, it's possible, but exceedingly unlikely—the hypothesis "the coin is fair" is probably not true, given the probability distribution obtained.[*] Later, we will calculate precisely *how* unlikely using a frequentist model—but, for now, an intuitive feel for small and large probabilities, for what results can largely be attributed to variability and what results cannot, is important.

---

[*] Here's a silly example. Suppose we have a hunch that the first name (or last name) of a person correlates with his or her profession; consider these a weak case of aptonyms. For instance, people with the name Dennis are likelier to be dentists, and those named Lawrence are more likely to be lawyers. The rare event rule approach to this problem would be as follows: Assume, going into such a study of first names, that those with names that sound like professions are *not* more likely to have those professions than anyone else. Then, collect data—lots of it. If the results belie the assumption, then it's probably the case that *name = profession* at a greater frequency than would be expected by chance alone. By the way, this idea of a name-profession link is called *implicit egotism*, and some studies have indeed shown a slight link.

ф

In the real world, as Stephen Jay Gould notes, "variation itself is the reality." Tasked with separating results of sampling due to variability (the chaff) from significant indicators (the wheat) is really a process of distinguishing noise from signal. The patterns in data help us to make decisions; but what appears to be patterns could be in fact be natural variability in the collection of the data. Consider this set of 60 rolls of an ostensibly fair die.

Simple Event	1	2	3	4	5	6
Frequency	7	17	8	11	8	9

Is the coin fair or not? Perhaps the coin is weighted toward 2, but it seems equally as likely that variability explains the uneven results. Weighting is not as apparent here as before; separating the chaff (inherent variability due to data collection) from the wheat (legitimate patterns in the data) is difficult. As Robert Hogg and Elliot Tanis point out in *Probability and Statistical Inference*, "In the face of variability, the statistician must still determine the best way to describe the pattern. Accordingly, statisticians know that mistakes will be made in data analysis, and they try to minimize those errors as much as possible and give bounds on the possible errors." Ideally, these error should be as small as possible; collecting more and more data, as long as the process of data collection is sound (i.e., randomized in some way), reduces the bounds on the error. Regardless, statisticians must accept the fact that "many decisions have to be made that involve uncertainty," and errors in both the analysis of data and the conclusions drawn from data sets are always possible.[*]

Example 1e.-1

You are asked to play two dice games. In Game A, if you roll more than 30% 6's, you win. In Game B, if you roll between 13% and 21% 6's, you win. For each game, you are given a choice: roll a fair die 20 times, or roll it 200 times. How many times should you roll the die for each game to give yourself the best chance of winning?

Solution. With fewer rolls comes the curse of more variability, and more noise. But with more rolls (trials), there is less variability and more signal—and you'll likely land the closer to your *expected* probability of 1/6, or around 17%. Therefore, roll the die 20 times for Game A and 200 times for Game B.

Example 1e.-2

Imagine yourself relaxing in a boat on Lake Placid, fishing rod in hand. You cast a reel into the

---

[*] In fact, the taking into account of the errors in measurements as a result of variability was, in many ways, an impetus to modern statistical methods, as we will see with multiple estimation methods later on.

lake, knowing that trout make up exactly 40% of the fish in the water, while salmon constitute the remaining 60%. After about an hour, you end up catching twenty fish: six trout (30%) and fourteen salmon (70%).

Throwing them all back into the lake, you go at it again. An hour later, another twenty fish. This time you catch eight trout (40%) and twelve salmon (60%), matching the overall percentages of fish in the lake.

You let the fish swim to freedom, and start over, one last time. Your final catch of the day: ten trout (50%) and ten salmon (50%).

What statistical concept best explains why your results varied from catch to catch, as well as why the results differed at all from the distribution of trout and salmon in Lake Placid?

Solution. Variability.

Example 1e.-3

Reconsider the fish swimming in Lake Placid. Fishermen Jeff and Frederick fish the lake for six months. Jeff captures ten fish per day from the lake, while Frederick captures 100 per day. The percentages of trout and salmon stay constant throughout the six months that they fish the lake: 40% trout and 60% salmon.

The number of days in which each fisherman captured more than 50% trout was recorded. Which fisherman, Jeff or Frederick, likely had more such days?

Solution. Since Jeff only captures ten fish per day, his daily trout-salmon distribution will vary much more widely than Frederick, who captures 100 fish per day—and probably hovers, on most days, somewhat close to a 40-60 trout-salmon split. So Jeff is likely to have more days in which more than 50% of his fish caught were trout.

To be fair, before going any further, we need to consider a second definition of variability that on the surface seems completely different than the first.

Definition 1e.-3

Variability. The spread of a data set.

Consider a data set of the weights (in grams) of ten fish captured by Jeff on a single day. If all these fish weights were plotted on some sort of graph, they will probably be scattered about. But if, instead, we plotted the weights of 100 fish captured by Frederick on a single day, many of his fish weights will be clustered together; hence, the variability, or spread, of Frederick's larger data set is less than Jeff's smaller data set. So, in general, we can say that the bigger the sample size, the less the variability.

Variability is the key link between probability and statistics. If a process or set of observations has at least some variability, then we'll make use of statistics to find their cause, trying to fit a distribution to best describe the data; but considering some process that's devoid of variability usually implies a theoretical, probabilistic study—a study to find effects, rather than causes. Zero variability implies that a process is perfectly predictable, i.e., it consists solely of signal; too much variability indicates a process that is haphazard and bordering on complete unpredictability, i.e., it consists solely of noise. The notion of variability is firmly planted in the real-world of data collection; the world of zero variability, or *antivariability*, is set in the abstract world of pure mathematical theory, or, alternatively, in the consideration of comprehensive census data for which further sampling is not necessary.

Definition 1e.-4

Antivariability. 1. A probabilistic process mapped out only in theory, in which real-world data collection is either impossible or not subject to consideration. 2. A reference to the bounds of error, which is zero (no error), on the examination of comprehensive census data—since every member of the population of interest is accounted for without further sampling.

For any process or set of observations, we can combine all of these aforementioned elements, and several additional germane elements, together on a spectrum, or axis; we might best term it the **axis of variation**.

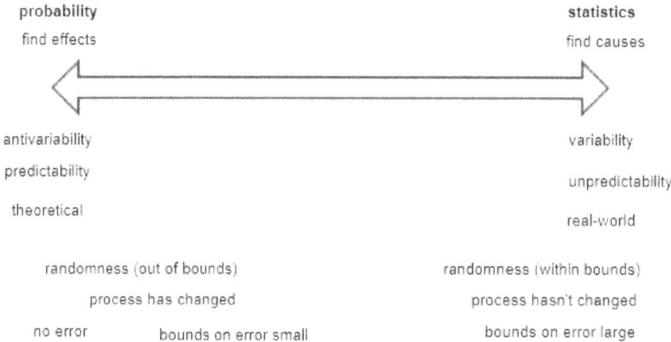

On the axis of variation, the higher the variability, the higher the bounds on error—until, ultimately, all gathered data is just error (or noise). But as the bounds of error decreases—ultimately to no error whatsoever—we have arrived at antivariability's second definition: there is zero error because the entirety of the population's data is extant.

Example 1e.-4

Consider the data set of the ages (in years) of all deceased U.S. Presidents at their death. Describe this data set's variability.

Solution. This data set has zero variability because it is, *ipso facto*, a census of a (small) population. No further sampling is necessary to obtain comprehensive information about the population. Thus, the dead presidents' data set exhibits antivariability.

Example 1e.-5

In a family with eight children, an equal number of boys and girls is less likely than seven kids of the same gender. Explain under what contexts variability and antivariability are exhibited here.

Solution. Considered theoretically—in other words, simply listing out possibilities in the sample space (and, more subtly, leaning on the Principle of Indifference to do so)—this process exhibits antivariability: there is complete predictability. But, instead, if many such families with eight children around the world were sampled, or if a computer was programmed to conduct many simulations of eight births, variability in the results would be present; it likely wouldn't be the case that the sampled families would match, in terms of proportions, the theoretical probability distribution.

But differences in variability may also signal that something has changed in a process.

Example 1e.-6

The Misery Loves Company (MLC) manufactures towels. To make sure that certain towels are being fabricated at the correct length—which is 14.1 inches—a quality control inspector takes a random sample of 25 such towels coming off of the factory's assembly lines. She finds the mean of this sample of towels to be 14.5 inches. Unsure if there is something wrong with the machines, which are set to cut at 14.1 inches, she nonetheless recalibrates the machines to cut at shorter lengths. She then obtains another random sample of 25 towels, now finding the towels' mean to be 14.3 inches. Has the production process of these towels truly improved, or can the second sample's decreased mean be attributed to variability?

Solution. Although we're not yet ready to look at the mathematics behind the procedure, the question must be reframed as follows: How far away—in the sense of variability's second definition, that of the spread of the data set—is what we actually observed from what we expected to observe? If there's not much of a difference between what we've observed with what we expected to observe, then any differences can be safely attributed to chance variation (for example, that the MLC machines aren't making the towels any shorter despite the smaller sample mean). But if the difference between what's observed and what's expected is sufficiently large, chance variation is unlikely to be an valid explanation—and we can claim that the process has changed in some way (for instance, that the MLC machines are in fact cutting the towels a bit shorter).

The axis of variation gives us quick rules of thumb to describe a data set. But we must consider another spectrum, or axis: the **axis of collection**. This axis, shown on the next page, centers on how the data were collected. Recall from earlier examples that if data were not collected using some random process, then usually the data are worthless.

In addition, mathematically, the bigger the sample size, the better—as long as the data are collected randomly. (Although this isn't always strictly true in the real world; it might be that gathering data beyond a certain size is prohibitively expensive or stretches other resources beyond their breaking point.)

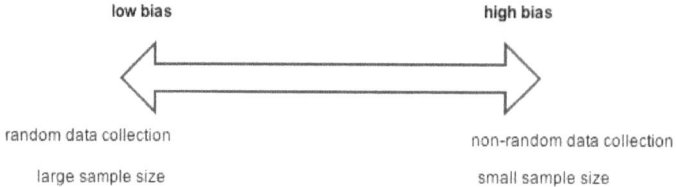

low bias   high bias

random data collection   non-random data collection

large sample size   small sample size

Which leaves us with a final spectrum, or axis, to delve into: the **axis of interpretation**. Once data is gathered and summarized, we need to be able to draw valid inferences. Sometimes, though, we make a mistake—without even realizing it (perhaps ever).

One of these potential mistakes is called a Type I error, more commonly called a *false positive*.

Definition 1e.-5

False positive (or *Type I error*). Our beliefs show or our results indicate the presence of a condition, signal, or pattern, when in reality there is no condition, there is noise, or there's only mere randomness.

Example 1e.-7

A roulette wheel has three colors a ball can land on: red, black, and green. (Not Russian roulette—which, come to think of it, is also a game of chance.) Red and black are equally likely. Suppose the ball has landed on black ten times in a row. Should you bet against black?

Solution. Betting against black would signify that you believe one of two possibilities: (1) That because black has come up many times in a row, another color is "due," or (2) You are suspicious that a streak of black might mean that the roulette wheel's ball is not equally likely to land in the thirty-eight slots of the roulette wheel.

But randomly occurring streaks are much more likely than we might initially suppose. The book *How We Decide*, by Jonah Lehrer,[*] relays an interesting anecdote.

> [P]eople are surprised when a shuffled song [in an electronic music player] repeats or when a flipped coin exhibits extended streaks of heads or tails. The most famous example of such a phenomenon occurred in a Monte Carlo casino in the summer of 1913 when a roulette wheel landed on black twenty-six times in a row. During the run, gamblers bet against black, since they as-

---

[*] Okay, okay—there were portions of this book, along with Lehrer's follow-up, that were fabricated. It all led to a journalistic scandal, covered well in *So You've Been Publicly Shamed* by Jon Ronson. Lehrer's books were withdrawn from circulation. Nonetheless, his description in the block quote above is cogent, concise, and accurate.

sumed that the randomness of the roulette wheel would somehow "correct" the imbalance and cause the wheel to land on red. The casino ended up making millions.

The innumerate Monte Carlo casino players in 1913 had a bad case of *gambler's fallacy*: Believing, in a game with *independent trials*, like roulette, that past outcomes affect or predict future outcomes. If you believe a particular color of the roulette wheel is "due," then you're also suffering from a case of gambler's fallacy.

Spotting patterns which signify nothing, such as noting extended streaks of balls landing on black on a roulette wheel, is human nature, even outside of circumscribed game-like conditions. For instance, look up at the sky on a cloudy day: Do you see any shapes, or objects, or cartoon characters? How about ghosts: Have you ever seen one? Pop psychology has embraced this notion of visual free association most clearly with the Rorschach test, where subjects analyze randomly generated inkblots for patterns (which may, in turn, reveal subjects' inner psychological states). Michael Shermer, historian of science, has even given this human proclivity to spot connections and patterns, irrespective of their connection to truth or falsity, a name: *patternicity*. As Bruce Poulsen notes in an article from *Psychology Today*, "Many people perceive faces in seemingly random places—such as in clouds, in patterns of dirt left on cars, or on the moon. We take such patterns a step further by ascribing meaning to them. People have seen the images of Jesus and Mary inside a halved orange; or the face of Jesus on a piece of toast…. Conspiracy theories, such as the belief that the twin towers of 9/11 were destroyed in a controlled demolition perpetrated by the government, are confabulations based on misperceived patterns."

Shermer was not the first to bundle up this neurological phenomenon of patternmaking into a catchall term. That distinction (sort of) goes to Klaus Conrad, half a century ago, with his *apophenia*.[*] This definition is especially important—after all, the term is in the title of this primer—so let's quote directly from *Merriam-Webster's*:

Definition 1e.-6

Apophenia. "[T]he tendency to perceive a connection or meaningful pattern between unrelated or random things (such as objects or ideas)."

When conjuring up apophenia, Conrad didn't have probability or statistics in mind[†] but, rather, mental illness. In her *Slate* article "It's All Connected," Katy Waldman explains Conrad's thinking:

---

[*] More precisely, Conrad coined *apophanie*, a meshing of two Greek words: *apo*, which means "away," and *phaenein*, which means "to show." The psychologist Peter Brugger later introduced the English version of the term we use here.

[†] As an adjacent notion, this one more closely tied to statistics, consider the *Texas sharpshooter fallacy*: focusing on a small segment of a data set for whatever reason, to the exclusion of all else—thus finding potentially non-representative patterns.

He was describing the acute stage of schizophrenia, during which unrelated details seem saturated in connections and meaning. Unlike an epiphany—a true intuition of the world's interconnectedness—an apophany is a false realization. Swiss psychologist Peter Brugger introduced the term into English when he penned a chapter in a 2001 book on hauntings and poltergeists. Apophenia, he said, was a weakness of human cognition: the "pervasive tendency … to see order in random configurations," an "unmotivated seeing of connections," the experience of "delusion as revelation." On the phone he unveiled his favorite formulation yet: "the tendency to be overwhelmed by meaningful coincidences."

Seeing order from randomness; perceiving delusion as revelation; letting cascading coincidences overwhelm one's self with meaning—Brugger's research has revealed that apophenia may result from "over activation of the right hemisphere," which is more responsible for connection-making, holistic awareness, and pattern-seeking; higher levels of dopamine may also trigger an excessive tendency toward pattern-making. Novelist Christopher G. Moore even claims that a touch of apophenia is necessary to filling in story "patter," the multivariate connections between the tinniest of narrative details. "Someone with apophenia makes these spontaneous connections throughout the day, in every setting, and out of all the unrelated people, events and objects that she has experienced," he says. "If your mind automatically switches into this method of assembly of people and events to tell a story, then you have the right mental stuff to be a writer." Psychologist Carl Jung might add an additional element to Moore's prototypical novelist: *synchronicity*, where there is no obvious causality behind what seem to be meaningful coincidences.

But what of statistics—should we resolve never to see patterns when presented with what looks to be random noise? Should we strive to not fall under the spell of apophenia (or synchronicity)? Should we completely disregard patterns in all aspects of life—such as while playing the stock market?[*] Steeling ourselves against finding even the weakest signal amongst lots of noise can sometimes be taking things too far.

Definition 1e.-7

False negative (or Type II error). Our beliefs show or our results indicate the presence of no condition, only noise, or mere randomness, when in reality there is a condition, signal, or pattern present.

Our brains are natural pattern detectors, which probably had significant survival benefits when our ancestors were prancing around the Serengeti. Instead of ignoring that rustling of leaves—which could very well be caused by a hiding lion, waiting patiently to catch some prey—our ancestors exited the scene, evading the potential danger as fast as possible (the *flight* in the fight-or-

---

[*] Burton Malkiel, in his classic *A Random Walk Down Wall Street*, describes the so-called technical analysis of the "chartists," that (small) group of stock analysts who use stock chart patterns to predict future stock price movements. Citing a definition of the *random walk hypothesis* (more on that later), Malkiel writes, "If the weak form of the random-walk hypothesis is valid, then, as my colleague Richard Quandt says, 'Technical analysis is akin to astrology and every bit as scientific.'" Apophenia indeed.

flight response). Early humans who didn't dismiss their suspicions, and who were paranoid at every turn, were selected to survive; those who were too nonchalant and didn't err on the side of caution, figuring the rustling of leaves was due to the wind—well, those humans by and large didn't make the cut. So we have all been left with a rich genetic legacy: more often than not seeing patterns where there are none, and finding meaning in coincidences that have no overarching connection.

Although it's not yet completely adopted into the common parlance, parapsychologist David Luke recently coined an antonym for apophenia, which has a much catchier name than "Type II error":

Definition 1e.-8

Randomania. The tendency to attribute to chance or merely noise that which actually is signal or pattern.

It's unfortunate that Luke's motivation behind coming up with the term seems so patently unscientific (he's a parapsychologist, after all). To wit: "Reclaiming the dream experience, if you were to work with, record, and study your dreams every day as I did for just 18 months, then you might actually discover, as did I, that on average 1 dream in 10 had some compelling precognitive component.... While such self-reports are not evidential, can the law of truly large numbers actually account for these rates of occurrence?" First, we'll be exploring *the law of large numbers*, first proved by Bernoulli, in a later §; and second, mathematicians such as Paulos in *Innumeracy* (among other places) have already explained the mathematics of the likelihood of coincidences in dreams. Nonetheless, his reasoning is rooted in a frequentist approach to probability, however skewed.

So let's not dismiss Luke's work out of hand. He continues: "Indeed, suggesting that such frequent occurrences are expected by chance is essentially the opposite of what psychiatrist Klaus Conrad (1958) somewhat oddly called apophenia, the discovery of patterns in (apparently) random data. Perhaps we should call this opposite phenomenon of attributing chance probability to (apparently) related phenomena randomania, as a label for believing that everything one cannot currently explain is just due to chance and coincidence." Luke's conceptual framework here is outstanding. But then it crumbles halfway through his next sentence: "One assumes that such a condition derives from a deep-seated rejection and fear of the paranormal...."[*] Ugh. Shades of Freud.

So, bundling up all these notions leads us to a graphic of the axis of interpretation, shown on the next page.

---

[*] Luke's excellent result—his new term for the opposite of apophenia—belies the wisdom of his choices, to paraphrase the master aphorist and two-time Super Bowl-winning head coach, Bill Parcells. "You may do something that turns out well, but it's still not the most prudent choice," he says in *Parcells: A Football Life*, while affixing an acronym to this idea: NATO, or Not Attached to Outcome.

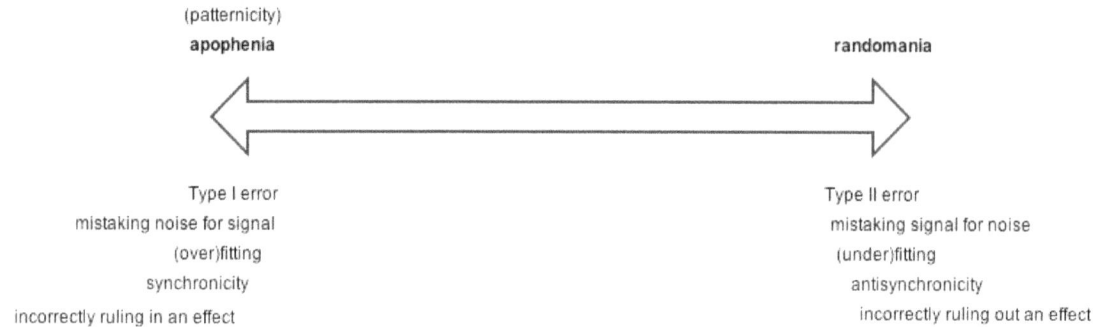

Realize that the "antidote" to apophenia of this primer's title does not imply that we should venture into randomania; that would be swapping one intellectual fallacy with another. Rather, we should take these three axes—of variation, collection, and interpretation—to serve as a background conceptual framework while formulating mathematical judgments, uncovering the principles behind and the methods of probability and statistics.

Part 1

φ

The Axis of Variation

| probability | statistics |
| find effects | find causes |

antivariability · variability

predictability · unpredictability

theoretical · real-world

randomness (out of bounds) · randomness (within bounds)

process has changed · process hasn't changed

no error · bounds on error small · bounds on error large

§2. *Data and Its Discontents*

φ

Some Basic Vocabulary. The Journey to Inference. Simple Graphs of Categorical Data Sets. Deceptions with Graphs. Contingency Tables. The Rise of Technology in Statistics. Simple Graphical Displays of Distributions. Measures of Center. Measures of Spread.

§2a. *Some Basic Vocabulary*

Let's set aside probability for now and focus in on statistics. Before we can begin solving statistics problems, we must become more familiar with some basic vocabulary and definitions.

Definition 2a.-1

Data. Information, such as numbers, that must be presented in some sort of context.

Example: Heights of students in a classroom. Note that no numbers have been presented here, like 65, 54, and 56 inches; yet the "heights of students in a classroom" has a proper context—so it is considered data.[*]

Definition 2a.-2

Individuals/Observational Units/Cases. A person or thing to which the variables are "attached" or assigned; they must be nouns.

Examples: Students; dogs; cats; cars; etc.

---

[*] Data are already antiseptic enough, even with context. As the journalist Paul Brodeur said, "Statistics are human beings with the tears wiped off."

Definition 2a.-3

Variable. A feature or characteristic of an individual that can change.

Examples: Ages of students; weights of dogs; colors of cats; MPG of cars; etc. A *constant*, unlike a variable, cannot change. For instance, the weight of Jenna's cat, named Chairman Meow, at the instant of measurement, is a single non-varying value.

Definition 2a.-4

Quantitative (or Measurement) variable. A variable that measures.

Examples: Ages of students; weights of dogs; lengths of cats; MPG of cars; etc.

Definition 2a.-5

Qualitative (or Categorical) variable. A variable that describes some sort of category.

Examples: Races of students; breeds of dogs; colors of cats; types of cars; etc.

Definition 2a.-6

Binary variable. A qualitative variable assigned to two categories.

Examples: Sexes of students, dogs, or cats; types of cars (big or small); side a coin lands on (heads or tails); etc.

It might be helpful to analogize the typology of variables with the videogame *The Sims*. The objective of the game centers on the effective maintenance of a simulation of human interactions; at its start, you are called upon to create a virtual human being by utilizing a number of digital building blocks. For instance, you can choose your Sims' body type (categorical variable), gender (binary variable), and a whole host of measurable characteristics such as "niceness" and "playfulness" (quantitative variables).

Definition 2a.-7

Univariate data set. The collection of data of a single variable from individuals.

Example: Heights of students in a classroom.

Definition 2a.-8

Bivariate data set. The collection of data of two variables from individuals.

Example: The heights and weights of students in a classroom.

Definition 2a.-9

Sample. A set of data that has been collected.

Examples: Thirty randomly selected students at Madison High School; voters surveyed in an upcoming election; etc. The most commonly used symbol to denote the size of a sample is n (although scholarly articles also use N). So, if our sample size is thirty, we would express that as $n = 30$. A random sample gives us the best chance of gathering representative data.

Sometimes sampling only one person—called an "N of 1" design—is all that's possible. Nonetheless, when collecting data one should always strive to capture many individuals rather than few, as much as resources reasonably allow. As far back as 1698, Charles Davenant, an English economist, argued that conclusions "must not [be] argue[d] from single instances, but from a thorough view of many particulars."

Definition 2a.-10

Population. The entire set from which samples can be drawn.

Examples: The entire student body at Madison High School; all potential voters in an upcoming election; etc.

Definition 2a.-11

Census. The gathering of data from every individual in the population.

Examples: Collecting data from every enrolled student at a local high school, or from all potential voters in an upcoming election. If censuses were always feasible or possible to conduct, then inferential methods for data analysis would have never been developed: drawing conclusions about a population from a representative sample would have been beside the point. But gathering data from every member of some large population is rarely easy. Suppose, for instance, that we wished to know the proportion of fish in a large lake that were salmon; how realistic would a conducting a census here be? What would be have do—drain the lake? Or suppose instead that we wanted to test the durability of plastic fruit-drink bottles. Is it productive or cost-effective to damage or destroy *every* bottle in the fruit-drink factory to achieve our aim?

Example 2a.-1

Is the U.S. decennial Census a *true* census?

Solution. This question is a corollary of the census definition. And the answer is no—the once-every-ten-years U.S. Census misses out on the homeless, the itinerant, the undocumented workers, the mole people (who live underground), ... the list goes on. As Charles Seife explains in his book *Proofiness: How You're Being Fooled by the Numbers*, billions are spent every decade on capturing a small segment of folks in the American population who can't or won't return their Census information cards. Simple random sampling of the population won't do the trick, since the problem lies in the U.S. Constitution. The requirement for a census, rather than just an estimated count of the inhabitants of the "several States," hinges on the word "enumeration," meaning "to count everyone." The insertion of "enumeration" into the Constitution was made for reasons of style, not substance, by the U.S. Constitution Committee on Style. Furthermore, Seife persuasively argues, head counts of millions and millions of people is in practice a process of estimation anyway—opening up the possibility of sampling some rather than attempting to capture all.

Throughout recorded history, censuses usually have had one of two objectives: taxation or conscription for the military. Written examples of censuses date at least as far back as the Old Testament, specifically in the section entitled Numbers (one of the Books of Moses), in which some of the people of Israel were counted—namely, men at least the age of twenty (no women or children)—to prepare for war.

Example 2a.-2

Identify all of the pertinent features of the following data collection processes.

(a) Whether or not an American child is a latchkey kid.
(b) The SAT math and SAT verbal scores an American high school student obtained.

Solution. For part (a), the population is all American children; the individuals are American children. The variable is binary: note the "Whether or not," denoting two options. Because the variable's binary, it's automatically categorical as well. And the data set collected is univariate.

For part (b), the population is all American high school students; the individuals are American high school students. The variables are the math and verbal scores from the SAT. Both variables are quantitative, because they perform a measurement, and the data set itself is bivariate, because two variables are being collected from each individual.

Since conducting a census on our population of interest is usually beyond the scope of possibility, we will need to obtain a sample to learn about the population. Much of the study of statistics is about (1) how to soundly obtain that sample of data, (2) the proper ways to analyze the data from the sample we've obtained, and (3) how to determine which conclusions we've made about our sample can be said to also apply to our population of interest.

Example 2a.-3

Determine whether each of the following data sets is a population or a sample:

    (a) The age of each state governor.
    (b) The speed of every third car passing a police speed trap.
    (c) The cholesterol levels of 20 patients in a very large hospital.
    (d) The number of pets in each U.S. household.
    (e) The number of televisions recorded from a survey of households.

Solution. The answers are (a) population, since the "each" implies "all"; (b) sample; (c) sample; (d) population; and (e) sample.

Example 2a.-4

Identify the population and the sample in each situation.

    (a) A study of 34,054 men and women in Greece looked for a link between smoking and lung cancer.
    (b) A recent survey of business owners found that 33% of Asian owners said they've been rejected for loans.

Solution. For part (a), the population is all men and women (or all men and women in Greece, if you believe there might be something fundamentally different about that group of people as compared to the rest of the world), while the sample is the 34,054 individuals. For part (b), the population is all business owners, and the sample is those business owners who were actually surveyed for the study.

§2b. *The Journey to Inference*

S everal more terms need to be defined before continuing.

Definition 2b.-1

Statistics. The science of data.[*]

---

[*] Recall John Tukey's take on statistics: "Statistics is a *science* in my opinion...."

Definition 2b.-2

Descriptive statistics. Methods of organizing data, whether in graphical or tabular form.

Definition 2b.-3

Inferential statistics. The process of drawing conclusions from data; also, generalizing about the population from which the sample data originated.

The year 1854 marked yet another mass cholera outbreak in the industrialized world, this time in Soho, London, England. As detailed in Stephen Johnson's book *The Ghost Map*, which traces the outbreak day by day, cities with poor sanitation facilities, like London, were hotbeds of disease.

> But the single most important factor driving London's waste-removal crisis was a matter of simple demography: the number of people generating waste had almost tripled in the space of fifty years…. Even with modern civic infrastructure, that kind of explosive growth is difficult to manage. But without infrastructure, two million people suddenly forced to share ninety square miles of space wasn't just a disaster waiting to happen—it was a kind of permanent, rolling disaster, a vast organism destroying itself by laying waste to its habitat…. [I]t was drowning in its own filth.

The dominant disease paradigm at the time was the theory of miasma, or "bad air," which, based on sensory experience, certainly isn't far-fetched: raw sewage smells badly, and—as a known now, thanks to the germ theory of disease (which supplanted miasma theory by the end of the nineteenth century)—presents bacteria with a fertile breeding ground to spread illness.

When the cholera outbreak hit London, John Snow, a prominent physician, was skeptical of miasma as the cause, and instead looked for patterns in the incidence of the disease. Snow's nascent epidemiology study ultimately, with a little luck and a lot of help from the community, traced the source of the outbreak to Broad Street pump, a central water supply avenue for residents. (Employees at a beer brewery, who were located near the pump, instead used water from a different homegrown source and as a result didn't contract cholera.)

Snow wrote, "On proceeding to the spot, I found that nearly all the deaths had taken place within a short distance of the [Broad Street] pump. The result of the inquiry, then, is, that there has been no particular outbreak or prevalence of cholera in this part of London except among the persons who were in the habit of drinking the water of the above-mentioned pump well." Snow constructed a dot map to illustrate the incidence of the disease around the pump. Generally, the farther away from the pump, the less dense the distribution of cholera.

As shown on the next page, Snow's dot map *graphically* organizes a complex data set—hence, this visual illustration is a prime example of descriptive statistics. Snow's ultimately recommendation—remove the pump's handle—shows inferential statistics, or the drawing of conclusions from data, in action. (The source of the outbreak, by the way, was an infant, whose diapers had leaked fecal bacteria into a nearby cesspit.)

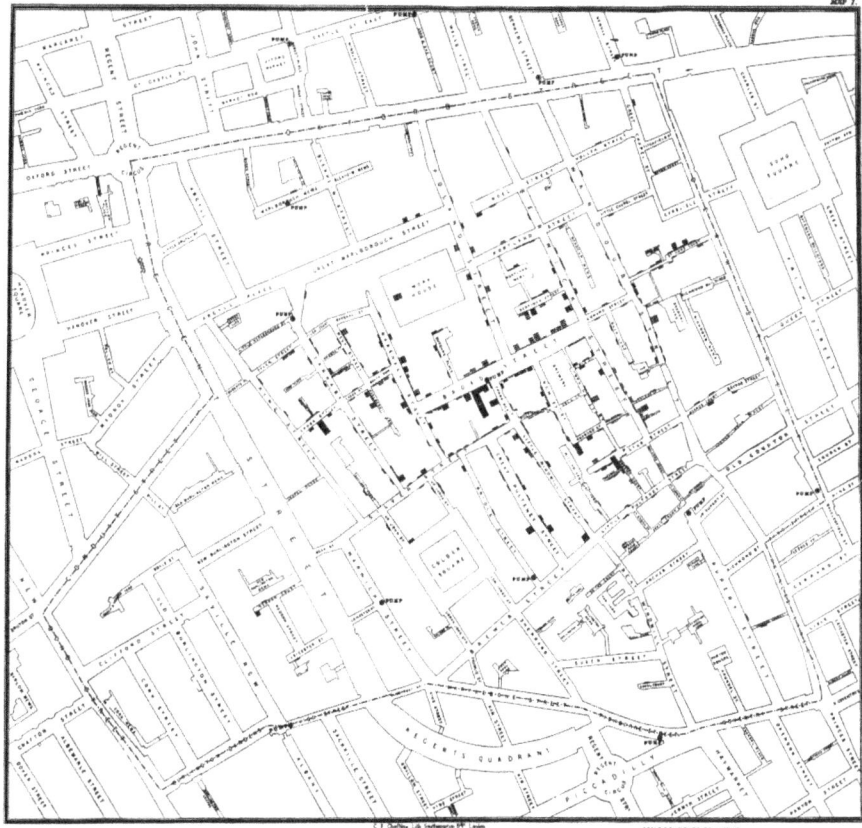

We need to travel back to an earlier plague in London—roughly two hundred years earlier—to find the origins of inferential statistics in the person of John Graunt, a tradesman draper. Graunt wrote perhaps the first text that can be legitimately classified as a modern statistics book, the *Natural and Political Observations Upon the Bills of Mortality*.

The *Observations* took Bills of Mortality data, some of which was either unreliable (e.g., in the listed causes of death) or incomplete (e.g., children not members of the state church were excluded), and, with a number of systematic calculations, produced a count estimating the total population of London at the time: 384,000. In addition, Graunt tracked population data over time, including through deaths by plague and increased births post-plague coupled with immigration. Gavin Kennedy notes that Graunt's two most statistically brilliant innovations were his use of *representative samples* to generalize to learning about a population and his construction of the first life table (which, in more advanced forms, was used by life insurance companies to price policies profitably).[*]

After Graunt's groundbreaking work, it was only a matter of time before his methods of ana-

---

[*] Edmund Halley, of Halley's Comet fame, extended Graunt's life tables by examining data not from England (as Graunt did) but Breslau, Germany, where records were meticulously kept, laying the path for the modern insurance industry; Halley assembled mortality data, showing risks of death within each age group, making actuarial calculation possible despite Halley ignoring Pascal's previous work on probability.

lyzing data would be taken up and improved by others. Take Sir William Petty, for instance. A contemporary and friend of Graunt's, Petty, a medical doctor, dabbled in a sort of proto-epidemiology, asking deep questions of data such as, "Whether of 1000 patients to the best physicians, aged of any decade, there do not die as many as out of the inhabitants of places where there dwell no physicians." If Petty had actually sought out the survival rates, he may have systematically provided the impetus to overhaul the medical profession centuries before William Osler's reformations, but Petty was more interested in describing the statistical methods and prospective data necessary for inferences rather than actually carrying out the cumbersome and messy data collection. (He did urge the English government to create a central statistical office.) Petty also dabbled in economics, and made a number of contributions to the dismal science.

§2c. *Simple Graphs of Categorical Data Sets*

We need to be able to quickly distinguish between counts, proportions and percentages. You probably already have some level of comfort with these ideas, so this should be review. One item of note, though. When talking about proportions in statistics, we almost always mean the decimal, not the fraction, form. Round your decimals to two or three digits. (Don't worry about significant digits.)

Example 2c.-1

A researcher in Pennsylvania wants to determine if there is a connection between blood type and hospital patients—in other words, Do more people of one blood type end up in the hospital than other blood types? So the researcher programs a hospital computer to capture a random sample of hospital patients' blood types. The following sample data is printed out on the hospital computer screen.

A A A A A A A A A A A A A A A A A A A A A A A A A A A A A A A A A A A A A A A A A A A A A A A A A A A A A A A

B B B B B B B B B B B B B B B B B B B B B B B B B B B B B B B B B B B B B B B B B B B B B B B B

AB AB AB AB AB AB AB AB AB AB AB AB AB AB AB AB AB AB AB AB AB AB AB

O O O O O O O O O O O

Here are the data again from the printout above, this time in tabular form:

A	B	AB	O
57	48	23	11

(a) Who/what are the individuals in this study?

(b) What is the variable being studied, and what *type* of variable is it?

(c) We wish to make a summary of the data. Making summaries of data is performing what kind of statistics: descriptive or inferential? Explain.

(d) Create a summary of the counts (called the "frequencies"), the proportions, and the percentages of the data.

(e) The researchers conclude that since blood type A has the highest frequency, people with a blood type of A are more likely to get injured or sick, landing them in the hospital. What's wrong with this conclusion?

(f) A *bar graph* is used to display a distribution of categorical variables. Create a bar graph of the proportions of the blood types.

Solution. The answers are as follows.

(a) Patients.

(b) Blood type; categorical.

(c) Descriptive statistics, since we are constructing visual summaries of data.

(d) Your tabular summary should look like this:

Blood Type	Frequency	Proportion	Percentage
A	57	0.41	41%
B	48	0.35	35%
AB	23	0.17	17%
O	11	0.08	8%
Total	139	1.00	100%

Note that the rounded proportions or percentages might not always quite sum to 1.00 or 100%. This is called *roundoff error*.

(e) A bit less than half of people in the population have blood type A overall, so on average we'd expect a relatively large proportion of blood type A patients.

(f) The bar graph is shown below.

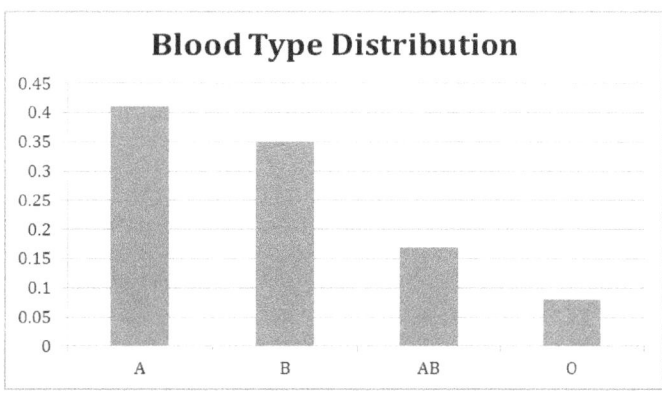

71

The *y*-axis scale need not go to 1.00, but rather as high as the biggest category proportionally.

A pie chart is very similar to a bar graph; it too can display the distribution of categories for one qualitative variable. The same set of blood type data can easily be transformed into a pie chart.

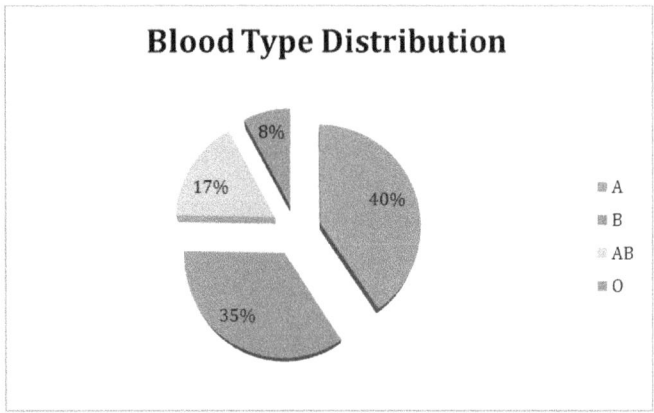

Scottish engineer William Playfair is generally considered to be the creator of the pie chart (among many other statistical graphs, such as line graphs, bar charts, and time-series charts, in the construction of which he used "lineal arithmetic"). But Florence Nightingale's "coxcomb" graph, essentially an exploding pie chart detailing the causes of mortality each month during the Crimean war, is perhaps the most famous use of Playfair's original design. Similar to the cholera outbreak, poor sanitation in barracks and hospitals was causing the bulk of the deaths in the war (as shown below in the thickest wedges protruding from the centers of the graphs below).

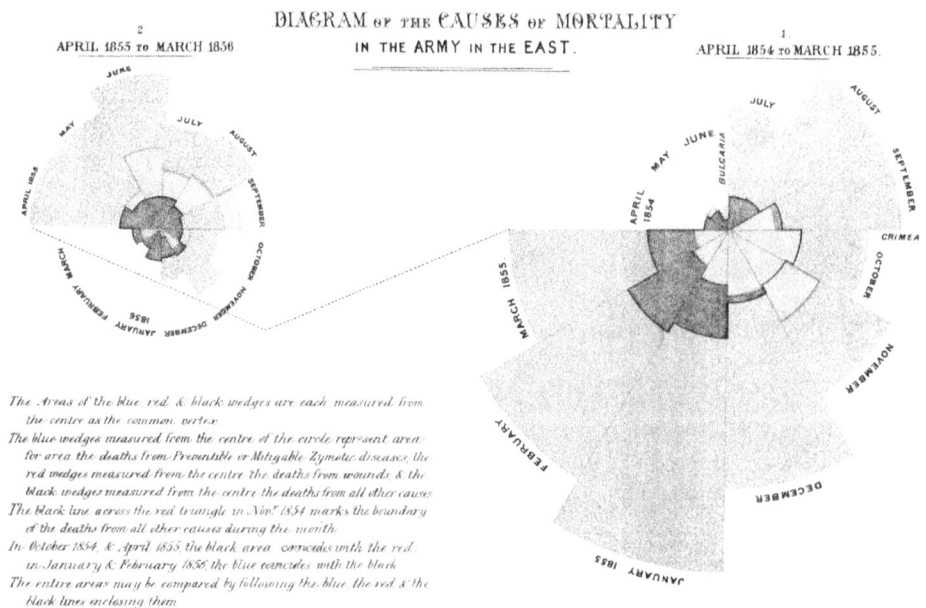

Nightingale presented the mortality data in graphical, instead of tabular, form to maintain Queen Victoria's interest as well as to press for the need for quick action to save lives.

§2d. *Deceptions with Graphs*

Graphs, such as pie and bar charts, may be presented misleadingly. Here is a set of data detailing average new car prices (in U.S. dollars) over a recent five-year period.[*]

Year	2011	2012	2013	2014	2015
Average MSRP	34,500	35,000	37,000	36,000	41,000

Suppose you'd like to present the data in a way that demonstrates that cars are quickly becoming unaffordable to the average consumer. A bar graph supporting this sort of narrative might look like this:

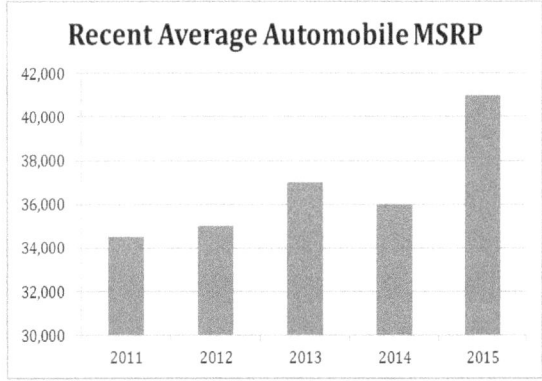

Now, instead, suppose you wish to spin a different narrative: that car prices have held relatively steady. Here's a bar graph supporting that conclusion.

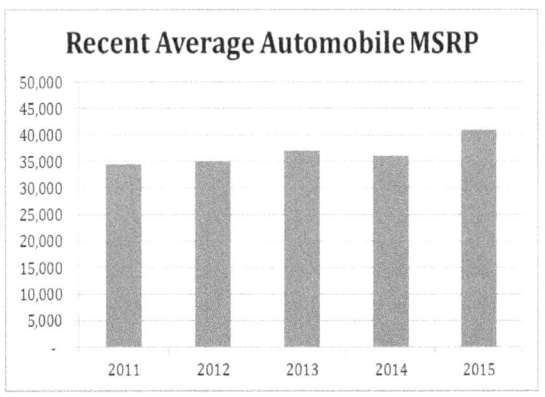

---

[*] Data come from the *WolframAlpha* website, querying the string "new car prices."

Startling, isn't it? These two bar graphs tell two different stories about the same data set by simply changing the scale of the *y*-axis, but neither graph is the "correct" (or the "incorrect") graph. The classic text *How to Lie with Statistics* has many more examples of these types of misleading bar graphs, in addition to displaying misleading pictographs, which are drawings of objects used to graphically depict relationships among data sets. Pictographs are usually presented as drawings in three dimensions, and because objects in three dimensions have volumes, serious distortions may arise.

§2e. *Contingency Tables*

One of the most common, and effective, ways to display qualitative (categorical) data in tabular form is to use a *two-way table*, also called a *contingency table*. Because the format of a contingency table is so user-friendly, this sort of visual display is widely used in different disciplines. First, two definitions are in order, however.

Definition 2e.-1

Explanatory variable. Also called the *independent variable*, the explanatory variable "explains" the other variable; usually the explanatory variable comes first in time.

Definition 2e.-2

Response variable. Also called the *dependent variable*, the response variable is "explained" by the other variable; usually the response variable does not come first in time.

Note: To make Examples easier to read, from this point forward, questions and answers will be integrated together within the Examples—but only where necessary.

Example 2e.-1

Suppose that 74 people in the U.S. were asked the following question: Should more tax dollars be spent on welfare? In addition to obtaining the answer to this question—either a "yes" or a "no"—those same 74 people were asked about their political identification: Independent, Democratic, or Republican.

(a) What are the two variables in this study, and what kind of variables (quantitative or qualitative) are they?

The two variables are "politics" and "opinion," and both are qualitative.

(b) Identify the explanatory and response variables in the study.

The explanatory variable is "politics," while the response variable is "opinion," since it's more likely your political leanings inform your beliefs about welfare spending than the other way around.[*]

A contingency table of the survey results is presented below. The total row and total column are called *marginal totals*; the "Total Total" cell is termed the *grand total*.

	Dem.	Ind.	Rep.	Total
Agree	22	12	4	38
Disagree	7	8	21	36
Total	29	20	25	74

(c) What proportion of people in this study are not Democrats?

$$\frac{(20+25)}{74} = \frac{45}{74} = 0.608$$

(d) What proportion of those who disagree are Republican?[†]

$$\frac{21}{36} = 0.583$$

(e) What percentage of those who are either Independent or Republican agree?

$$\frac{(12+4)}{(20+25)} = \frac{16}{45} = 0.356 \ , \text{or } 35.6\%$$

(f) Republicans are how many times more likely to disagree than are Democrats?

$$\frac{\frac{21}{36}}{\frac{7}{36}} = \frac{21}{7} = 3$$

(g) What proportion of people are either Democrats or those who agree (or both)?

---

[*] Although I suppose that there's an ready-made argument for causation in the other direction, but let's hold off on a discussion about causal direction for now.

[†] The group of possible people who are Republicans in this question has shrunk as compared to part (c): from everyone in the survey to just those who disagree.

$$\frac{(29+38-22)}{74} = \frac{45}{74} = 0.608$$

Before proceeding, we need to call attention to two things:

(1) In part (f), the "how many times more" calculation is called a *relative risk*; relative risk is usually most useful when examining disease incidence and other health outcomes.

(2) In part (g), there are 22 people who are both Democratic and agree (called the "overlap"), and they are being double-counted, so they must be subtracted out; treated as a probability problem, the *addition rule* accounts for the overlap (a rule we'll fully explore in a later §).

The cleanest, clearest way to present data from a two-way table graphically is with a *segmented bar graph*. Unlike a "regular" bar graph, which displays either frequencies or relative frequencies (or percentages) by category, a segmented bar graph splits each category up by relative proportions given by the other variable. In order to construct a segmented bar graph, however, a *conditional distribution* must be constructed.

Definition 2e.-3

Conditional distribution. A distribution of the relative frequencies of one value of a variable for all possible values of another.

When constructing a segmented bar graph, the proportions of the values of the variable must be "stacked" so that each bar sums to 1.000 (or 100%). Several examples will make this clear.

Example 2e.-2

Reconsider the welfare question survey data from the previous example.

(a) Create a conditional distribution of opinion for each political inclination. Then, construct a segmented bar graph of the conditional distribution.

	Dem.	Ind.	Rep.
Agree	0.759	0.600	0.160
Disagree	0.241	0.400	0.840
Total	1.000	1.000	1.000

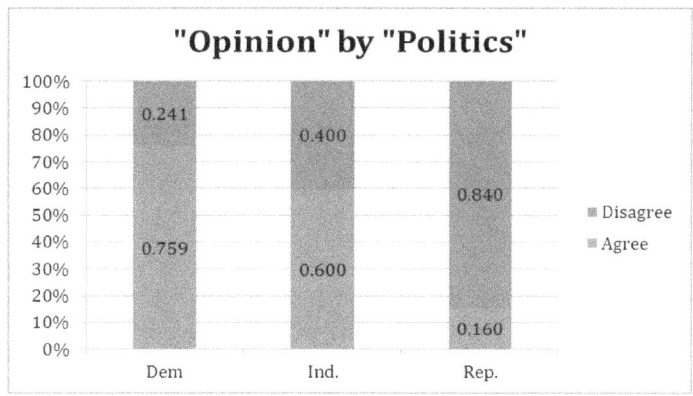

(b) Create a conditional distribution of political inclination for each opinion. Then, construct a segmented bar graph of the conditional distribution.

	Dem.	Ind.	Rep.	Total
Agree	0.579	0.316	0.105	**1.000**
Disagree	0.194	0.222	0.583	**1.000**

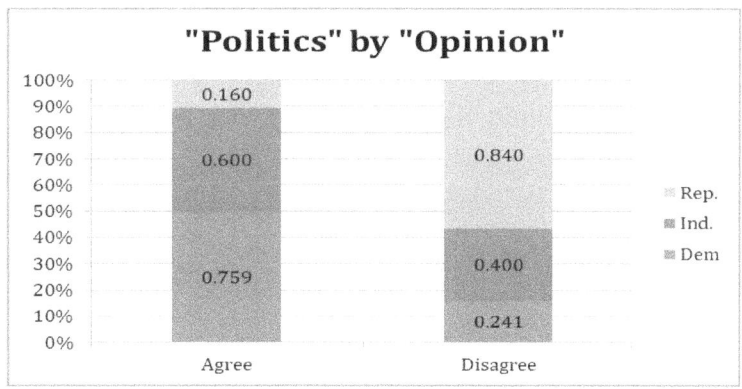

(c) What inferences can we draw from the data?

The more conservative you are, the less likely you are to believe that more tax dollars should be spent on welfare.

Definition 2e.-4

Independence. A property of a two-way table in which the conditional distributions are identical—i.e., the breakdown proportions amongst the categories are the same for all values of the two variables.

When variables are independent, there is no association between them. Because the two conditional distributions of the welfare question survey data are not identical row-wise or column-wise, the variables of political inclination and opinion are not independent.

Example 2e.-3

Suppose we asked 21 people about Misery Loves Company (MLC) towels—specifically, whether they like the towels or not. The collected survey data are organized into the following contingency table:

	Like	Dislike	Total
Male	2	12	14
Female	1	6	7
Total	3	18	21

(a) Create a conditional distribution of gender for each preference.

	Like	Dislike
Male	0.667	0.667
Female	0.333	0.333
Total	1.000	1.000

(b) Create a conditional distribution of preference for each gender.

	Like	Dislike	Total
Male	0.143	0.857	1.000
Female	0.143	0.857	1.000

(c) Based on your conditional distributions, are the two variables independent? Explain.

Yes, the variables are independent, because the conditional distributions are identical by proportion for each variable.

Realize that, mathematically, if one conditional distribution exhibits independence, there's no need to construct a conditional distribution in the other direction—it will also show independence as well, as both conditional distributions in the example above do. What follows is a proof of this property for a 2x2 contingency table, which can also be called a *fourfold table*.

Consider the following fourfold table:

	1	2	Total
1	a	b	$a + b$
2	c	d	$c + d$
Total	$a + c$	$b + d$	$a + b + c + d$

Suppose that $\dfrac{a}{a+c} = \dfrac{a+b}{a+b+c+d}$. Then, using cross-multiplication, we see that

$$a(a+b+c+d) = (a+b)(a+c)$$

$$a^2 + ab + ac + ad = a^2 + ac + ab + bc$$

$$ad = bc$$

Thus, $a = \dfrac{bc}{d}$. We wish to demonstrate that $\dfrac{a}{a+b} = \dfrac{a+c}{a+b+c+d}$ given that $a = \dfrac{bc}{d}$. Substituting $\dfrac{bc}{d}$ for a on the left side of the equation gives us $\dfrac{c}{c+d}$; that same quotient is obtained on the right side. Note that this result could be generalized for ever-bigger *n*-by-*m* two-way tables simply by synthesizing more unique variables. Therefore, we have shown that if a conditional distribution in one direction exhibits independence, it does so in the other direction as well.

In addition to recognizing when data in a contingency table exhibits independence, and what the independence means in context, you should also be able to construct a two-way table that is independent *given* a set of marginal totals.

Example 2e.-4

Based on the marginal totals shown below, fill in the blank cells so that the variables are independent.

	Like	Dislike	Total
Male			22
Female			11
Total	24	9	33

Solution. Remember: the proportions by value for each variable are identical, both row-wise and column-wise. So, for instance, the proportion of males who "Dislike" is equivalent to

$$\frac{22}{33} = \frac{2}{3}$$

Thus, to determine the number of males in the dislike category, we need two-thirds of 9, which is 6. With that particular cell populated, the rest of the cells can be filled in using simple arithmetic:

	Like	Dislike	Total
Male	16	6	22
Female	8	3	11
Total	24	9	33

Given this set of marginal totals, this is the *only* solution.

When working with contingency tables, perhaps the most counterintuitive concept is called *Simpson's paradox.*

Definition 2e.-5

Simpson's paradox. Describes a trend that has an opposite direction when comparing the parts with the whole.

Consider the overall numbers of cars and trucks sold by two dealerships this past month, as shown in the two-way table below.

OVERALL	Dealer 1	Dealer 2
New Vehicles Sold	66	26
Used Vehicles Sold	64	29
Total	130	55
% New Sold	**51%**	**47%**

It's clear that, when considering the percentage of new cars sold, Dealer 1 did better—slightly—than Dealer 2.

But let's now see those same numbers, disaggregated by type of vehicle: car or truck (that's an exhaustive list). Take a look at cars first:

CARS ONLY	Dealer 1	Dealer 2
New Vehicles Sold	6	15
Used Vehicles Sold	14	25
Total	20	40
% New Sold	**30%**	**38%**

Even though Dealer 1 sold a higher percentage of new cars overall, Dealer 2 sold the higher percentage of new cars. You might suspect that Dealer 1 *must* sell the higher percentage of new trucks, since overall Dealer 1 sold the higher percentage of vehicles, and there are only two types of vehicles we're considering: cars and trucks. Let's find out.

TRUCKS ONLY	Dealer 1	Dealer 2
New Vehicles Sold	60	11
Used Vehicles Sold	50	4
Total	110	15
% New Sold	**55%**	**73%**

Even though Dealer 1 sold the higher percentage of new vehicles overall, Dealer 2 sold *both* the higher percentage of new cars *and* new trucks!

There are several examples of the Simpson's paradox phenomenon appearing in real life. Bill James, the sports statistician, noticed that the paradox occasionally occurs when comparing baseball players' batting averages from the first and second halves of a season. And the University of California, Berkley, was sued for gender discrimination in graduate school admissions— except that when the admissions data were disaggregated by department, the bias vanished, and so did the lawsuit.

§2f. *The Rise of Technology in Statistics*

After Girolamo Cardano, Blaise Pascal, and Pierre de Fermat made use of gambling games to affix a great deal of mathematical rigor to the notions of probability more than three hundred years ago, others turned their attention to using data to predict mortality.

Demographer John Graunt, along with the astronomer Edmund Haley, made the first actuarial life tables; and, later, insurance companies—such as Lloyd's of London, the oldest insurance company in the world[*]—would lean on these vast aggregations of data to price insurance products profitably.

The problem in the industry of pricing risk, however, was that massive amounts of data needed to be collected and sorted by hand. Women were usually employed as "computers"—the original meaning of the term was not an electronic but a human tabulator of information—especially by the time of World War II, when many men were off fighting.

Besides the abacus, automated analog calculating machines and computers were first conceptualized by Blaise Pascal. By the early 1800s, Charles Babbage sketched out the Difference Engine, a mechanical computer which was, in theory, capable of solving algebraic equations. More impressive still was his Analytical Engine, a programmable device utilizing a punched card system lifted from the state-of-the-art Jacquard loom sewing machine. Despite receiving funding from the British government, Babbage's contraptions were never fully built,[†] although his epistolary collaborator Ada Lovelace, daughter of Lord Byron, in her *Notes* addendum to an article

---

[*] Initially centered on the shipping industry, this famous insurance company was started by Edward Lloyd in 1687 as a small coffee house (coffee houses were hotbeds of rumor and gossip about trade at the time). With the expansion of trade came the expansion of the insurance industry.

[†] Until recently, where they are now on display in the London Science Museum.

about Babbage's Analytical Engine, rigorously worked through precisely how to program the machine, including explicating the first programing loop. Lovelace viewed programming as a "poetical science," a nexus between science and art.

Although the U.S. Census in the late nineteenth century made use of some automated calculating devices, the first true ancestor to the personal computer we make use of today—in that the machine was electronic, digital, and programmable—was the ENIAC, constructed in the late 1940s. The theoretical work of mathematicians such as John von Neumann and especially Alan Turning paved the way for that electronic monstrosity of searing-hot vacuum tubes and wires.

The vacuum tube was terribly inefficient, so it didn't take long for scientists to look for a more efficient alternative. AT&T's Bell Laboratories, that hotbed of innovation, birthed the transistor, but not without controversy. As detailed in Jon Gertner's *The Idea Factory*, the definitive institutional history of Bell Labs, physicists Walter Brattain and John Bardeen, under the loose direction of William Shockley, made the breakthrough; Shockley, upset at being left out of the day-to-day process and envious of his underlings reaping all the credit, secretly improved the design, ultimately leading all three physicists to be awarded the Nobel for the invention of the transistor.

The transistor, by using semiconductor materials, was a stunning leap over the vacuum tube. But there was still room for improvement: namely, the microchip, which could fit many transistors onto one small circuit. Jack Kilby of Texas Instruments stitched together the first one, although it was "ugly" and "impractical" compared to the (slightly) later, cleaner printed circuit of Robert Noyce's at Fairchild Semiconductor, according to *The Innovators*, Walter Isaacson's history of the people behind the computer's creation.

Eventually, Kilby was recruited by TI's brass to construct a handheld calculator that "can do the same tasks as the thousand-dollar clunkers that sit on office desks. Make it efficient enough to run on batteries, small enough to put into a shirt pocket, and cheap enough to buy on impulse...." Although the calculator initially could only employ the four basic arithmetic functions, "[a] new market had been created for a device people had not known they needed," and Moore's Law,[*] which anticipated ever-smaller and more powerful computers, cleared the path for multifunction graphing calculators.

In addition to its faculty at traditional forms of math, data sets can be manipulated and analyzed relatively easily with graphing calculators. Casio and Hewlett Packard developed the first graphing calculators in the early 1980s. The oldest of Texas Instruments' graphing calculators is the TI-81.

The integrated circuit, along with some effective statistical software (and the graphing calculator), has allowed statistics to flourish from the second half of the twentieth century onward. Data can be analyzed seamlessly with computers. Interestingly, decades before electronic computers were invented, many statistical and probabilistic methods were developed to avoid or shortcut massive amounts of manual calculations; today, those calculations are no longer necessary (although statistics in practice has had a hard time catching up with the technology). Current statis-

---

[*] Coined by the low-key founder of Intel named Gordon Moore, who in a 1965 paper noted, "The complexity for minimum component costs has increased at a rate of roughly a factor of two per year. Certainly over the short term this rate can be expected to continue...."

tics computer software, such as SPSS, PSPP,* Minitab, Fathom, SAS, R, and even Excel, are all powerful statistical programs designed to analyze masses of data quickly and efficiently using sophisticated methods, such as predictive analytics. Data mining, a contemporary process that assists statisticians in discovering patterns in large data sets—and is used in business, science, engineering, sports, and education—would be almost impossible without these computer programs. (The Victorians, who obsessively measured anything and everything they could get their hands on in the noble furtherance of "science"—skewed then, such as it was, with now long-discredited ideologies—would have had a field day with our modern statistical equipment and methods. Perhaps we're currently in the midst of a similar sort of measurement fever, complete with our own set of latent ideological agendas, driving the pursuit of science; after all, as Stephen Jay Gould notes, science is not conducted in a vacuum, and "quantitative data are as subject to cultural constraint as any other aspect of science, [thus] they have no special claim upon final truth.")

The statistician John Tukey was involved with the development of modern computers at AT&T's Bell Labs in the 1960s. He coined the term "software," and was a proponent of using computers and technology to make data analysis faster and easier—although ironically he rarely made use of computers himself.[†] Tukey wrote a statistics book in 1977 entitled *Exploratory Data Analysis* in which he developed new types of visual displays, quickly able to be constructed on computer screens, to concisely summarize data sets' main characteristics, such as boxplots and stemplots (which we will make use of in this primer).

To make data analysis easier, sometimes you'll want to enter data sets into a graphing calculator to analyze; the TI-83/84 series is the many-years-running gold standard of calculators. Data sets that are entered into this calculator series are referred to as "lists."

So let's make use of what Jack Kilby of Texas Instruments wrought. Consider a data set of ATM withdrawals (in dollars):

35	10	30	25	75	10	30	20	20	10	40
50	40	30	50	60	25	40	10	60	20	90
40	25	20	10	20	25	30	50	60	20	

To store these numbers in the calculator,[‡] go to your Home screen (the screen in which your calculator starts when you turn it on; to get there, type 2ⁿᵈ and MODE). Then open a brace by typ-

---

* A free open-source program similar SPSS, hence the name. One may be reminded of *2001*'s evil computer HAL, an acronym, in each character, one letter away from IBM—the company which, coincidently, produces SPSS.

† In the early 1960s, Tukey presciently said, "Today, software and hardware together provide far more powerful factories than most Statisticians realize, factories that many of today's most able young people find exciting and worth learning about on their own. Their interest can help us greatly, if statistics starts to make much more nearly adequate use of the computer."

‡ I make use of the approach to data entry detailed in the *Workshop Statistics* (Rossman et al.) family of textbooks;

ing 2nd and $($. You should see a { on your screen. Now, type in each number above, using a comma to separate the numbers. Next, close your list by typing 2nd and $)$. You should see a } appear on your screen.

We need to store these numbers. Find the $\boxed{\text{STO→}}$ button on the calculator (it's on the bottom left of your keypad) and press it. This function tells your calculator to store the list of data. But you need to label the data with a name, so turn on the Alpha pad (which allows you to type letters instead of numbers) by typing 2nd and $\boxed{\text{ALPHA}}$. We'll call this list ATM, so type in those three letters. (List names can be a maximum of five characters in length.) Before pressing the $\boxed{\text{ENTER}}$ button, make sure that this is what's showing your screen:

```
{35,10,30,25,75,
10,30,20,20,10,4
0,50,40,30,50,60
,25,40,10,60,20,
90,40,25,20,10,2
0,25,30,50,60,20
}→ATM
```

After hitting $\boxed{\text{ENTER}}$, the ATM list's contents should repeat right below the word ATM on the calculator screen. Use the left and right arrows to scroll through the numbers of the list, and make sure that they're all there.

Yet viewing the data this way is really inconvenient. To see the data in the Stat Editor, a Microsoft Excel-type format—in other words, to see the numbers organized vertically in cells—do the following:

1. Hit the $\boxed{\text{STAT}}$ key, bringing up a statistics menu.
2. Select option 5, SetUpEditior, by scrolling down to it and hitting $\boxed{\text{ENTER}}$; the word SetUpEditor should appear on your Home screen.
3. The calculator is expecting a list name. In order to pull up the list name ATM, type 2nd and $\boxed{\text{STAT}}$ and scroll down the list names until finding ATM. Then select it by hitting $\boxed{\text{ENTER}}$.
4. On your Home screen should be the following: SetUpEditor $_L$ATM. You are instructing the calculator to set up the list editor to show the ATM list. Press $\boxed{\text{ENTER}}$.
5. Now hit the $\boxed{\text{STAT}}$ key again and select the first option, Edit... You are now in the Stat Editor. Scroll up and down using the cursor keys to see the contents on the ATM list.
6. In the Stat Editor, you can easily edit numbers in a list, delete them by using the $\boxed{\text{DEL}}$ key, or add data points. Right now, though, don't change any of the ATM data points.
7. You can return to the Home screen at any time by typing 2nd and $\boxed{\text{MODE}}$.

---

in my view, although idiosyncratic, *Workshop* offer the best approach.

Sometimes we need to sort data in lists in order to allow for proper analyses. We have two choices: we can either sort a list in ascending order (smallest numbers to largest numbers) or descending order (largest numbers to smallest numbers). Let's sort our list in ascending order.

1. Go to STAT and select option 2, SortA(.
2. Select the ATM list from the list menu. On your screen should be SortA($_L$ATM
3. Close the parentheses by typing) and hit ENTER.
4. Confirm that the list is in order by viewing it in the Stat Editor.

Next, entering data from a table and make calculations based off of the data in that table bears close examination. Consider the following table, which contains the grades of a single student, by assignment, for an entire quarter in high school.

assignment	score	max points	assignment	score	max points
Free Write	15	20	Test on Ch. 1	79	100
"Runner" Quiz	23	25	"Gatsby" Report	85	100
Report on "K.B."	41	50	Fitzgerald Essay	36	50
Class Demo	7	10	Quiz on Gatsby	18	25
Group Work	32	43	Test on Ch. 2	67	100

1. Create two lists, one called SCORE and the other called MAX, that contain the student's scores and maximum points shown in the table, respectively. Recall that list names can be a maximum of five characters in length. (Besides the method mentioned earlier, in which you'd enter the lists using the {}'s and store it into a list name, you could set up your editor to show the blank lists SCORE and MAX and then type in the numbers in the Stat Editor directly.)
2. A common mistake is to not realize that the score and maximum points data is broken up into two columns in the table above. So, before you continue, make sure you have actually typed all ten numbers in for the SCORE list, and all ten numbers in for the MAX list.
3. Let's set up the Stat Editor to show the SCORE and MAX lists *side by side*. To do this, type in SetUpEditor $_L$SCORE,$_L$MAX and press ENTER.
4. Pull up the Stat Editor. The Stat Editor should now show both lists next to each other.

Let's create a third list that calculates the average of *each assignment*. We will construct this third list based off of the other two lists; we will need to *divide* each assignment's score by the maximum number of points for that assignment to get a list of the student's individual assignment averages. To do this,

1. Open a set of parentheses, and pull up $_L$SCORE, the SCORE list.
2. Press the division key and pull up $_L$MAX, the MAX list.

3. Close the parentheses. To find a percentage you multiply the quotient by 100, so press the multiplication key and then type 100.
4. Store this calculation into a list called GRADE.
5. Before pressing ENTER, make sure that this is what you've typed on-screen: (_LSCORE/_LMAX)*100→GRADE
6. Hit ENTER and set up your Stat Editor to show the SCORE, MAX, and the new GRADE lists next to each other by typing SetUpEditor _LSCORE,_LMAX,_LGRADE
7. Pull up the Stat Editor. The Stat Editor should now show all three lists next to each other. The GRADE list on the right shows the student's average on each individual assignment.

And, finally, suppose you wish to view the three lists in order of the highest to lowest assignment averages. When sorting more than one list, place the list you wish to sort by after the open parenthesis *first*; also, each of the related lists must be sorted in the command also—otherwise one list's order changes while the others do not, corrupting the lists.[*] Thus, to sort all three lists in order from highest to lowest assignment average, type SortD(_LGRADE,_LSCORE,_LMAX) and press ENTER. In your Stat Editor, the lists should now be arranged in descending order by highest to lowest assignment average.

§2g. *Simple Graphical Displays of Distributions*

Our ultimate objective with data isn't merely to collect it; instead, we seek to understand what it tells us. Toward that end, as Mario Triola in the text *Elementary Statistics* suggests, when we examine data sets, it is critical to identify the following characteristics:

- *Center.* The "middle" of the data set.
- *Variation.* (or *spread* or *variability*) The spread of the data set.
- *Distribution.* The nature or shape of the distribution (symmetrical, bell-shaped, uniform, skewed).
- *Outliers.* Observations that are set apart from the rest of the data.
- *Time.* Characteristics of a data set that alter over time; this doesn't apply to every data set.

Triola groups these bulleted points into an acronym: CVDOT. But there are other characteristics related to those enumerated above we need to be aware of, such as peaks (high points), clusters (distinct pockets of data), and gaps (perhaps indicating data from different populations or homogenous subsets of a single population); unimodal (one mode—one peak at the mode), bimodal (essentially, two peaks), or multimodal (more than two peaks); and granularity (data values that

---

[*] And, just like with the actions you take in your life, there's no "undo" button on this series of graphing calculators. So always be careful before hitting ENTER.

only appear at constant intervals). But the first four of the five Triola characteristics (CVDO from CVDOT) are most important to begin our analyses with.

Let's try our hand at identifying the characteristics of distributions of quantitative data sets. But we can't do that just yet: it is not enough simply to look at a list of the values in a data set; we need to have some way to display the distribution of data values. *Dotplots* are one such way.

Definition 2g.-1

Dotplot. A simple visual display of a quantitative data set, consisting of a dot representing each observation plotted above a horizontal scale.

Example 2g.-1

Reconsider the ATM withdrawals data set. Construct a dotplot of the data.

Solution.

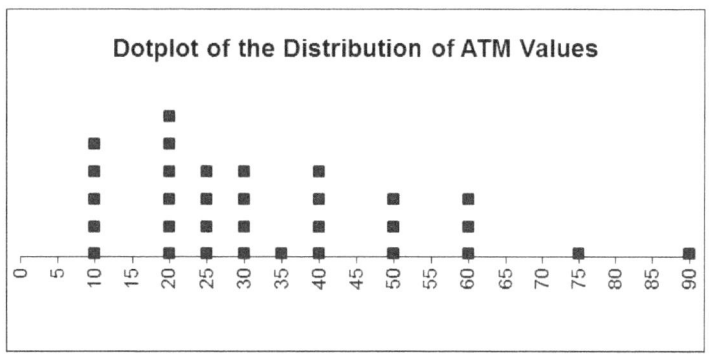

We're almost ready to analyze the characteristics of the ATM data set, but first we need to sketch out what these characteristics might look like in any distribution of data.

Example 2g.-2

Construct dotplots that satisfy the listed criteria.

(a) The center is at about 50, spread is from 10 to 70, and there are two clusters of data.
(b) The center is at about 30, spread is from 20 to 90, and there is an outlier on the high end.
(c) The center is at about 30, spread is from 0 to 60, and the data is symmetrical and mound shaped (bell shaped), with a single peak at 30 (unimodal).
(d) The center is at about 50, spread is from 10 to 90, and the data is skewed to the right.

Solutions.

(a) Think of the "center" of a data set, informally, as the place in which an imaginary fulcrum could be placed that would "balance" the entire dotplot—if every dot "weighed" an identical amount. The "spread," or "variability" simply reports the lowest and the highest data points. And clusters are distinct pockets of data.

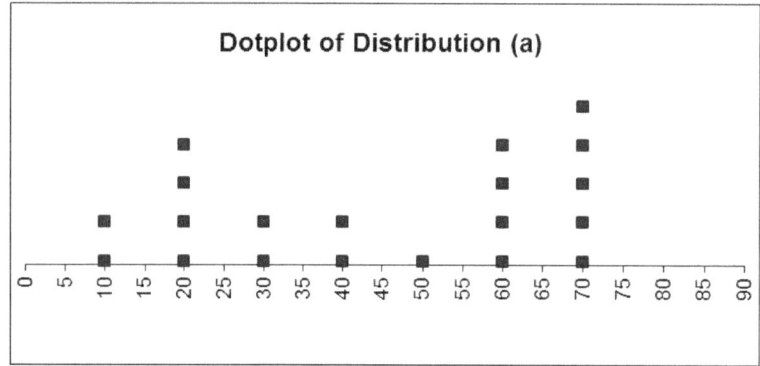

(b) An outlier is a data point separated from all of the rest, either at the high end or the low end of the scale.

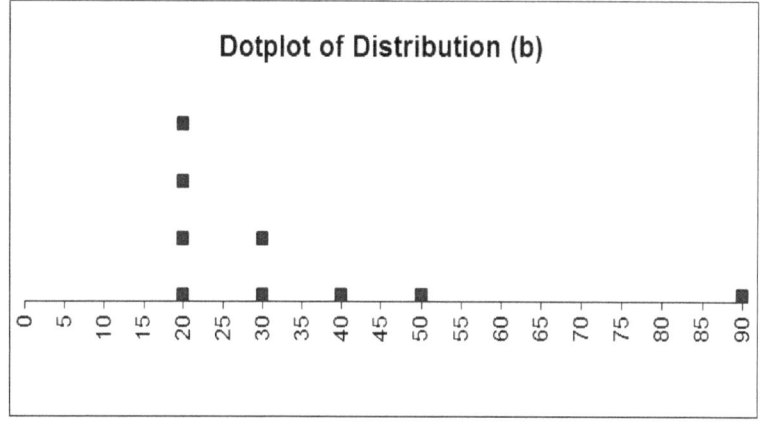

(c) Recall the classic bell curve shape when considering mound-shaped, symmetrical data sets.

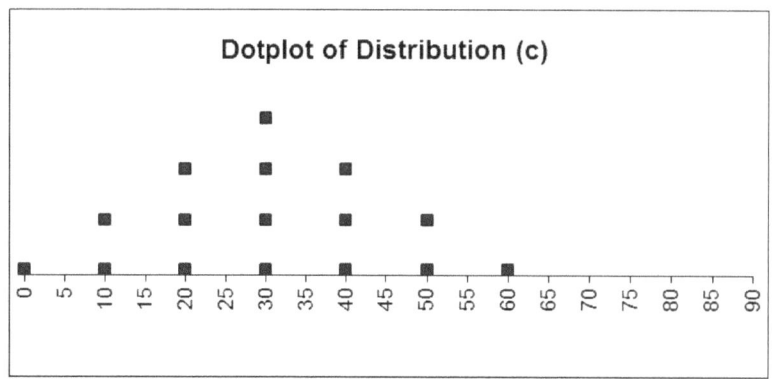

(d) A right-skewed distribution means that there's a "tail" of data on the right side of the plot, with the majority of the data on the left side. Sometimes, distributions skewed to the right are termed "positively skewed." Unsurprisingly, left-skewed, or negatively skewed, data sets are mirror images of right-skewed data.

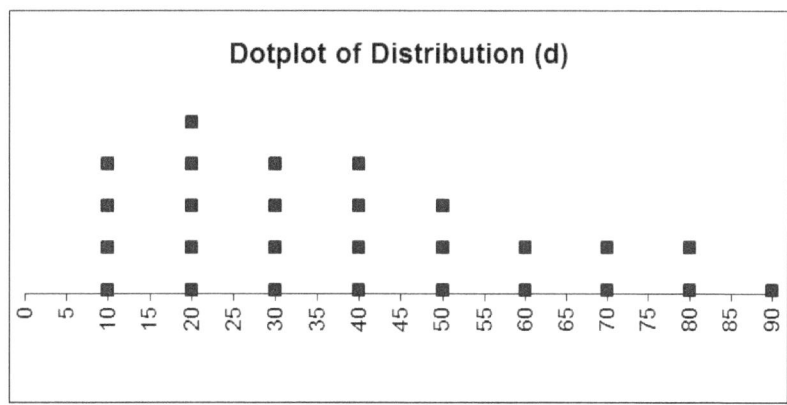

Example 2g.-3

Look again at the ATM dotplot,[*] and analyze the key characteristics of the ATM distribution.

Solution. The center is at about 35, the spread is from 10 to 90, and the data set is skewed to the right (positively skewed). Perhaps 90 can be considered an outlier, although it's unclear right now just how far apart an observation needs to be from the remainder of the data to be classified as such.

    There are many more ways to summarize data visually. The best data displays reveal heretofore hidden or non-obvious characteristics of a data set. We are not simply making new types of graphs for their own sake; rather, we are trying to tease details out of data that will help us to more fully understand a data set.

    In addition to dotplots and bar graphs, other common data displays are histograms, frequency polygons, ogives (pronounced "oh-jives"), Pareto charts, pie charts, time-series plots (data examined through time), stemplots, and scatterplots (also called scatter diagrams). Although all are important and can reveal interesting features of a data set, we won't spill ink about them all here, just several fundamental ones.

---

[*] By the way: Although bar graphs are straightforward to make in Microsoft Excel using the "Charts" feature, constructing dotplots requires a bit of jury rigging. First, fill in a column of data points, starting in cell H3, in ascending order. Then, one column over, input the following "vertical position of dot" formula into cell I3: =IF(H3=H2,I2+0.1,0.01) Next, copy this formula cell straight down, ensuring that formulas are populated in all cells adjacent to the data points. Finally, after highlighting all of the populated data and formulas' cells, set the Chart type to *Scatter* and remove the vertical axis.

Like a dotplot, a *stemplot* (also called a *stem-and-leaf plot*, first made by John Tukey) is a simple graphical display that can give us a quick picture of the shape of a distribution of quantitative data. Stemplots are best made with small data sets.

To make a stemplot, each observation (each case) must be separated into a stem, consisting of all but the final digit, and a leaf, the final digit. Arrange the numbers in order. Finally, make sure to construct a key that contextualizes the data set.

Example 2g.-4

Consider, yet again, the ATM data set. Construct a stemplot of the data.

Solution.

```
0 |
1 | 0 0 0 0 0
2 | 0 0 0 0 0 0 5 5 5 5
3 | 0 0 0 0 5
4 | 0 0 0 0
5 | 0 0 0
6 | 0 0 0          [ 7 | 5 = 75 dollars from the ATM]
7 | 5
8 |
9 | 0
```

If you turn this stemplot 90 degrees counterclockwise, you (sort of) have a dotplot. Thus, all of the characteristics of data sets we identified from dotplots—such as center, skewness, and the like—can be obtained from stemplots as well.

Sometimes, to help read a stemplot better, a *split stemplot* is constructed. The *split stems* of this type of plot divide each stem in half: one stem for leaves 0 to 4, and another stem for leaves 5 to 9. For instance, if we were to make a split stemplot of the ATM data, the "2" stem would appear as follows.

```
2 | 0 0 0 0 0 0
2 | 5 5 5 5
```

We can also compare distributions using a *back-to-back* or a *side-by-side stemplot*, which locates the stems in the middle of the plot with leaves flanking each side, from lowest (closest to the middle) to highest (farthest out).

Example 2g.-5

Consider the annual salaries (in thousands of dollars) of a randomly selected group of teachers in California and Pennsylvania shown below.

California salaries. 50, 53, 56, 59, 59, 60, 63, 63, 69, 70, 75, 75, 75, 76, 77, 77, 77, 77, 83, 85, 85, 87, 89, 94, 94, 94, 95, 102, 104, and 111.

Pennsylvania salaries. 46, 48, 50, 51, 52, 56, 60, 60, 61, 61, 65, 66, 66, 67, 69, 69, 69, 72, 74, 76, 77, 93, 94, and 100.

(a) Create a side-by-side stemplot of the annual salaries. Put California's on the left side of the plot and Pennsylvania's on the right side.

```
               | 4 |  6 8
      9 9 6 3 0 | 5 |  0 1 2 6
        6 3 3 0 | 6 |  0 0 1 1 5 6 6 7 9 9 9
  7 7 7 7 6 5 5 0 | 7 |  2 4 6 7
      9 7 5 5 5 3 | 8 |
        5 5 4 4 | 9 |  3 4
            4 2 |10 |  0          [ 4 | 9 | 3 = $94,000 for CA; $93,000 for PA]
              1 |11 |
```

(b) Which state has a greater number of salaries in the $60,000 range?

Pennsylvania.

(c) What is the center of each data set?

California: approximately $70,000; Pennsylvania: approximately $60,000.

(d) Describe the shapes of both distributions of data.

California is skewed to the right (there are more lower salaries, with a tail of higher salaries). Pennsylvania is approximately symmetric.

(e) Which state *tends* to have higher salaries? How can you tell?

Since there are more leaves at the bottom of California's plot than Pennsylvania's, California tends to have the higher salaries. It's not true in every case, of course: not *every* California salary is higher than every Pennsylvania salary. Also, since California tends to have higher salaries, it also has the higher mean salary.

Definition 2g.-2

Histogram. A visual display similar to a stemplot but that can be produced for very large data sets; essentially, it is a bar graph for quantitative data.

Unlike a stemplot, which shows every single data point, a histogram breaks a range of values of a variable into groups and displays only the count or percent of the data points that fall into each group. Before making a histogram, we need to be familiar with some associated terms.

Definition 2g.-3

Class. A grouping of data into a single interval; each class should have an *upper* and *lower limit* (endpoints), and each should be the same size.

Definition 2g.-4

Class mark. The midpoint of the upper and lower limit of each class.

Definition 2g.-5

Frequency. The count or amount of observations.

Definition 2g.-6

Frequency distribution table. A tabular display of the counts within each class.

Definition 2g.-7

Relative frequency distribution table. A tabular display of the percentage of observations within each class.

Example 2g.-6

The data set below details the selling prices of a random sample of 30 houses sold in a city during the past year.

67,500	72,000	54,600	38,900	87,400	28,300
105,000	91,600	46,500	62,800	139,600	81,200
59,900	76,200	107,100	64,400	25,500	51,800
88,000	63,900	105,000	70,500	28,200	59,500
108,700	81,100	42,600	57,300	77,700	64,800

We wish to make a histogram of this data using six classes from $20,000 to $140,000.

(a) Define the upper and lower limits of all six classes.

First, we need to calculate the size of each class. To make things simpler, we'll work in thousands of dollars.

$$\frac{140-20}{6}=20$$

So each class should be size 20 (or $20,000). Thus, the six classes are 20 to 40, 40 to 60, 60 to 80, 80 to 100, 100 to 120, and 120 to 140.

(b) Find the class marks of all six classes.

Class marks, recall, are the midpoints of the classes. The midpoints of the classes are 30, 50, 70, 90, 110, and 130.

(c) Construct a frequency distribution table of the data.

Class	Frequency
20 to < 40	4
40 to < 60	7
60 to < 80	9
80 to < 100	5
100 to < 120	4
120 to < 140	1

The frequencies were found by counting the number of homes in each class. For instance, there were four homes with prices between $20,000 and $40,000 exclusive (hence the < symbol).

(d) Which class interval has the greatest frequency?

The $60,000 to $80,000 interval.

(e) Construct a frequency histogram of the homes' data set.[*]

---

[*] Short of using a statistical software package, such as SPSS (it's very expensive but somewhat user-friendly) or R (it's free but not easy to use—you need to hard code in every statistics command), the Internet has a cornucopia of websites that will produce histograms to your specifications. For the histogram here, I made use of the site http://www.imathas.com/stattools/

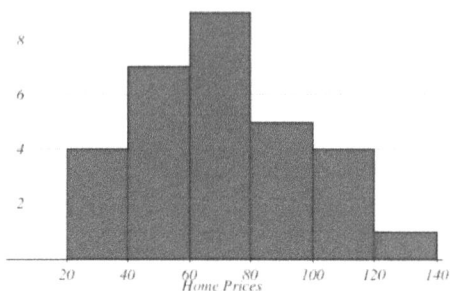

The *x*-axis need not be the class limits; instead, you could insert tick marks through the center of each bar. For instance, the 20 to 40 bar could instead be labeled as

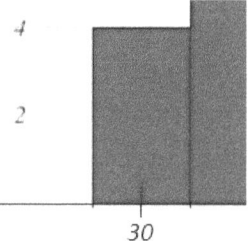

with all the other classes following suit.

Instead of a frequency histogram (i.e., a "regular" histogram), you could make the rectangle heights correspond to the *proportions* or *percentages* (called the relative frequencies) of observations in the classes.

(f) Construct a relative frequency histogram of the homes' data set.

First we will add relative frequencies to the table in part (c):

Class	Frequency	Rel. Freq.
20 to < 40	4	0.13
40 to < 60	7	0.23
60 to < 80	9	0.30
80 to < 100	5	0.17
100 to < 120	4	0.13
120 to < 140	1	0.03

Next, we need to alter the scale on the *y*-axis, and then build the histogram:

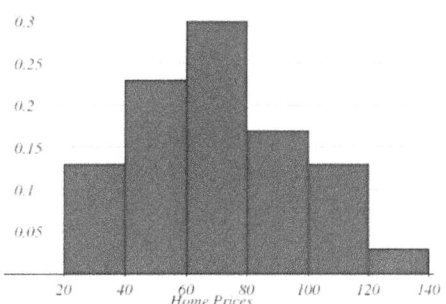

(g) Are the shapes of the frequency and the relative frequency histograms identical?[*]

Yes.

(h) Describe the shape, center, and spread of the histograms. Are there any outliers?

The center is at about 70, the spread is from 30 to 130, the shape is right skewed, and there are no outliers.

By the way, if an outlier *were* to appear in a histogram, it would be represented by a low-height bar set apart from the other bars. Like this:

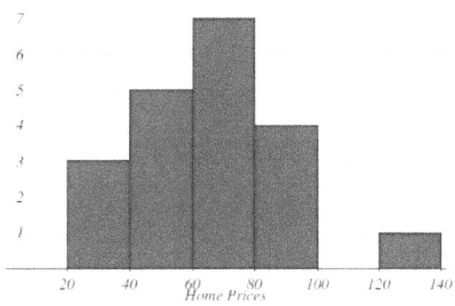

(i) What percentage of homes are priced less than $100,000?

$$13\% + 23\% + 30\% + 17\% = 83\%$$

(j) What proportion of homes are priced between $40,000 and $120,000?

$$0.23 + 0.30 + 0.17 + 0.13 = 0.83$$

(k) Using *only* the relative frequency histogram, can you determine the proportion of homes priced less than $70,000? Why or why not?

---

[*] Frequency and relative frequency are not the only types of histograms that can be constructed; there are also *cumulative* varieties of frequency and relative frequency histograms, which sum up the counts of ascending classes.

No, you cannot. Even though 70 is a class midpoint, you can't simply assume that half of the data inside the interval is less than $70,000 and the other half greater.

The TI-83/84 graphing calculator can also construct histograms.

- First, you must input all of the data into a list. Input the homes data set into the calculator; call the list HOMES.
- Now, go into the STAT PLOT menu by typing 2nd $\boxed{Y=}$ and select the first plot. Turn the plot ON and select the third type of plot (the icon that looks like a histogram). For the "Xlist," select the ∟HOMES list. For "Freq," type in the number 1.
- Hit the $\boxed{ZOOM}$ key and select option 9—which is "ZoomStat." After hitting $\boxed{ENTER}$, you should see the histogram. Next, press the $\boxed{TRACE}$ button and scroll left and right to see the values.
- The class width of a histogram can be adjusted by pressing the $\boxed{WINDOW}$ key, altering the number next to the "Xscl=" row, and then hitting the $\boxed{GRAPH}$ key. Experiment with different class widths.

Also try constructing a histogram of the ATM data set using the calculator. The left screenshot below shows the setup for the plot; the right screenshot below shows the calculator's default histogram of the data set.

 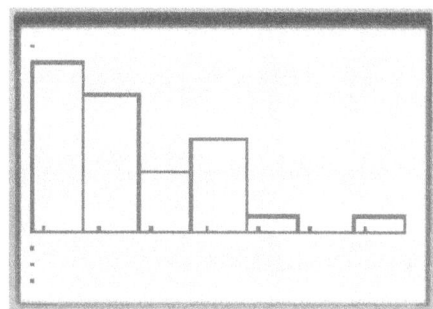

Both histograms[*] and stemplots can show key characteristics of quantitative data: symmetry, gaps, clusters, outliers, and even peaks. Although both types of graphs can be constructed by hand, it's usually most efficient to let computers do the grunt work.

---

[*] There are other varieties of histograms not covered here—such as the *density histogram*, in which the areas of the bars themselves are in proportion to the frequencies of the classes.

§2h. *Measures of Center*

N ow let's use measures of descriptive statistics to describe the center of a distribution. If the observations of a quantitative data set are plotted on a dotplot, an obvious choice for the center would be at the location an imaginary fulcrum underneath the dotplot (i.e., the "seesaw") could be placed so that the dots to the left side and the right side of the fulcrum *balance*.

We need to be more mathematically rigorous when describing the center, though. Bearing that in mind, consider the following definitions.

Definition 2h.-1

Measure of center. The value found at the middle of a data set.

Definition 2h.-2

Mean. The sum of the numbers divided by the number of numbers. The formula is $\bar{x} = \dfrac{\sum x_i}{n}$

Before we continue, note that sample size is denoted by n and population size is given by N. The mean of a population is represented by μ, a Greek letter pronounced "mu," with the mean of a sample specified by $\bar{x}$, pronounced "x-bar."

Example 2h.-1

We wish to find the average amount of money spent per visit among those who shop at a certain store at the mall. After looking through a series of randomly selected sales receipts, we find the mean to be $56.12. Report the results using the correct symbols.

Solution. The sample mean is $\bar{x} = \$56.12$, whereas the population mean μ represents the true mean amount of money *all* shoppers at the store spend per visit.

Definition 2h.-3

Median. The middle value of the data set, denoted by $\tilde{x}$ or M or *Med*.

Definition 2h.-4

Mode. The observation that occurs the most.

Roughly, there are three classifications of mode: unimodal (one mode), bimodal (two modes), or multimodal (more than two modes).[*] Mode is the only measure of center applicable to categorical data. For instance, calculating a mean zip code won't find you an "average" location.

Definition 2h.-5

Midrange. A value halfway between the minimum and maximum values of a data set. The formula is $\dfrac{max + min}{2}$

Realize that the midrange is not necessarily equal to the median, since the midrange ignores all values except for the extremes. For example, the data set 1, 2, 3000 has a median of 2, but a midrange of 1500.5.

Definition 2h.-6

Trimmed mean. A certain percentage of the highest and lowest observations of a data set are first removed (or "trimmed"); then, the mean of the remaining observations is calculated.

For instance, if a 10% trimmed mean is requested, then the lowest 10% *and* the highest 10% values in a data set are discarded before the mean is calculated.

Definition 2h.-7

Resistant. A measure not affected, or mostly unaffected, by the presence of outliers.

A quick word of caution: the term "average" doesn't necessarily refer only to the mean—there is ambiguity. Darrel Huff, in *How to Lie with Statistics*, devotes an entire chapter to how one can be deceived by uses of the term "average," and how the interpretation of averages can be suspect.

Your graphing calculator can find most, but not all, of the commonly used measures of center listed above automatically. Consider the data set of ATM withdrawals yet again. The 1-Var Stats calculator function will relay common descriptive statistics of a list stored in the calculator, as ATM should be (if you've been following all of the steps so far).

1. At your Home screen, press STAT and scroll over to the menu on the top middle of the screen called CALC.
2. Select the first option, called 1-Var Stats.
3. Now pull up your list by pressing 2nd and STAT and highlighting it.

---

[*] There are no formal names for trimodal, etc., because, when collecting real-world data, more than two distinct modes are a rarity. Distributions with multiple modes "usually indicate inhomogeneity in the system, or, in plainer language, different causes for different modes," according to Stephen Jay Gould.

4. After pressing ENTER, your screen should show look like the screenshot below on the left; scroll downward, and your screen should appear as the screenshot below on the right.[*]

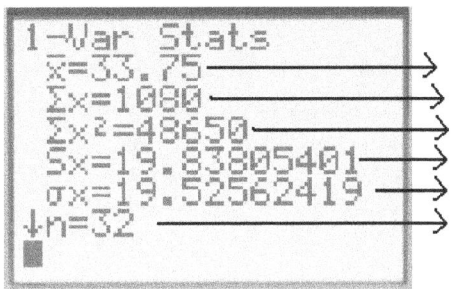 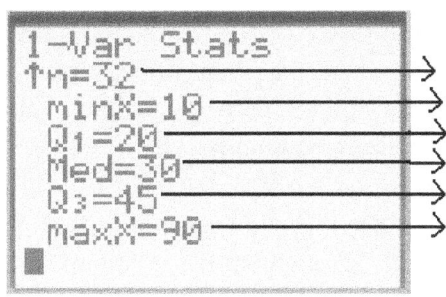

Note the arrows coming from each summary measure. Here's what they represent (you may wish to write them in on this page, if that will make it easier for you to remember them going forward):

- On the screenshot on the left, the $\bar{x}$ represents the mean, the $\sum x$ represents the sum of all of the observations in the data set, the $\sum x^2$ represents the sum of each observation squared, the Sx represents the sample standard deviation (a measure of spread we'll explore in subsequent pages), the σ_x represents the population standard deviation, and n represents the sample size.
- On the screenshot on the right, Q1 represents the first quartile—the median of the lower half of the data—and Q3 represents the third quartile—the median of the upper half of the data. The other items on-screen are self-explanatory.

From the screenshot above (on the right), the midrange can be found: $\dfrac{90+10}{2}=50$

The mean, median, and mode are the most commonly used measures for the center of a distribution of data. We need to do more than simply calculate these measures of center—we need to explore and discover their properties. Answering the following questions will help us get started.

Example 2h.–2

(a) Does the mean *have* to be one of the numbers in the data set? Fully justify your answer.
No, but it could be. For instance, the "data set"[†] 1, 2, 3 has a mean of 2, whereas the data

---

[*] You might instead, when selecting 1-Var Stats, be presented with a menu. If that's the case, then make sure that your list is pulled up, leave the frequency at the default value (which is 1), and scroll down and press ENTER to execute the command.

[†] The scare quotes are a reminder that numbers must have some context to be properly construed as a data set, which

set 1, 2, 3, 4 has a mean of 2.5.

(b) Does the median have to be one of the numbers in the data set? Fully justify your answer.

If *n* is odd, then yes. But if *n* is even, only if the middle two numbers are the same. (Because, to get the median of an even number of numbers, you simply take the mean of the middle two numbers.)

(c) Suppose that the largest number in a data set, which is not an outlier, is increased significantly enough so that it becomes an outlier.

   a. How does this outlier affect the mean and the median of the data set?

   If the data set is 1, 2, 3, then the mean and median are both 2. But if the data set is changed to 1, 2, 3000, then the median stays the same, but the mean jumps considerably—to 1001.

   b. So, which measure of center is *resistant*: the mean or the median? Explain.

   The median is resistant since it stays locked in place, whereas the mean is "attracted" to the outlier.

(d) Suppose the mean and the median salary at a company is $50,000 and all employees (1) get a $1,000 raise; or (2) get a 10% raise; or (3) receive a 5% reduction in salary. Find the new mean and median salary in each case.

For (1): if every employee gets a $1,000 raise, then everybody's salary goes up by $1,000. Imagine a dotplot of salaries—with the raise, every employee's "dot" would have shifted to the right by an equal distance. Therefore, both the mean and the median go up—by the same distance on the scale, $1,000.

   For (2): If everyone gets a 10% raise, then employees who aren't making much money are getting a smaller raise, money-wise, then employees who are making a lot. Thus, the dotplot stretches outward; specifically, it becomes 10% lengthier. To calculate the new mean and median, take $50,000 and multiply it by 1.10 (i.e., a 110% increase).

   For (3): Same logic as (2), except reversed. The dotplot shrinks in length by 5%. To get the new mean and median, multiply $50,000 by 0.95 (i.e., you want to calculate 95% of the original value; think of it this way: if an item in a store was 5% off, you'd be paying 95% of its original value).

   Here, the mean and median reacted the same way to the transformations. In question

---

these do not. But we'll put that aside for now, and I'll stop using the scare quotes.

(c), however, the mean and median reacted differently.

(e) An ice cream shop offers two flavors: vanilla and chocolate. If vanilla is the mode of ice cream sold, is it definitely the case that more than 50% of customers purchased vanilla? Explain.

Yes. With only two options and one mode, the mode must be the majority.

(f) A different ice cream shop offers three flavors: vanilla, chocolate, and strawberry. If chocolate is the mode of ice cream sold, is it definitely the case that more than 50% of customers purchased chocolate? Explain.

No. For instance, chocolate: 40%; vanilla: 30%; strawberry: 30%. Even though chocolate is the mode, it's not the majority.[*]

(g) Is it possible for 90% of employees at a company to earn less than the mean salary? Explain.

Yes. Recall that the mean isn't resistant, so it's attracted to outliers. The outliers in the salaries of employees at a company usually are top-paid management. So, not only is it possible that 90% of employees earn less than the mean salary, that sort of thing is actually quite common. (If the question were reworded to ask about the *median* salary, though, the answer would be: impossible.)

(h) Will an outlier added to a data set affect the mean more if $n = 10$ or $n = 100$? Justify your answer.

$n = 10$, since there are fewer numbers to "anchor" the mean in place. For instance, a freshman who has taken few classes can affect her GPA more going forward than a senior who has completed many classes. (Of course, in general, with more data points, there's a greater chance of capturing extreme values.)

(i) Create a data set of 9 integers, 0 to 10 inclusive, which has the following properties: the mean, median, and mode are all equal, and not every number in the data set is the same.

There are many possible answers. Here's one: 1, 2, 2, 3, 3, 3, 4, 4, 5.

---

[*] Real-life example: In the 1992 U.S. presidential election, there were three major candidates: George H.W. Bush (Republican), Bill Clinton (Democrat), and H. Ross Perot (Independent). The popular vote percentages were 37.4%, 43.0%, and 18.9%, respectively. Even though Clinton didn't obtain a majority, he easily won the election, dominating Bush electorally.

(j) Create a data set of 9 integers, 0 to 10 inclusive, which has the following properties: the mean and median are equal, and the data set is bimodal.

Again, there are many possible answers, but one might be: 1, 1, 1, 1, 5, 9, 9, 9, 9.

(k) We will mostly use the mean and median to describe the center of data sets. There is no "right" measure of center to use, but if a data set has extreme values, which measure of center might be preferable?

The median, since it is resistant to outliers.

(l) Draw the placement of the mean, median, and mode in a distribution skewed to the right (positively skewed). Then, give a real-world example of a distribution skewed right.

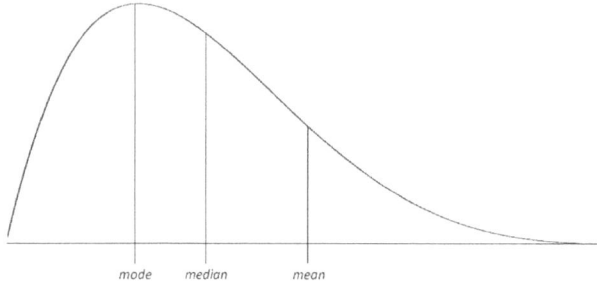

The mode, which occurs the most, is the peak; the median splits the area under the curve in half; and the mean follows the tail (where the extreme values lie). The salaries of employees at a company is one real-world example; prices of new cars is another.

(m) Draw the placement of the mean, median, and mode in a distribution skewed to the left (negatively skewed). Then, give a real-world example of a distribution skewed left.

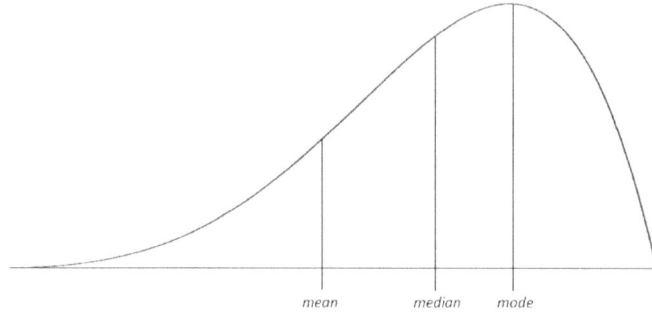

Real-world examples include scores on an easy exam and the years minted among all coin money in circulation.

(n) Draw the placement of the mean, median, and mode in a symmetrical distribution (zero skew). Then, give a real-world example of a symmetrical distribution.

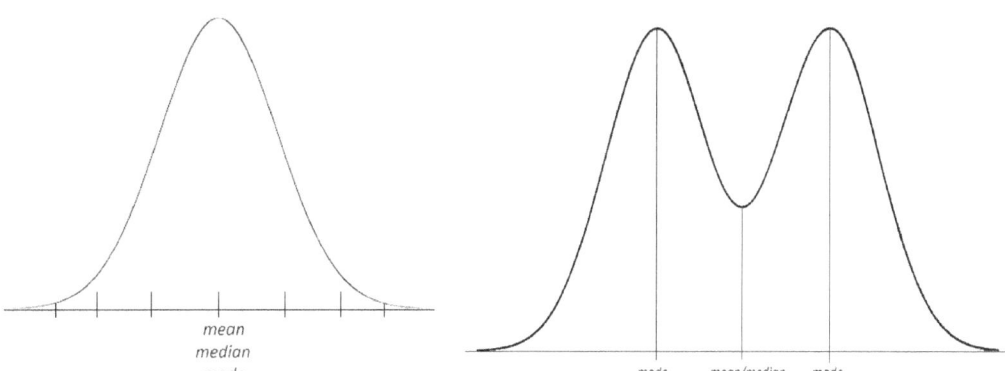

On the left lies the classic bell curve. The mean, median, and mode of this mound-shaped symmetrical distribution are located at the center. Heights of adults males in a population and IQ of students are two examples. Biological characteristics generally tend to assume bell-shaped distributions since, typically, these characteristics are calculated off of the averages of many factors. For instance, the heights of adult males are determined by the interplay between genes, nutrition, and prior health, among other things. Same with blood pressure measurements, stock market percentage returns over time from very diversified portfolios, and chance outcomes in games, such as with the sum of rolling two dice or coin flips. In addition, distributions of errors also tend to be bell shaped.[*]

On the right lies a bimodal distribution, where the two modes are presented as peaks. The heights of males and females grouped together into one data set is a possible example.

(o) In 2011, two reported measures of center for NBA players were $5.15 million and $2.33 million, one of which was the mean and the other the median. Which one was which?

The mean was likely the higher number, $5.15 million, because the so-called superstars make exceptional sums of money, dragging the mean up (since the mean is attracted to the extreme values). The data set of NBA salaries is probably skewed to the right.

The real-world consequences of the different types of measures of center (or averages) can be found in the Stephen Jay Gould essay "The Median Isn't the Message" (a play on McLuhan's famous observation that the medium is the message). After being diagnosed with abdominal mesothelioma, which he learned had a "median mortality of only eight months after discovery,"

---

[*] Nutrition is a powerful predictor of height. For instance, compare the heights of North and South Korean adult males. Those in North Korea are around three inches shorter than in South Korea. Although genetically similar, nutritionally North Koreans suffer from a variety of politically-manufactured sanctions.

Gould notes how understanding statistics lifted his attitude and "played a major role in saving [his] life."

Instantly realizing that rather than believing he will be dead in eight months—as the median figure may suggest to the layperson—the notion of variability rears its head: "[A]ll evolutionary biologists know that variation itself is nature's only irreducible essence.... I had to place myself amidst the variation." He knew half the people diagnosed with abdominal mesothelioma would live longer than eight months—but what were his chances of landing in that fortunate half?

He was relatively young and otherwise healthy, the cancer was caught early, and he lived in a developed country: all factors increasing his odds of survival. But, beyond that, the distribution of survival years was skewed right, meaning the "upper (or right) half [tail of the distribution] can extend out for years and years, even if nobody ultimately survives," unlike the left side of the survival distribution, which extended only so far as zero years (i.e., immediate death). Gould ended up living another twenty, mostly healthy, years after his diagnosis.

We can obtain the approximate mean of a data set given only a frequency distribution if we multiply each class's frequency by the midpoint of that class, sum the results, and divide by the sum of the frequencies.

Example 2h.-3

Refer back to the frequency distribution we constructed of the selling prices of homes. Using *only* this frequency distribution, find the mean selling price of the homes and record this value.

Solution. Work through this formula:

$$\bar{x} = \frac{\sum (f \cdot x)}{\sum f} = \frac{30 \cdot 4 + 50 \cdot 7 + 70 \cdot 9 + 90 \cdot 5 + 110 \cdot 4 + 130 \cdot 1}{30} = 70.67; \ \$70,670.$$

> We've multiplied each class midpoint by the frequency of each class, and divided by the total of the frequencies.

With that example complete, we now turn to the *weighted mean*, also called the *weighted average*. Oftentimes we have to find a mean that is not directly calculated as simply the sum of the observations divided by the number of observations, because different values have different weights. When different values have different weights, we have to find a weighted average of the data.

Example 2h.-4

    (a) Suppose you received a 60% for the first marking period, an 85% for the second marking period, and a 95% for the midterm. Also, each marking period is worth 40% of your semester grade, with your midterm counting for the remaining 20% of your semester grade. What is your average for the semester?

	Score	Weight	Score × Weight
1st Marking Period	60	0.40	24
2nd Marking Period	85	0.40	34
Midterm	95	0.20	19

Your average for the semester is $24 + 34 + 19 = 77$. Any number of categories in a weighted average calculation can be accounted for in this way, as long as the weights of all the categories sum to 1.

Sometimes the weights (in percentage or proportion form) of the observations in a data set will not be given. When this occurs, after multiplying, you will need to divide by the total number of observations in order to obtain the weighted average.

(b) A teacher gives a test in the two sections of her class. Her first class' test mean is 55. Her second class' test mean is 85. Her first class has 20 students, and her second class has 30 students. Calculate the mean test score among both sections of the teacher's class.

$$\frac{55 \cdot 20 + 85 \cdot 30}{50} = 82$$

This calculation is similar to the homes example above.

Tally charts are used to systematically organize categorical data. Usually, the categories in tally charts take the form of numbers, such as points possible during a game. The "tally" represents the count of each category in the data set.

Example 2h.-5

Suppose the following tally charts show the number of points possible during some sort of game.

(a) Look at this tally chart:

Points	6	8	10	12	14
Tally	2	1	0	5	1

    a. Find the mode.

        The mode is 12.

    b. Find the mean (hint: you will need to find a weighted average).

$$\frac{6 \cdot 2 + 8 \cdot 1 + 10 \cdot 0 + 12 \cdot 5 + 14 \cdot 1}{9} = 10.44$$

    c. Find the median.

       The median is the fifth number: 12.

(b) Now, look at this one:

Points	2	7	15	20	21
Count	7	4	2	0	1

    a. Find the mode.

       The mode is 2.

    b. Find the mean.

       The mean is 6.64.

    c. Find the median.

       The median is 4.5.

    d. Is this data set skewed left, right, or is it symmetrical? Is there an outlier?

       The data set is skewed right, since most of the numbers are on the low side (if plotted).

(c) Finally, examine this one (the < 3 captures any number less than 3):

Points	< 3	4	5	6
Tally	2	2	0	3

    a. Find the mode.

       The mode is 6.

    b. Find the median.

       The median is 4.

    c. Can you find the mean? If so, do it. If not, explain why.

No. It's not clear which numbers are in the < 3 category.

In 2009, the British magician Derren Brown quite amazingly predicted—or at least *seemed to* predict—the Midweek Lotto Numbers live on England's Channel 4. The odds[*] of guessing the numbers correctly? About fourteen million to one. Brown explained his method this way:

> I gathered a panel of 24 people who wrote down their predictions after studying the last years' worth of numbers. Then they added up all the guesses for each ball and divided it by 24 to get the average guess. On the first go they only got one number right, on the second attempt they managed three and on the third they guessed four. By the time of last week's draw they had honed their technique to get six correct guesses, and these were the numbers drawn on the TV program.

Although it's extremely unlikely Brown actually obtained his lottery numbers using this method—rather than statistical analysis, trickery is the more probable explanation—using the aggregate of many guesses has a long pedigree, stretching as far back as Sir Francis Galton.

Galton, a heavyweight in the history of statistics and science (and first cousin of Charles Darwin), arrived at two important statistical ideas—correlation and regression toward the mean—as well as piloting innovative ways to survey people. The dichotomous "nature versus nurture" argument stems from him as well; in fact, he coined the phrase.[†]

In 1906, at England's Plymouth country fair, Galton chanced upon a strange game at a livestock fair. A crowd of people gathered around the slaughtering of an ox; observing carefully, each person was summoned to guess the animal's weight. Although nearly a thousand people participated, not one landed on the correct value. Galton nonetheless noticed something interesting: "[T]he middlemost estimate expresses the *vox populi* [the voice of the people], every other estimate being condemned as too low or too high by a majority of the voters." In other words, the median hit the mark (as did the mean, which Galton also checked). Although no one person had the whole story, many revealed small segments of the truth, all of which pointed toward the ox's weight; but those who were completely off the mark, not having any clue of the ox's true weight, tended to cancel each other's random guesses out.

Many years later, this idea—that the mean of many individual responses may cluster around the true mean in some situation—has been termed *the wisdom of the crowd* (after the title of a book by James Surowiecki), or *crowdsourcing*. Netflix used crowdsourcing to help build a better algorithm to predict their users' ratings. And the technique has even been leveraged by social

---

[*] Odds are subtly different than probability; more on that later.

[†] Galton loved measuring things. For example, he wrote a book on fingerprinting, which ultimately led to its adoption by police departments. But Galton also dabbled in eugenics, or the selective breeding of human beings, and was thus unsurprisingly one of Hitler's intellectual heroes. (And an inspiration to dog breeders everywhere.) Nonetheless, Galton's contributions to statistics, as we will see, are undeniable. As Peter L. Bernstein notes, "Galton moves us into the world of everyday life, where people breathe, sweat, copulate, and ponder their future," rather than testing theories at gambling tables or by stargazing.

media to diagnose illness, as this recent article from *Time* explains:

> Crowdsourcing isn't a new phenomenon is health care—experts have long relied on mass media surveillance to monitor flu outbreaks, for example—but…the rise of social media [has increased its prevalence]. When doctors or patients are stumped by symptoms, Facebook, Twitter and medical networks allow them to tap the wisdom of the crowd with a few clicks. "The potential is huge," says Aydogan Ozcan, a UCLA professor who showed that a group of gamers could diagnose malaria from microscopic slides almost as well as an experienced pathologist.

Sometimes, it seems, it's wise to follow the crowd.[*]

§2i. *Measures of Spread*

Measures that describe how a data set is distributed can provide us with useful information.

Example 2i.-1

The highway mileages of 20 cars, as measured by *Consumer Reports* (May 2015), arranged in increasing order, are

6	11	20	25	27	29	29	30	30	31
31	32	33	35	35	37	38	40	56	86

(a) Store this data into your calculator as a list called HMPG. Using 1-Var-Stats, find both the mean and the median of the highway miles per gallon data set.

The mean is 33.05, and the median is 31.

(b) Based only on the mean and the median, can you determine the skew/symmetry of the highway miles per gallon data set? Explain.

Since the mean is a little larger than the median, and distributions where the mean > median tend to be skewed right (there are exceptions, however), the data set is likely skewed right.

---

[*] Although Burton G. Malkiel, who spills much ink in *A Random Walk Down Wall Street* describing "the madness of crowds" with respect to market speculations throughout history—such as the tulip-bulb craze of seventeenth century Holland, the South Sea Bubble of eighteenth century England, and the stock market crash of twentieth century America—would likely disagree.

The median and the mean are of course both measures of center, but a measure of center alone can be very misleading. Two companies might have the same median salary, but be distributed very differently at the extreme ends of the salary scales. Two neighborhoods, A and B, may have the same mean property values—with neighborhood A containing properties of mostly commensurate, medium valuations, and neighborhood B laying claim to mostly low-value properties with a few extremely high-value properties tugging up the mean. Thus, knowing the *spread* or *variability* of a data set will give us a more complete picture of the distribution.

One way to measure spread is to calculate the *range*, which is simply the difference between the largest and smallest observations.

(c) Calculate the range of the highways miles per gallon data set.

The range is 6 subtracted from 86, or 80 mpg.

(d) Is the range a resistant measure of spread? Justify your answer.

No. The range could be dramatically affected by outliers, since the minimum or maximum (or both) might be an outlier.

Yet range is not a useful measure of spread because we don't know how the observations *between* the minimum and maximum values of the data set are distributed. Let's break a data set into four equal parts: by chopping it in half (at the median) and in two halves again. The first quartile (or lower quartile) Q1 is the median of the first half of the observations. The third quartile (or upper quartile) Q3 is the median of the second half of the observations.

(e) Using your calculator, find the upper and lower quartiles of the highway miles per gallon data set.

The lower quartile is 28, and the upper quartile is 36.

By using the quartiles, the median, and the minimum and maximum values, we can better describe the spread of a data set. In addition, we can define *percentiles*—the percentage of data at or below a certain observation—to get a sense of the variability. (The 50th percentile is at the median since 50% of the data in any data set lies at or below the median. Percentiles are usually used to analyze standardized test results. If you score at the 0th percentile on a standardized test, though, that means that you're *at best* tied for the lowest score.) A percentile is calculated by finding the percentage of observations less than that particular observation in the distribution.

(f) Find the percentile of the Volkswagen Beetle, which gets 32 mpg.

$$p = \frac{11}{20} \cdot 100 = 55th$$

Now consider the diagram on the top of the next page, which can be applied to any data set.

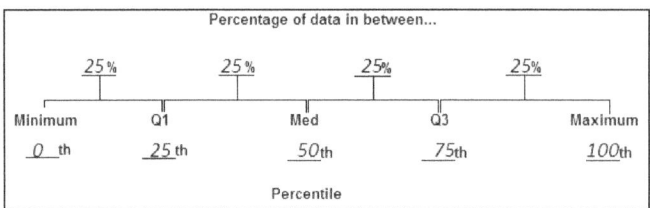

(g) Based on the diagram, what percentage of data of *any* data set is between the lower quartile and the upper quartile?

The middle 50%.

The difference between the upper quartile and the lower quartile is called the *interquartile range*, or IQR. The IQR represents the *middle 50%* of a data set (unlike the range, which represents 100% of a data set).

(h) Find the IQR of the highway miles per gallon data set. Also determine if the IQR is a resistant measure of spread.

The IQR of the data set is 8. The IQR is resistant, since the first and third quartiles are not at the extremes.

Statistician John Tukey bundled together the minimum, the first quartile, the median, the third quartile, and the maximum into what he called *the five-number summary*, which can give you a quick and dirty[*] feel for the distribution.

(i) Report the five-number summary of the highway miles per gallon data set below.

Min: 6; Q1: 28; Med: 31; Q3: 36; Max: 86.

The five-number summary leads us to another visual representation of a distribution, called a *boxplot* (or *box-and-whisker plot*). A boxplot is made by drawing a box to indicate where the quartiles lie and "whiskers" to extend the box to the minimum and maximum values of the data set. The boxplot was introduced by Tukey in 1977.

(j) Draw a boxplot of the highway miles per gallon data set. Be sure to label the values of the five-number summary on the plot.[†]

---

[*] Yes, this is a real term, essentially meaning "crude" or "approximate."

[†] I used the following website to construct the boxplot: http://www.imathas.com/stattools/

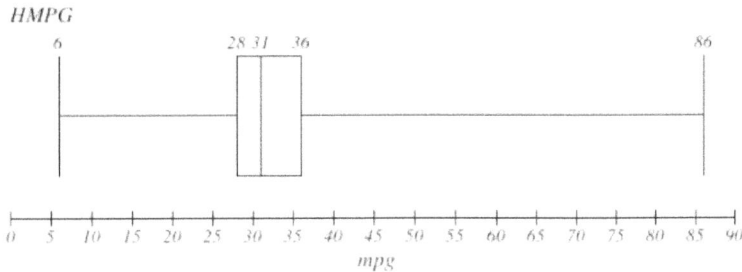

Your graphing calculator can also construct this boxplot. Go into the STAT PLOT menu by typing $\boxed{2^{nd}}$ $\boxed{Y=}$ and select the first plot. Turn the plot ON and select the fifth type of plot (the icon that looks like a boxplot, without the dots extending from it; the boxplot with the dots next to it is called a *modified boxplot*, which will show outliers as stars). For the "Xlist," select the $_L$HMPG list. For "Freq," type in the number 1. Hit the $\boxed{ZOOM}$ key and select option 9—which is "Zoom-Stat." After hitting $\boxed{ENTER}$, you should see the boxplot. Next, press the $\boxed{TRACE}$ button and scroll left and right to see the values of the five-number summary.

We have not had, up to this point, a formal way to test if an observation is an outlier. Tukey came up with a rule (actually, he came up with several rules).

Example 2i.-2

Reconsider the highway mileages from the previous example.

(a) Which two observations above appear to be outliers?

Perhaps 56, and probably 86.

(b) Does the "regular" or "skeletal" boxplot that we constructed in the previous example allow you to determine if these two predicted observations are actually outliers? Why or why not?

No. You're unable to tell the spread of the data *within* the whiskers (and boxes) themselves.

We need to have a way to formally test for an outlier; we also need a way to indicate on a boxplot diagram that an observation or observations are in fact outliers. We will construct a *modified boxplot* to do so.

Here is the rule to determine if an observation is an outlier: see if it lies more than 1.5 IQRs away from the nearer quartile. This is called the *1.5 x IQR rule* (or *the outlier test*).[*] Let's put this rule into practice.

---

[*] Tukey also made an allowance for "extreme outliers," or those observations more than three IQRs away from the nearer quartile. We will ignore this distinction.

(c) Calculate 1.5 x IQR.

$8 \cdot 1.5 = 12$

(d) Add your result from part (c) above to Q3, the upper quartile. Are any observations above that number? If so, which one(s)? These are the outliers.

$36 + 12 = 48$. Therefore, according to the outlier test, both 56 and 86 are outliers.

(e) Next, take your result from part (c) above and subtract it from Q1, the lower quartile. Are any observations below that number? If so, which one(s)? These are also outliers.

$28 - 12 = 16$. Thus, 6 and 11 are also outliers.

To construct a modified boxplot, or a boxplot which shows outliers (unlike a skeletal boxplot), use a * for observations classified as outliers, and extend the boxplot's whiskers to the most extreme non-outlying value(s).

(f) Construct a modified boxplot of the data below.

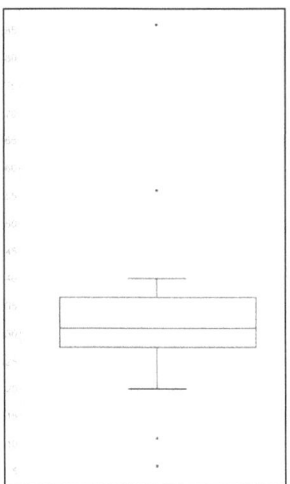

Your graphing calculator can also create modified boxplots. Go into the STAT PLOT menu by typing 2nd Y= and select the first plot. Turn the plot ON and select the fourth type of plot (the icon that looks like a boxplot, but with the dots extending from it). For the "Xlist," select the ∟HMPG list. For "Freq," type in the number 1. Hit the ZOOM key and select option 9—which is "ZoomStat." After hitting ENTER, you should see the boxplot. Next, hit the TRACE button and scroll left and right to see the values of the five-number summary, along with any outliers (see the screenshot on the top of the next page).

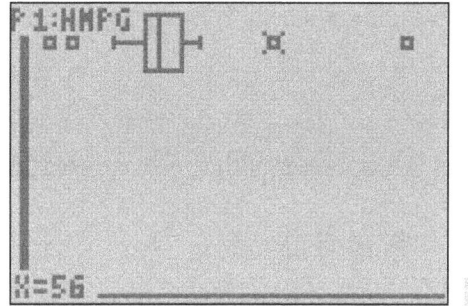

Comparative boxplots (or *side-by-side* or *parallel boxplots*) display at least two boxplots on the same scale. Usually, comparative boxplots compare and contrast distributions of data which are related in some way. We have to not only be able to construct comparative modified boxplots, using appropriate scales, but we also need to be able to analyze the relationships between these boxplots.

Example 2i.-3

Take a look at the following skeletal (i.e., not modified) boxplots.

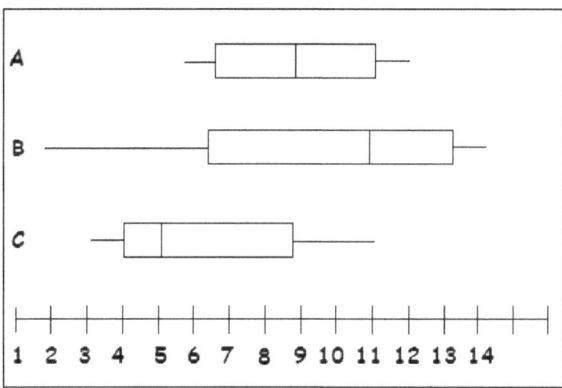

(a) Which boxplot represents the symmetrical data set? How can you tell?

Boxplot A, since the whiskers and boxes are the same size.

(b) Which boxplot represents the data set skewed to the left? How do you know?

Boxplot B. The median is shifted rightward, and the left whisker is longer—meaning that the first half of the distribution is spread out over a larger area than the second half.

(c) Which boxplot represents the data set skewed to the right? How do you know?

Boxplot C. Opposite reason of part (b).

(d) Which boxplot represents the data set with the greatest range? Calculate this range.

Boxplot B. The range is about 12.

(e) Which boxplot represents the data set with the smallest IQR? Calculate this IQR.

Boxplot A. The IQR is approximately 4.5.

(f) You can actually find the *mean* of the data set represented by boxplot A above, despite not having the original set of data. Explain how this is possible, and then state the mean.

If a data set is perfectly symmetrical, then the mean and the median are equal. So the mean of the data set is 9.

Example 2i.-4

Consider the following dotplots.

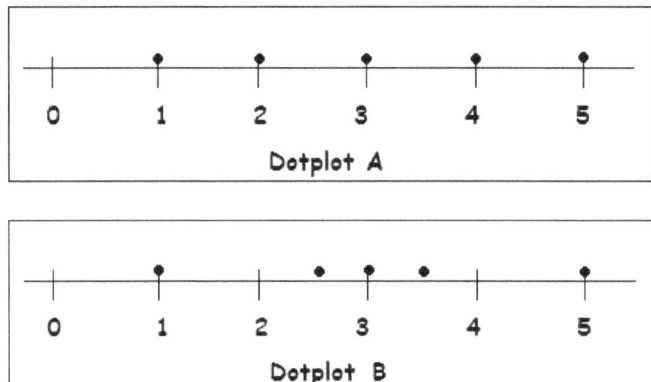

(a) Which dotplot has data that's more *spread out*? How can you tell?

It's tough to tell. But perhaps dotplot A, since the dots inside the distribution are farther away from each other, relatively speaking.

We know how to get the center of the data sets displayed in dotplots A and B. We've also been able to describe the variability (or the range) of the data shown in dotplots. But how can we come up with a way to measure spread—a way to measure how things are spread out *within* the range of the data?

(b) Find the range of the data of dotplots A and B. Are they the same?

For both, the range is 4. Since that's the answer for A and B, this measure doesn't help us differentiate between the data sets at all.

(c) Find the mean of the data of dotplots A and B. Are they the same?

For both, the mean is 3. Again, since that's the answer for A and B, this measure doesn't help us differentiate between the data sets.

Suppose we looked at each dot's *distance* from the mean. Distance (by definition) must be positive.

(d) Find each dot's distance from the mean of dotplot A's data set. You should have 5 distances, since there are 5 dots.

The distances are 2, 1, 0, 1, and 2.

(e) Find each dot's distance from the mean of dotplot B's data set. You should have 5 distances, since there are 5 dots.

The distances are 2, 0.5, 0, 0.5, and 2.

For each dotplot, let's take the mean of these distances from the mean. This is a measure of the spread of the data called *Mean Absolute Deviation*, or MAD.

(f) Find the Mean Absolute Deviation (MAD) of dotplot A's data set.

$$MAD_A = \frac{2+1+0+1+2}{5} = 1.2$$

(g) Find the Mean Absolute Deviation (MAD) of dotplot B's data set.

$$MAD_B = 1.0$$

(h) Explain what MAD tells you about the distribution of the data.

MAD tells the mean of the data points' distances to the mean. It describes how the data is distributed within the range.

Although MAD is a relatively easy and straightforward measure to find, statisticians don't normally use it. The problem with the calculation is that MAD uses absolute values—but absolute values are not part of the set of *algebraic operations*.[*] In addition, the sample MAD is not an *unbiased estimator* of the population MAD, meaning the sample value doesn't target the intend-

---

[*] Algebraic operations include addition, multiplication, taking roots, and raising to powers.

ed measure well. And, finally, using absolute values would cause us complications in statistical methods later on, so we need to avoid them.[*]

But if we're not going to use MAD, then how can we describe the spread of the values of a data set accurately? We will use a measure called *standard deviation*; when calculating standard deviation by hand, we'll need to find the *variance* first. First, let's define both of these terms (and one more).

Definition 2i.-1

Standard deviation. The average deviation around the mean.

The symbol for the population standard deviation is σ, a Greek letter pronounced "sigma," while the symbol for the sample standard deviation is s. The formula for the standard deviation of a sample is given as

$$s = \sqrt{\frac{\sum(x-\bar{x})^2}{n-1}}$$

which we will unpack shortly. (Although, for now, note that the units of standard deviation are the same as the data set's—so, for instance, the standard deviation of a distribution of heights in inches will have a standard deviation with the units also being in inches.)

Definition 2i.-2

Variance. A measure of variation equivalent to the square of the standard deviation.

The symbol for the population variance is, unsurprisingly, σ^2, while the symbol for the sample variance is s^2. The formula for the variance of a sample is given as

$$s^2 = \frac{\sum(x-\bar{x})^2}{n-1}$$

which we will also unpack shortly. (Please observe that variance is given in *squared* units.)

Example 2i.-5

Let's find the standard deviation (and the variance) of a data set by hand. Six randomly selected students were surveyed about their television-viewing habits—in terms of how many hours of

---

[*] The source for this paragraph's ideas: Mario Triola's *Workshop Statistics*.

TV they watch per week. Here are the results:

10	5	7	14	4	8

(a) Find the sample mean.

$\bar{x} = 8$

Besides just the mean, we'll need to find what's called the "deviation from the mean"—the difference between any individual data point and the mean of all of the data—and the absolute and squared deviations.

(b) Fill in all of the cells in the following table.[*]

	Hours	Deviation from the Mean	Absolute Deviation	Squared Deviation
	10	10–8 = 2	2	4
	5	5–8 = –3	3	9
	7	7–8 = –1	1	1
	14	14–8 = 6	6	36
	4	4–8 = –4	4	16
	8	8–8 = 0	0	0
Total	**48**	**0**	**16**	**66**

(c) Using the table, calculate the Mean Absolute Deviation (MAD).

To calculate MAD, recall we need the mean of the *distances* to the mean. The distances are found in the "Absolute Deviation" column. (The units will be in hours.)

$MAD = \dfrac{16}{6} = 2.667$

(d) Using the table, calculate the variance. (The units will be in squared hours.)

$s^2 = \dfrac{66}{5} = 13.2$

Before continuing, we need to take stock of a couple of things. First, note that you can't simply square the sum of the "Absolute Deviation" column to obtain the sum of the "Squared Deviation" column—the sum of the squares is arithmetically not the same as the square of the sum.

---

[*] This table was inspired by a similar exercise in Rossman's *Workshop Statistics*.

Second, although it might seem a bit strange that the denominator of the variance calculation has 5 in it instead of 6, there's a good reason. Look at the "Deviation from the Mean" column's sum: zero. No matter what numbers are in a data set, that sum will *always* equal zero. Knowing that, suppose we accidently deleted one of the rows of the table—let's say the row of the student who watches 14 hours of television per week. We could repopulate that entire row because we always have the sum of the "Deviation from the Mean": zero. Thus, since one individual's data point is effectively unnecessary, we really have five pieces of independent data, not six—hence the denominator's value.[*]

Another way to conceptualize this is to imagine a huge dotplot of *every* student's television-watching habits per week (not just these six students). There would be lots of dots at zero—for those who don't watch TV—and many millions of dots scattered about, with a maximum of 168 hours (the total number of hours in a week). A small sample culled from these millions upon millions of data points is unlikely to be *as spread out* as the population's dotplot; so, to adjust for this lack of "spreadness," the sample variance is artificially raised a bit by dividing by a smaller number—hence, the $n-1$ in the denominator of sample variance formula.

(e) Find the standard deviation. (The units will be in hours.)

$s = \sqrt{13.2} = 3.633$, which is reasonably close to the MAD.

(f) Use the graphing calculator to confirm the calculation of standard deviation.

Running the 1-Var Stats calculator function does confirm the value.

The standard deviation is a powerful measure of variation. The standard deviation informally represents the "typical" distance from any point in the data set to the mean. Roughly speaking, the standard deviation gives you the average distance of all the data points to the mean.[†] The smaller the standard deviation, the better the mean represents the "typical" values of the data set—since the observations tend to be clustered together when the standard deviation is relatively small—and vice versa.

Example 2i.-6

(a) Can standard deviation ever be negative? Why or why not?

---

[*] This is usually called "*n* minus one degrees of freedom"—one less than the sample size. (It seems like grammatically I should have written "one *fewer* than the sample size," but since degrees of freedom are not restricted to natural numbers, *less* is in fact the correct word to use.)

[†] The qualifier "roughly speaking" is necessary since the exact "average distance of all the data points to the mean" is the MAD.

No. Distance can't be negative. Also, standard deviation is found by taking the square root of the variance, and the radicand can't be negative (no imaginary numbers permitted here).

(b) What is the smallest the standard deviation can be? Give an example of a data set with the smallest possible standard deviation.

Zero. An example: 1, 1, 1, 1. Or any data set with every observation having the same value.

(c) Suppose the mean salary at a company is $50,000, with a standard deviation of $500. All employees get a $1,000 raise. What happens to the standard deviation of the salaries?

The standard deviation remains $500. Here's why: Imagine a dotplot of salaries. If everyone gets a $1,000 raise, all of the dots shift over by an equal amount—leaving the spread between the dots the same.

(d) Suppose the mean salary at a company is $50,000, with a standard deviation of $500. All employees get a 10% raise. What happens to the standard deviation of the salaries?

The new standard deviation is $550. The dotplot stretches in length by 10%, increasing the spread of the data. So the new standard deviation is found by taking 500 and multiplying it by 1.10.

(e) Is the standard deviation a resistant measure? Why or why not?

Since the calculation of standard deviation relies on the mean, and the mean is not resistant, neither is the standard deviation.

In his textbook *Elementary Statistics*, author Mario Triola presents a quick rule of thumb to determine if an observation from a data set is "unusual" or not: if that observation is more than two standard deviations away from the mean of the data. In addition, Triola presents a quick estimate for standard deviation: one quarter of the range.

Example 2i.-7

Reconsider the ATM withdrawals data from earlier (min = $10, max = $90, mean = $33.75, and $s = \$19.84$).

(a) Estimate the value of the standard deviation, then compare it to the true value.

$s \approx \dfrac{90 - 10}{4} = 20$, which is very close to the actual sample standard deviation. (How close the value ends up depends largely on the shape of the distribution.)

(b) Determine if an ATM withdrawal amount of $90 would be considered an "unusual value." Justify your answer with appropriate calculations.

An unusual value is more than two standard deviations from the mean. So, since

$$33.75 \pm 2 \cdot 19.84 = (-5.93, 73.43),$$

a $90 withdrawal amount is unusual.

Measures of spread factor into data forensics. Steven Levitt is an American economist best known for his work on crime analysis. Recall that he is the originator of the very controversial abortion-crime rate hypothesis: that since 1973, when *Roe v. Wade* was overturned (making abortion legal throughout the U.S.), the crime rate has gone down—because there is a shortage of potential criminals who would have been born.

But Levitt also examined possible standardized test cheating by teachers in the Chicago Public School (CPS) system. As he explains in his popular book *Freakonomics*,

> To catch a cheating teacher, it helps to think like one. If you were willing to erase your students' wrong answers and fill in correct ones, you probably wouldn't want to change too many wrong answers. That would clearly be a tip-off. You probably wouldn't even want to change answers on every student's test—another tip-off.
>
> So what you might do is select a string of eight or ten consecutive questions and fill in the correct answers for, say, one-half or two-thirds of your students. You could easily memorize a short pattern of correct answers, and it would be a lot faster to erase and change that pattern than to go through each student's answer sheet individually. You might even think to focus your activity toward the end of the test, where the questions tend to be harder than the earlier questions. In that way, you'd be most likely to substitute correct answers for wrong ones.

And that's exactly what Levitt discovered in Chicago. CPS gave Levitt masses of data, making available a database of the test answers of all third grade to seventh grade students from 1993 to 2000. The data were organized by classroom for math and reading tests. Levitt constructed a computer algorithm to tag suspicious blocks of answers in students' testing sheets according to the criteria he detailed in the block quote above.[*]

Another advanced method of analysis of standardized tests is called *erasure analysis*. Pennsylvania, New Jersey, New York, and Georgia, among numerous states, do erasure analysis; such analyses have been frequently reported on by the media. Here's how it works: the mean percentage of erasures on students' answer sheets is calculated statewide. Then, the percentage of eras-

---

[*] For example, as he shows in the book, he found a *d-a-d-b-c-b* string of correct answers, all in a later (and more difficult) section of a standardized test, among a number of students whose other answers weren't correlated.

ures of answer sheets for each school is examined. If that percentage is (let's say) more than two standard deviations from the mean (too many erasures, or too few), the school is "flagged" for further investigation.[*] Erasure analysis assumes that the distribution of erasures theoretically forms a symmetrical mound shape, which may or may not be true—there is disagreement in the literature.

---

[*] This is a bit of a simplification. The results of erasures, such as wrong-to-right, are also examined. In addition, what constitutes an erasure, rather than just a stray mark or other artifact, isn't agreed upon, bringing to mind the "hanging chad" fiasco of the 2000 presidential election.

§3. *What is Normal, Anyway?*

☖

Cumulative Graphs. Standardized Scores. Simple Transformations of Data.
The Uniform Distribution. The Empirical Rule.
Chebyshev's Inequality. The Normal Distribution.

§3a. *Cumulative Graphs*

Cumulative frequency graphs start at a base value, and, class by class, sum up the frequencies in ascending order. Let's start by examining histograms of the cumulative variety.

Example 3a.-1

Look again at the homes' prices data set.

(a) Add two new columns which account for cumulative frequencies and cumulative relative frequencies.

Class	Frequency	Rel. Freq.	Cumul. Freq.	Cumul. Rel. Freq.
20 to < 40	4	0.13	4	0.13
40 to < 60	7	0.23	11	0.36
60 to < 80	9	0.30	20	0.66
80 to < 100	5	0.17	25	0.83
100 to < 120	4	0.13	29	0.96
120 to < 140	1	0.03	30	1.00

(b) Construct histograms for both the cumulative frequencies and the cumulative relative frequencies.

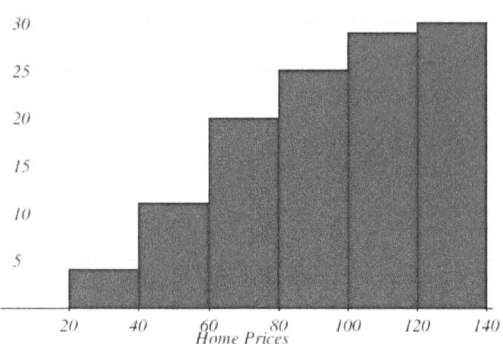 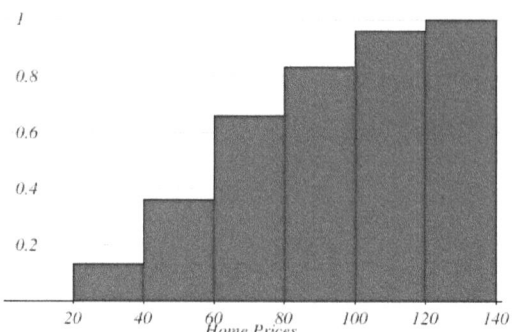

A cumulative relative frequency histogram's *y*-axis must always reach up to 1.

Definition 3a.-1

Ogive. A line graph of the cumulative frequency or cumulative relative frequency of each class in ascending order.

Example 3a.-2

(a) Construct an ogive of the relative frequencies of the home prices data set.

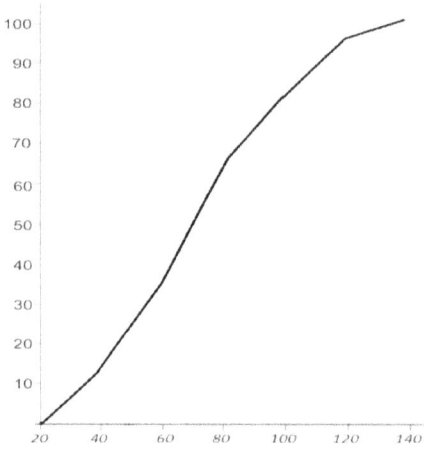

(b) Based on your cumulative relative frequency graph, complete the five-number summary table as best you can. (Remember that, for example, Q1, the lower quartile, is some number which 25% of the data fall below.)

Min	Q1	Median	Q3	Max
20	50	70	90	140

(c) What is the interquartile range of the internet usage times?

Take 50 from 90—the IQR is 40.

(d) Approximately what house price represents the 30th percentile? What does this mean?

55, or $55,000. This means that if the house price is $55,000, the house costs more than about 30% of homes in the sample.

(e) Approximately what internet usage time represents the 90th percentile? What does this mean?

110, or $110,000. This means that if the house price is $110,000, the house costs more than about 90% of homes in the sample.

(f) What percentage of houses cost less than $100,000?

Looking at the graph, it appears that approximately 80% of homes cost less than $100,000, since $100,000 is at the 80th percentile.

It turns out that the shape of the cumulative frequency graph reveals the skew of the data. Look at the ogives below; the first is left-skewed, the second is symmetrical, and the third is right-skewed.

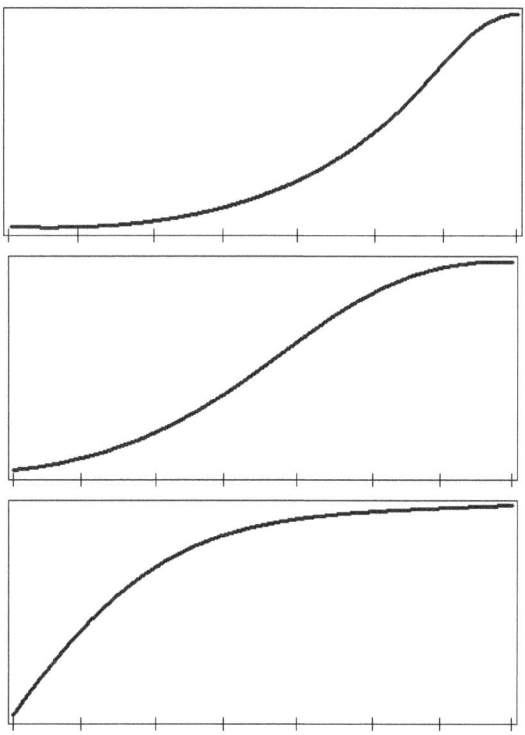

§3b. *Standardized Scores*

Percentiles are one way to compare one observation from a data set to all others. Percentiles reveal the relative position of observations. Standardized scores, also called *z-scores*, are another way to make such comparisons.

Definition 3b.-1

Z-Score. Measures how many standard deviations above (or below) the mean a particular "value of interest" falls in a distribution.

A *z*-score is found using the following formula, where *x* represents the "value of interest":

$$z = \frac{x - \mu}{\sigma}$$

More useful than using z-scores to compare observations in the same data set, however, is using z-scores to compare the position of individuals in different distributions.

Example 3b.-1

As of 2015, the mean of the top ten all-time homerun hitters in Major League Baseball (MLB) is 658, with a standard deviation of 66. The mean of the top ten all-time homerun hitters in minor league baseball is 216, with a standard deviation of 77.

(a) Babe Ruth, of the MLB, hit 714 homeruns, while Mike Hessman, of the minors, has hit 425. Find each player's *z*-score and compare your results.

$$z_{BR} = \frac{714 - 658}{66} = 0.85$$

$$z_{MH} = \frac{425 - 216}{77} = 2.71$$

Even though Ruth hit 289 more homeruns than Hessman, Hessman has the much higher *z*-score—Hessman's homerun total is nearly three standard deviations above the mean.

(b) Recall that an unusual value will be at least two standard deviations above the mean. Who had the "unusual" number of homeruns?

Hessman, since his total was more than two standard deviations above the mean.

(c) Hessman is the all-time homerun leader in the minor leagues. In MLB, Barry Bonds has the most, with 762 homeruns. If Babe Ruth had hit enough homeruns to have the same *z*-score as Hessman, would Ruth (instead of Bonds) have been the all-time homerun leader in the MLB?

Set up a *z*-score, and then solve for *x*:

$$2.71 = \frac{x - 658}{66}, x = 837$$

Yes, Babe Ruth would have been the MLB leader with 837 homeruns, 75 more than Barry Bonds.

§3c. *Simple Transformations of Data*

We wish to consider what happens to the common summary measures of descriptive statistics—mean, median, range, standard deviation, quartiles, and interquartile range—when we either add (or subtract) a constant to each observation or multiply (or divide) each observation by a constant.

Example 3c.-1

Here are summary statistics from a statistics exam.

Summary Statistics	Score	Summary Statistics	Score
Mean	62	First Quartile	48
Median	59	Third Quartile	68
Range	45	Interquartile Range	20
Standard Deviation	9		

(a) Suppose an error was made—one of the questions on the statistics exam was unanswerable. Since each question was worth 5 points, 5 points is added to everyone's exam score. What will the new summary statistics be?

Summary Statistics	Score	Summary Statistics	Score
Mean	67	First Quartile (Q1)	53
Median	64	Third Quartile (Q3)	73
Range	45	Interquartile Range	20
Standard Deviation	9		

The measures of center all increase by 5 points. But the measures of spread don't change, since all of the data points are relatively the same distances from one other.

(b) Suppose that Jenna's test, before the error was noticed, was located at the 75th percentile. After 5 points is added to everyone's exam (including hers), what is her new percentile?

Jenna's percentile has not changed, since her position relative to her peers is the same.

(c) Now suppose that, instead of adding 5 points to each person's exam, each exam's score is *increased* by 5%. Calculate the new summary statistics.

Summary Statistics	Score	Summary Statistics	Score
Mean	65.10	First Quartile (Q1)	50.40
Median	61.95	Third Quartile (Q3)	71.40
Range	47.25	Interquartile Range	21.00
Standard Deviation	9.45		

These values were obtained by multiplying each summary measure by 1.05—in effect reporting 105% of each measure.

(d) Consider one more possibility: if all exam scores had *instead* been lowered by 10%, what would the new summary statistics be?

Summary Statistics	Score	Summary Statistics	Score
Mean	55.80	First Quartile (Q1)	43.20
Median	53.10	Third Quartile (Q3)	61.20
Range	40.50	Interquartile Range	18.00
Standard Deviation	8.10		

These answers were found by multiplying each summary measure by 0.90.

If a set of data needs to be transformed in both an additive way (like adding the five points to everyone's exam, as we did in the previous example) and a multiplicative way (like increasing everyone's score by a certain percentage), then the transformations to the summary measures (e.g., the mean, median, etc.) should be applied in the same order as they were to the data points themselves.

Standardizing a data set is a common additive-multiplicative transformation. Here, the *z*-score of each observation is taken, creating a new data set—simply of *z*-scores. The mean of *any* set of standardized data is zero, and the standard deviation of any standardized data set is one.

Here's why. Suppose x_1, $x_2, \ldots, x_n$ constitute a data set. The mean of the data is $\bar{x}$, with a standard deviation of s. Note the mean is equal to $\bar{x} = \dfrac{x_1 + x_2 + \cdots + x_n}{n}$ and the standard devia-

tion is equal to $s = \sqrt{\dfrac{(x_1 - \bar{x})^2 + (x_2 - \bar{x})^2 + \cdots + (x_n - \bar{x})^2}{n-1}}$. To find the *z*-score of, let's say, x_1,

we would subtract the mean and divide by the standard deviation:

$$z_1 = \frac{x_1 - \bar{x}}{s}$$

But we need to find the *z*-score of not just the first data point x_1, but of all them. Then, we need to find the mean of all the *z*-scores to get the mean of any standardized score.

$$\bar{x}_z = \frac{\left(x_1 - \dfrac{x_1 + x_2 + \cdots + x_n}{n}\right) + \left(x_2 - \dfrac{x_1 + x_2 + \cdots + x_n}{n}\right) + \cdots + \left(x_n - \dfrac{x_1 + x_2 + \cdots + x_n}{n}\right)}{n}$$

$$\bar{x}_z = \frac{\left(\dfrac{nx_1}{n} - \dfrac{x_1 + x_2 + \cdots + x_n}{n}\right) + \left(\dfrac{nx_2}{n} - \dfrac{x_1 + x_2 + \cdots + x_n}{n}\right) + \cdots + \left(\dfrac{nx_n}{n} - \dfrac{x_1 + x_2 + \cdots + x_n}{n}\right)}{n}$$

$$\bar{x}_z = \frac{\left(\dfrac{2x_1 - x_2 - \cdots - x_n}{n}\right) + \left(\dfrac{2x_2 - x_1 - \cdots - x_n}{n}\right) + \cdots + \left(\dfrac{2x_n - x_1 - x_2 - \cdots - x_{n-1}}{n}\right)}{n}$$

$$\bar{x}_z = \frac{\left(\dfrac{2x_1 - x_2 - \cdots - x_n + 2x_2 - x_1 - \cdots - x_n + \cdots + 2x_n - x_1 - x_2 - \cdots - x_{n-1}}{n}\right)}{n}$$

$$\bar{x}_z = \frac{\left(\dfrac{0}{n}\right)}{n}$$

$$\bar{x}_z = 0$$

On to standard deviation. When altering a measure of spread—as a *z*-score will do, by dividing by the standard deviation of the entire data set—adding (or subtracting) doesn't have an effect. So, for instance, when finding the *z*-score of the first data point, x_1, we can ignore any instances of where the mean shows up:

$$z_1 = \frac{x_1}{\sqrt{\dfrac{(x_1)^2 + (x_2)^2 + \cdots + (x_n)^2}{n-1}}}$$

So the standard deviation of all of the *z*-scores is

$$s_z = \sqrt{\frac{\left(\dfrac{x_1}{\sqrt{\dfrac{(x_1)^2 + (x_2)^2 + \cdots + (x_n)^2}{n-1}}}\right)^2 + \cdots + \left(\dfrac{x_n}{\sqrt{\dfrac{(x_1)^2 + (x_2)^2 + \cdots + (x_n)^2}{n-1}}}\right)^2}{n-1}}$$

$$s_z = \sqrt{\frac{\left(\dfrac{(x_1)^2}{\dfrac{(x_1)^2 + (x_2)^2 + \cdots + (x_n)^2}{n-1}}\right) + \cdots + \left(\dfrac{(x_n)^2}{\dfrac{(x_1)^2 + (x_2)^2 + \cdots + (x_n)^2}{n-1}}\right)}{n-1}}$$

$$s_z = \sqrt{\frac{\dfrac{(x_1)^2 + (x_2)^2 + \cdots + (x_n)^2}{\left(\dfrac{(x_1)^2 + (x_2)^2 + \cdots + (x_n)^2}{n-1}\right)}}{n-1}}$$

$$s_z = \sqrt{\frac{\dfrac{1}{\left(\dfrac{1}{n-1}\right)}}{n-1}}$$

$$s_z = \sqrt{\frac{n-1}{n-1}}$$

$$s_z = \sqrt{1}$$

$$s_z = 1$$

Through all of these tedious steps we have shown that the mean of any set of standardized data is zero, and the standard deviation of any standardized data set is one, as expected.

§3d. *The Uniform Distribution*

If enough observations from a population are displayed on some sort of graph, the distribution may be predictable enough to be represented by a smoothed curve, called a *density curve*.

Definition 3d.-1

Density curve. A smooth curve that describes a distribution; the area underneath it must be equal to one.

Definition 3d.-2

Uniform distribution. Graphed as a rectangle, the distribution describes some process that has an equal probability of occurring at any value over a set interval.

A graph of a uniform distribution allows us to calculate proportions (or percentages) of observations by simply finding the areas underneath the density curve at given values. Those areas are easy to calculate because of the rectangular shape of the graph.

Example 3d.-1

Suppose the probability of the length of time a library patron at the Huntingdon Valley Library actually stays in the library building on any given visit is given by the following uniform probability distribution.

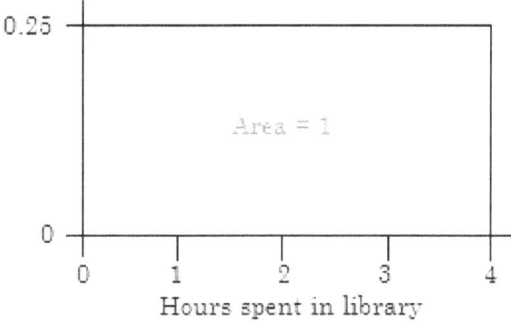

(a) Find the proportion of patrons who spend fewer than 2 hours in the Huntingdon Valley Library.

Take 2 multiplied by 0.25 to get the answer: 0.5.

(b) Find the percent of patrons who spend more than 3 hours in the Huntingdon Valley Library.

25%.

(c) Find the proportion of patrons who spend between 1 and 4 hours in the Huntingdon Valley Library.

0.75.

(d) Calculate the mean and median of the Huntingdon Valley Library density curve.

Since the mean and median of any symmetrical distribution are located at the center, both the mean and median of this data set is 2.

§3e. *The Empirical Rule*

Of all possible density curves, the *normal curve* is the most significant, mainly because many real-world distributions tend to be normally distributed. All normal curves are symmetrical and bell-shaped (or mound-shaped), and the mean, median, and mode all reside in the center. And you won't have to actually *see* a normal curve to describe it; rather, you'll need only two pieces of information: its mean and standard deviation. The mean describes the curve's location on an axis, and the standard deviation describes the curve's spread (or variability)—the bigger the standard deviation, the wider (and shorter) the normal curve will be.

An important rule of thumb to estimate the percentage of data between set values within normal distributions is called *the Empirical Rule*, or the *68-95-99.7 Rule*. The Empirical Rule will help us calculate *approximately*[*] what percentage of data lies between two observations. Specifically,

- Approximately 68% of the observations lie within 1 standard deviation of the mean μ;
- Approximately 95% of the observations lie within 2 standard deviations of the mean μ;
- Approximately 99.7% of the observations lie within 3 standard deviations of the mean μ.

If the curve is normal, these three percentages will always hold—no matter if the curve is low, stout, and spread out (called *platykurtic*) or toweringly tall and thin (called *leptokurtic*).

Example 3e.-1

An IQ test is designed to measure intelligence, although what it measures isn't so clear-cut; for this reason, the test isn't necessarily considered valid, although it is reliable.[†] The test was origi-

---

[*] Not *exactly*. We'll see how to obtain exact values in a subsequent §.

[†] Validity = Does the test measure what it purports to measure? Reliability = Does the test give consistent results under similar conditions? In *The Mismeasure of Man*, Gould warns of using a single measure, such as IQ, to "convert abstract concepts [like intelligence] into entities," calling this tendency "reification": "We recognize the importance of mentality in our lives and wish to characterize it, in part so that we can make the divisions and distinctions among people that our cultural and political systems dictate. We therefore give the word 'intelligence' to this wondrously complex and multifaceted set of human capabilities. This shorthand symbol is then reified and intelli-

nally developed in part by Alfred Binet in the last century.[*] For children, its calculation is simple: mental age divided by chronological age times 100.

To use the Empirical Rule to solve problems, we must know three things: if the data is mound-shaped (i.e., if the data follow a normal distribution), and the values of the mean and standard deviation of the data set. For IQ scores, $N(100,15^2)$.[†]

(a) Draw the normal density curve of IQ scores below, showing the values one, two, and three standard deviations from the mean. Also label the inflection points with dots.[‡]

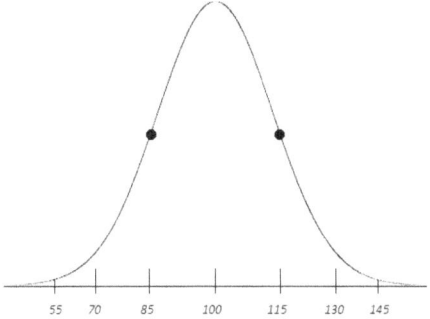

(b) What percentage of people have an IQ greater than 100?

No calculations necessary—it's 50%, as shown in the graph.

(c) What percentage of people have an IQ between 85 and 115? between 70 and 130? between 55 and 145?

Because we know that IQ scores are normally distributed—the test is normalized[§]—we can use the Empirical Rule to answer the questions: 68%, 95%, and 99.7%.

---

gence achieves its dubious status as a unitary thing." Reification goes hand in hand with ranking—people are still ranked and grouped using IQ as the measurement of choice.

[*] Again, see Gould's *The Mismeasure of Man* for a comprehensive treatment on the history and uses of IQ testing.

[†] A compact notation relaying three things: the letter tells you the type of distribution (here, N for normal), the first number inside the parentheses gives the mean, and the second number gives the variance. Like circles, which are uniquely defined by only two measures (length of the radius and location of the center), unique normal distributions exist for every mean and variance (or standard deviation).

[‡] The inflection points are at the locations where the curve's concavity changes. On a normal curve, underneath the inflection points lies exactly one standard deviation from the mean. A bit of calculus: the second derivative at an inflection point is equal to zero.

[§] Interesting tidbit about IQ: The Flynn effect, first observed by James Flynn, notes the increase in intelligence scores since such measurement began—so IQ has been going up over time, and the test has to be repeatedly rescaled

(d) What percentage of people have an IQ between 85 and 130?

There's 68% of the data between 85 and 115; thus, between 100 and 115, there's half of that—or 34%. And there's 95% of the data between 70 and 130, so half of that, or 47.5%, is between 100 and 130. Add up the two percentages—34% and 47.5%—to get the total between 85 and 130: 81.5%. See the diagram below.

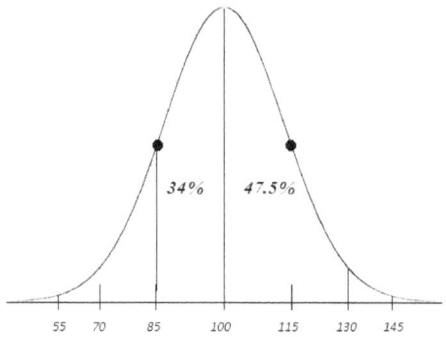

(e) What percentage of people have an IQ greater than 145?

Examine the tail at the far right of the normal curve. The entire normal curve—from the far left end to the far right end[*]—contains 100% of the data. Traveling within three standard deviations of the mean, though, will net you 99.7%. Subtract the two percentages, and you'll get the percentage in *both* tails: 0.3%. Since the normal curve is perfectly symmetric about the center, taking half of 0.3% will give you the percentage is one tail: 0.15%.

Interestingly, this result implies that the percentage of people with an IQ less than 55 and with an IQ greater than 145 is the same, meaning that, theoretically, you should be able to pair such people up, with a one-to-one correspondence. (Although how, exactly, that would work in practice seems open to debate.)

(f) What percentage of people have an IQ between 55 and 85?

---

to account for the change. Why might this be? Perhaps there's a kind of scaffolding taking place, or a cultural inheritance, a "cultural ratchet." As Vincent Deary explains in *How We Are*, "What distinguishes us from the pride of lions is pinpointed…in the analogy between individual habit and group custom…. The later we are born, the more knowledge we inherit about how to live. This has also been called by cognitive scientists the 'cultural ratchet'. While each new generation of lions is back to a kind of Year Zero of lion life, back to solving the problems of how to live again," humans have evolved a method (i.e., language and culture) of transmitting such knowledge through the generations.

[*] The normal distribution function is asymptotic to the axis, meaning that the curve gets closer and closer but technically never actually *touches* the axis, in both directions.

Less than 100 gives us 50%. All that's needed is to subtract 34% and 0.15% from 50%. The answer is 15.85%. See below.

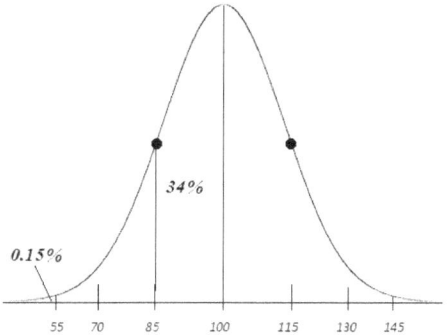

(g) Out of 1000 randomly selected people, how many have an IQ less than 55?

There are 0.15% of people with an IQ lower than 55. In a group of 1,000 people, as long as they're randomly selected, we'd expect $1000 \cdot 0.0015 = 1.5$, or about 2 people, to have an IQ lower than 55.

   If the problem did not specify random selection, however, we would not have confidence in the answer. (For instance, what if those thousand people were students at Harvard University? Then it's unlikely to find any with an IQ anywhere near that low, even if you're not sold on IQ measuring raw "intelligence" per se.)

(h) If you have an IQ of 55, how many standard deviations below the mean are you?

Three. Take a look at the scale at bottom:

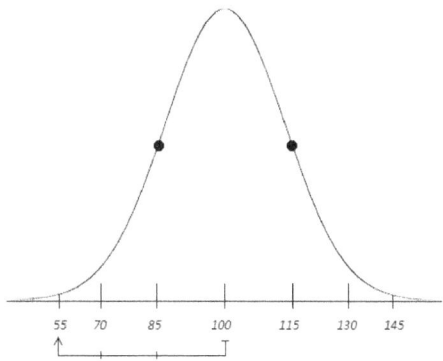

(i) If you have an IQ of 80, how many standard deviations below the mean are you?

Because an IQ of 80 lands in between two tick marks, this problem is best solved using a *z*-score.

$$z = \frac{80-100}{15} = -1.33$$

(j) Marilyn vos Savant has world's officially recorded highest IQ, at 240. How many standard deviations above the mean is Marilyn's IQ?

$$z = \frac{240-100}{15} = 9.33$$

(k) Samara takes an IQ test three times. Her scores are 98, 101, and 100. What is the most likely explanation for her different scores?

Variability—which means that even if we attempt to gather data repeatedly under conditions which are similar, the results won't necessarily be identical—best explains the differences in Samara's scores.

§3f. *Chebyshev's Inequality*

The Empirical Rule only applies to normal curves. Is there a rule that applies to *any* distribution (of any shape)?

There is, but the rule is limited. It's called *Chebyshev's Inequality*, named after the mathematician Pafnuty Chebyshev. With Chebyshev's Inequality (also sometimes called Chebyshev's Rule), if you have the mean and standard deviation of the data set (regardless of the shape), you'll be able to glean the percentage of data lying *at least* k standard deviations from the mean.

As long as k is more than zero, the formula is given as

$$P(|x - \mu| \geq k\sigma) \leq \frac{1}{k^2}$$

Usually the formula is more easily thought of as $1 - \frac{1}{k^2}$, which works as long as $k > 1$. Let's take a look at some examples.

Example 3f.-1

According to Chebyshev's inequality, *at least* what percentage of the data lie within two standard deviations of the mean of any data set? How about three standard deviations?

Solution. Using the formula, we obtain

$1 - \dfrac{1}{2^2} = 0.75$, or at least 75% for two standard deviations from the mean, and

$1 - \dfrac{1}{3^2} = 0.89$, or at least 89% for three standard deviations from the mean.

Example 3f.-2

The mean age of Ukraine's residents is around 43 years old, with a standard deviation of about 10 years.

(a) What can you learn if you apply Chebyshev's Inequality to this data set using $k = 2$?

At least 75% of Ukraine's residents are between the ages of 23 and 63 years old (those ages are two standard deviations from the mean age).

(b) Suppose we managed to bus in one million 85-year-olds into Ukraine. Would the results you obtained in the previous question using Chebyshev still hold?

No, since the mean and standard deviation of the age distribution would change.

The reason why these results have, at best, only limited utility lies with the "at least" portion of their interpretations. Even though we know that at least 75% of the data is located within two standard deviations of the mean of *any* data set, no matter the shape, the percentage of a particular data set—say, one dramatically skewed left— satisfying that condition might be nearly 100%. Chebyshev is a one-size-fits-all formula, unlike the normal-curve-only the Empirical Rule, and, as such, is inherently limited, the way one-size-fits-all clothes will fit a few people well but most other people poorly.

§3g. *The Normal Distribution*

Unfortunately, we cannot use the Empirical Rule to obtain the percentage between *any* two observations in a normally distributed data set. The density curve (or, as we'll call it in later pages, the *probability density function*) of the normal distribution function[*] is

$$f(x) = \frac{1}{\sigma\sqrt{2\pi}} e^{\frac{-1}{2}\left(\frac{x-\mu}{\sigma}\right)^2}$$

---

[*] Actually, it's not just one curve, but a family of curves; the area under each curve is equal to one, with the shape and position of the curve determined by the standard deviation and mean, respectively.

Note the *z*-score calculation in the exponent of *e*. The normal distribution has its origin in the investigation of errors in astronomical and geodesic survey calculations, especially by the mathematician Carl Frederick Gauss.[*] (Sometimes the normal distribution is instead called the *Gaussian distribution*.)[†] Eventually, the normal curve was shown to have great utility modeling phenomena besides error distributions.

The Empirical Rule gives us quick and dirty estimates for percentages of data lying underneath the curve. Generally, when we wish to obtain the area underneath mathematical functions that are not constructed of simple geometric shapes (such as squares, rectangles, trapezoids, circles, or the like), integration—whether by numerical or symbolic methods, like Taylor Series—is our only option.

To make the integration easier, mathematicians don't consider the whole class of normal curves, but integrate over a so-called *standard normal distribution*, where $N(0,1)$, as shown here:

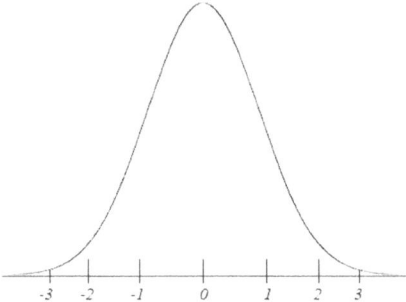

From this "standardized" normal curve, we can integrate using a much simpler formula, where the mean and standard deviation are set to 0 and 1, respectively:

$$y = \frac{1}{\sqrt{2\pi}} e^{\frac{-1}{2}x^2}$$

The coefficient $1/\sqrt{2\pi}$ ensures that the area underneath the curve is equal to one.

Using the *z*-score formula standardizes values, placing them on the same scale as the standard normal distribution. A table of values, residing in the Statistical Tables section at the back of this primer, converts the *z*-score to a proportion, percentage, area underneath the curve, or probability, depending on how the results are interpreted in context.

---

[*] Although Gauss, who headed a Bavarian geodesic survey (measuring land curvatures), wasn't the first to describe its properties, nor was Laplace. It was Abraham de Moivre, friend of Isaac Newton, inveterate gambler, and pioneer of probability theory, who cast the mathematical parameters of the normal distribution.

[†] Gauss wasn't the only mathematician taken by the normal distribution. Francis Galton, whom we met earlier, constructed a Quincunx "bean machine," or Galton box, that let hundreds of marbles loose down a chute, only to encounter resistance in the form of little pegs on their way to the bottom of the box. (Pascal's triangle can be used to calculate the multitude of different paths the marbles can take.) You can still see variants of the bean machine today, most notably in the pricing game Plinko on the TV game show *The Price is Right*.

Example 3g.-1

Let's reexamine IQ. Recall that the mean IQ is 100, the standard deviation is 15, and the data distribution is mound-shaped and symmetrical.

(a) Construct a graph of IQ, plotting the inflection points and the scale.

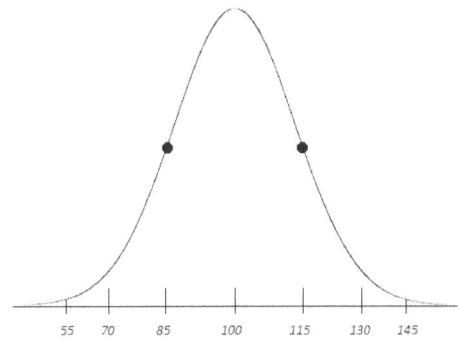

(b) Based on the graph, approximately what proportion of people have an IQ lower than 100?

Approximately 0.5.

(c) Find the *z*-score for an individual with an IQ of 100.

$$z = \frac{100 - 100}{15} = 0$$

(d) Use the standard normal distribution table to convert the *z*-score to a proportion. So, according to the table, what proportion of people have IQs below 100? Does this agree with your answer to part (b)?

Looking up a *z*-score of 0 lands you at the following (circled) spot on the table:

z	0	0.0
0.0	0.5000	0.50
0.1	0.5398	0.54
0.2	0.5793	0.58

This value, 0.5000, matches the answer to part (b).

Notice that the normal distribution table gives you the area underneath the normal curve to the *left* of the z-score you look up; this area represents the proportion or probability *less than* that particular z-score. Recall the total area underneath the normal curve is equal to one (the uniform distribution's area was also equal to one).

(e) Using a z-score and the table, calculate the proportion of people who have an IQ lower than 115 points.

$$z = \frac{115 - 100}{15} = 1$$

0.7	0.7580	0.7611	0.7
0.8	0.7881	0.7910	0.7
0.9	0.8159	0.8186	0.8
1.0	0.8413	0.8438	0.8
1.1	0.8643	0.8665	0.8
1.2	0.8849	0.8869	0.8

(f) Now, use the Empirical Rule to find the proportion of people who have an IQ lower than 115 points. Does your answer closely match part (e) above?

Take 50% + 34% = 84%. Yes, this result is close to the answer to part (e).

(g) If an American were selected at random, approximately what is the *probability* that he or she would have an IQ lower than 115 points?

Probability = proportion here, since the total area underneath the normal distribution is equal to one. Therefore, the answer is 0.8413.

(h) If 1,000 Americans were randomly selected, approximately how many of those 1,000 would have an IQ lower than 115 points?

$1000 \cdot 0.8413 \approx 841$ people.

(i) Suppose, though, that you wish to find out what proportion of Americans have an IQ higher than 115. How would you go about doing that, and what is the answer?

Since the entire area underneath the normal curve is equal to one, all we need to do is take $1 - 0.8413 = 0.1587$. So anytime we need a "more than," "greater than," or "at least" proportion (or probability), we'll simply find a z-score, look it up on the table, and subtract the table's value from one.

(j) Now suppose that you wished to find out the proportion of Americans who have an IQ between 100 and 115. First, return to your graph of the normal curve and shade in the area you are looking for. Next, use the proportions you calculated in parts (d) and (e) and subtract them. What is this proportion?

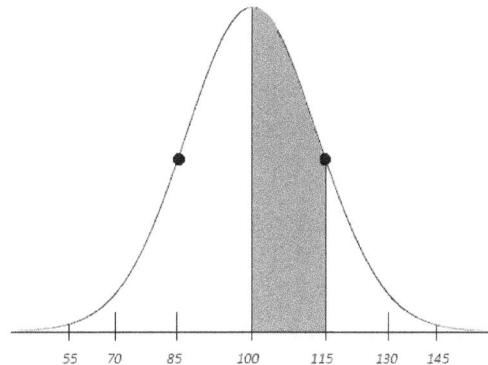

According to the Empirical Rule, the proportion should hover around 34%. Finding *z*-scores and looking them up on the standard normal distribution table, the result is $0.8413 - 0.5000 = 0.3413$. This isn't the *exact* answer—the values on the table are rounded to the ten thousandths place—but the result is closer to the true value than the Empirical Rule gives us.

(k) Calculate the proportion of Americans who have an IQ between 85 and 120 points.

Find two *z*-scores:

$$z = \frac{85 - 100}{15} = -1$$

$$z = \frac{120 - 100}{15} = 1.33$$

The first *z*-score is easy to look up: 0.1587. The second one, however, requires a new technique—traveling along the top row, which accounts for the hundredths places of *z*-scores:

			Standard Norm	
z	0	0.01	0.02	0.03
1.3	0.9032	0.9049	0.9066	0.9082
1.4	0.9192	0.9207	0.9222	0.9236
1.5	0.9332	0.9345	0.9357	0.9370

Finally, take the difference of the two looked-up values to obtain 0.7495.[*]

---

[*] Also note that the proportion at any particular *z*-score—let's say 1—is equal to zero, since a line has zero area. For instance, the proportion of people with an IQ of exactly 115 is zero, as is the proportion at any other number. That seems strange, but, mathematically, it's true. As we'll examine later, the issue lies with the normal distribution being

(l) Let's say we know that exactly 90.99% of people have an IQ below x. Find x.

In order to find out what x is, you have to work "backward": first, search the cells in the table for the z-score that has a proportion of 0.9099.

			Standard Normal Probab

z	0	0.01	0.02	0.03	0.04	0
1.3	0.9032	0.9049	0.9066	0.9082	0.9099	0.
1.4	0.9192	0.9207	0.9222	0.9236	0.9251	0.

Then, use the z-score formula to solve for x.

$$1.34 = \frac{x-100}{15}, \ x = 120.1$$

(g) Suppose we know that Max has an IQ at the 30th percentile. What is Max's IQ?

Again, let's look up the percentage (or proportion) "backward." In this case, find the closest value you can in the table to 0.3:

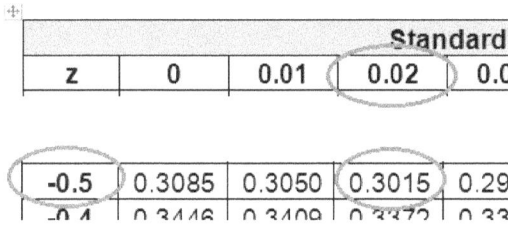

			Standard

z	0	0.01	0.02	0.0
-0.5	0.3085	0.3050	0.3015	0.29
-0.4	0.3446	0.3409	0.3372	0.33

And solve for x:

$$-0.52 = \frac{x-100}{15}, \ x = 92.2$$

We've now used the Empirical Rule to roughly check results obtained from the standard normal distribution; how about deriving the three Empirical Rule percentages by using the standard normal table?

---

continuous (lying over an interval), whereas single values, like 115, are *discrete* (finite, or countable).

Example 3g.-2

Use the standard normal table to match the Empirical Rule.

Solution. According to the Empirical Rule, approximately 68% of the data lies one standard deviation from the mean. If we look up these two z-scores, 1 and -1, and subtract the proportions found on the table, we get: $0.8413 - 0.1587 = 0.6826$.

Likewise, for two and three standard deviations from the mean, the proportions off of the standard normal table are 0.9544 and 0.9975, respectively.

All three proportions calculated here hew closely to the Empirical Rule.

Using notation when querying about the normal curve brings additional clarity and consistency to the proceedings, as the following example illustrates.

Example 3g.-3

Suppose that a data set, which is normally distributed, has a mean $\mu = 10$ and standard deviation $\sigma = 5$.

(a) If we wish to find the area under the normal distribution—or the probability—to the left of, let's say, $x = 15$, then we can simply ask: Find $P(x < 15)$.[*] Calculate this probability now.

$$z = \frac{15-10}{5} = 1.0, \ 0.8413$$

(b) Now, suppose we want to calculate the probability that an observation is greater than a certain value—say 11. Simply ask: Find $P(x > 11)$. Calculate this probability now.

$$z = \frac{11-10}{5} = 0.2, \ 1 - 0.5793 = 0.4207$$

(c) We might also want a probability between two observations. Calculate the following: $P(7 < x < 13)$.

$$z = \frac{7-10}{5} = -0.6, \ 0.2743$$

---

[*] The *P* notation refers to "probability," although it is equally valid to think of it as "proportion."

$$z = \frac{13-10}{5} = 0.6, \ 0.7257$$

$$0.7257 - 0.2743 = 0.4514$$

Sometimes you already have the z-score and simply wish to get the area underneath the normal curve. You then only need to consider standard normal distribution.

(d) Calculate this probability: $P(z < 1.54)$.

   0.9382.

(e) Please note that we could have also asked the question of $P(z < 1.54)$ in this way: Find $z < 1.54$. So, now, find the following: $z > -1.15$.

   $1 - 0.1251 = 0.8749$

You should be comfortable solving more complex, multi-step problems with normal distributions.

Example 3g.-4

The braking distance of the average Chevrolet Corvette on a dry surface is about 100 feet, with a standard deviation of 10 feet.

(a) Recall that an "unusual value" is more than two standard deviations away from the mean. Zoe has a Corvette that take 112 feet to stop on a dry surface. Is this unusual?

   No, since $100 \pm 2 \cdot 10 = (80, 120)$.

(b) Find the middle 50% of the braking distances. (Hint: This is the same as finding the values at the 25th percentile, P_{25}, and the 75th percentile, P_{25}.)

   Another way to look at this is by using quartiles—the 25th percentile is Q1, and the 75th is Q3 (see the sketch at the top of the next page). To get Q1, we'll have to look up 0.25 "backward" on the standard normal distribution table; thus, Q1 has a z-score of –0.67.[*]

---

[*] When looking up percentages or proportions "backward," always ask yourself this question: What percentage is less than the value I want? Since we want Q1, and 25% of the area underneath the curve is less than Q1, we looked up 0.25 "backward."

Likewise, looking up 0.75 gives us the *z*-score for Q3, which is 0.67.[*]

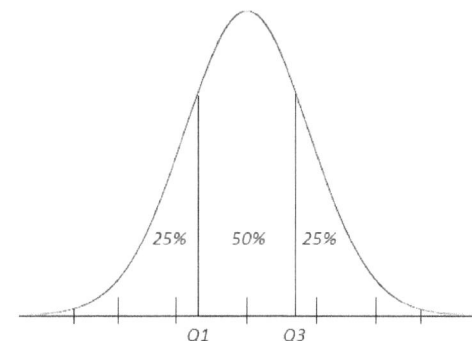

$$-0.67 = \frac{Q1 - 100}{10} = 93.3$$

$$0.67 = \frac{Q3 - 100}{10} = 106.7$$

(c) Find the interquartile range (IQR) of braking distances.

$$106.7 - 93.3 = 13.4 \text{ feet.}$$

Not every distribution that data originate from is normal. Sometimes, given just a mean and a standard deviation, normality can be shown to be impossible. Take salaries at a company, for instances. Salaries tend to be skewed right—since a few people make a lot, and a lot of people make a little.

Example 3g.-5

The salaries at a local company have a mean of $120,000, with a standard deviation of $67,000. Explain why this distribution cannot be normal.

Solution. In a normal distribution, approximately 95% of the observations lie two standard deviations from the mean. But that's impossible here, since $120000 - 2 \cdot 67000 = -14000$, and only a fool would agree to work for a negative income.

For inferential methods later on, it will be important to test a set of data to see if the population from which the data came from is (likely) normally distributed.[†] Examining a histogram of the

---

[*] Since Q1 and Q3 are equidistant from the normal curve's midpoint (the mean), by using symmetry you should have been able to deduce Q3's *z*-score without looking it up.

[†] We can never ascertain normality for sure unless we have collected every piece of data from the host population.

sample data is one way to tell; a second way is to look at a modified boxplot; and a third way is to study a *normal probability plot* (NPP).

Definition 3g.-1

***Normal probability plot* (*NPP*).** A scatterplot in which the ordered sample data are plotted against the *z*-scores of the corresponding data points' percentiles.

Example 3g.-6

Reconsider the ATM data set presented much earlier. Does the data set come from a normal population? Construct and interpret a histogram, a modified boxplot, and an NPP.

Solution. Using a graphing calculator, first, view the histogram.

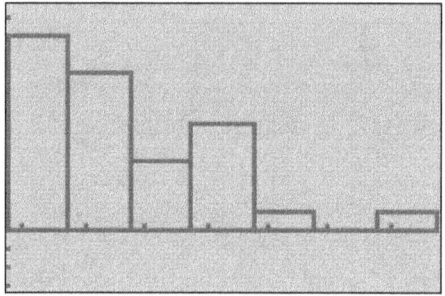

The ATM data set looks very skewed right; it also does not closely follow the Empirical Rule, which we would expect from normally distributed data. Therefore, we have little evidence to claim that the data set comes from a normal population, although we always have to keep the concept of variability in mind.

The modified boxplot of ATM supports our conclusion:

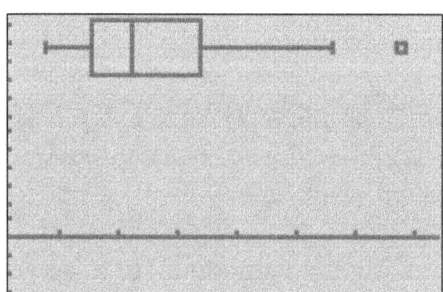

Again, observe the skewness. In addition, the presence of an outlier most likely disqualifies the data set from population normality, since the chance of sampling a data point that far away from the mean *in a normally distributed population* is remote.

In fact, before we examine the NPP, let's calculate the probability of finding an outlier in a normal distribution. Recall that any observation that's more than 1.5 IQRs away from the nearer quartile is considered an outlier (the process of checking for outliers this way is called *the outlier test*). The IQR of a standard normal distribution is $0.67-(-0.67)=1.34$, since the upper and lower quartiles have *z*-scores of 0.67 and –0.67, respectively, and $1.5\cdot1.34=2.01$. Any observation, then, with a *z*-score of less than $-0.67-2.01=-2.68$ or more than $0.67+2.01=2.68$ would be considered an outlier. Looking up –2.68 on the standard normal table, we get 0.0037. Looking up 2.68 on the standard normal table, and subtracting that proportion from one (since we want area to the *right* of that *z*-score—that's high outlier territory), we obtain 0.0037 again. Therefore, the probability of randomly selecting an observation in a normally distributed population that's an outlier is $2\cdot0.0037=0.0074$, or less than a 1% chance. If a sample of data has very few observations—say, less than twenty—then the chance of an outlier being among them, given that the host population is normal, is very small.

Finally, we'll use a normal probability plot to analyze the data set. To construct an NPP using your calculator, do the following:

- Go into the STAT PLOT menu by typing 2nd $\boxed{Y=}$ and select the first plot.
- Turn the plot ON and select the sixth type of plot (the icon that looks like a line plot).
- For the "Data List," select the ʟATM list.
- For "Data Axis," select Y.
- Hit the $\boxed{\text{ZOOM}}$ key and select option 9—which is "ZoomStat."
- After hitting $\boxed{\text{ENTER}}$, you should see the normal probability plot.

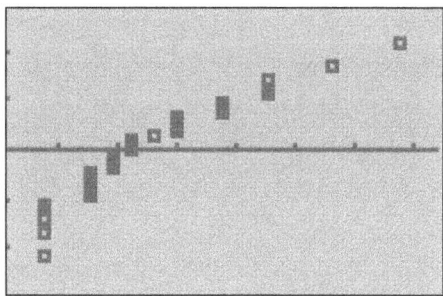

A perfectly normally distributed data set would form a straight, diagonal line in an NPP. Although you shouldn't overreact to portions of the plot that aren't straight, the curvature of the ATM NPP suggests clear departures from normality—and reconfirms our suspicion that the ATM data set is not from a normally distributed population.

§4. *They Come in Pairs*

ϙ

Scatterplots. Quantifying Relationships. Lines of Best Fit. Residuals.
The Coefficient of Determination. There's No Substitute for Looking. Computer Output.
Linear Transformations of Nonlinear Data.

§4a. *Scatterplots*

World Wrestling Entertainment (WWE) holds a pay-per-view contest every January called the *Royal Rumble*. In it, thirty superstars (the industry's catchall term for wrestlers) enter the ring, one by one, in two minute intervals. Their objective: outlast every other superstar by not getting thrown over the top ropes, which eliminates superstars from contention. The sole survivor punches his ticket to the main event of *WrestleMania*, in effect the Super Bowl of wrestling. Notwithstanding that the matches in professional wrestling are staged,[*] let's examine the entry and elimination spots in one of the most controversial[†] *Royal Rumbles*:

Wrestler	Entered	Eliminated	Wrestler	Entered	Eliminated	Wrestler	Entered	Eliminated
The Miz	1	1	Fandango	11	10	Damien Mizdow	21	16
R-Truth	2	2	Tyson Kidd	12	8	Jack Swagger	22	20
Bubba Ray Dudley	3	3	Stardust	13	15	Ryback	23	19
Luke Harper	4	4	Diamond Dallas Page	14	9	Kane	24	26

---

[*] Not fake; "fake" would imply that no one gets hurt, which is certainly not the case. Injuries happen on a nightly basis. In addition, even though move sets are designed to inflict a minimum amount of damage, wrestlers' bodies still experience the same sorts of wear and tear that professional athletes in any major sport have to deal with.

[†] The data we will examine come from the 2015 event, which was supposed to be a coronation for Roman Reigns, the heir apparent (and real-life relative) of movie star Dwayne "The Rock" Johnson; but Philadelphia's raucous crowd turned on the superstar, nearly booing him out of the building after his victory.

Wrestler	Entered	Eliminated	Wrestler	Entered	Eliminated	Wrestler	Entered	Eliminated
Bray Wyatt	5	24	Rusev	15	28	Dean Ambrose	25	25
Curtis Axel	6	--	Goldust	16	14	Titus O'Neil	26	17
The Boo-geyman	7	5	Kofi Kingston	17	13	Wade Barrett	27	21
Sin Cara	8	6	Adam Rose	18	12	Cesaro	28	22
Zack Ryder	9	7	Roman Reigns	19	30	The Big Show	29	27
Daniel Bryan	10	11	Big E	20	18	Dolph Ziggler	30	23

Before we analyze the data, notice that Curtis Axel,[*] who entered sixth, was never eliminated—he never even made it to the ring—so we'll ignore his data point. In addition, Roman Reigns, the winner, is considered "eliminated" thirtieth, even though, technically, he wasn't eliminated.

The individuals (also called the observational units, or cases) here are the wrestlers, since we're collecting the data from them. There are two quantitative variables: "Entered" and "Eliminated." Since the entry rank is likely to predict the elimination spot—since it's hard to last a long time in the ring, the probability of winning the *Rumble* increases the later a wrestler enters the contest—"Entered" is the explanatory variable, and "Eliminated" is the response variable.

Constructing a scatterplot is one of the easiest ways to visualize the relationship between a set of bivariate quantitative data. The explanatory variable is plotted on the *x*-axis, with the response variable affixed to the *y*-axis. Here's a trick to help you remember: On the *x*-axis goes the "*x*-planatory" variable.

A scatterplot of elimination versus entry[†] ranks is shown below.

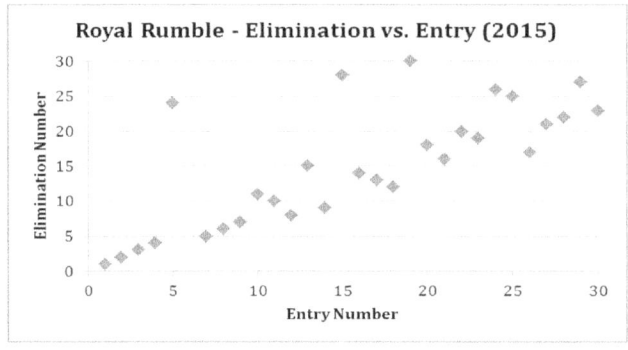

The plot clearly shows a relationship between the two variables, and one that we anticipated: in general, the later the entry number, the later the elimination. The plot also looks relatively linear

---

[*] Son of the legendary late professional wrestler Curt Hennig, a/k/a Mr. Perfect.

[†] Note: when titling a scatterplot, always use a *y* vs. *x* convention.

(i.e., straight; modeled by a line); some other plots in subsequent pages will have curvilinear shapes.

Also notice the outlier, Bray Wyatt.[*] Although he entered fifth, he lasted until late in the *Rumble*—he was eliminated twenty-fourth (having thrown a number of other wrestlers out of the ring, including fan-favorite Daniel Bryan, to boot). Outliers in scatterplots can be spotted simply by looking for observations that are departures from the patterns. But how can we best describe the pattern of the data?

Recall the idea of the slope of linear equations. If the slope of the scatterplot trends positive, then there is a positive association between the variables—higher values of one variable go with higher values of the other. If the slope of the scatterplot trends negative, then there is a negative association between the variables—higher values of one variable go with lower values of the other. If there doesn't appear to be any clear positive or negative slope to the scatterplot, then there is no association.

If, in the *Royal Rumble*, the entry rank predicted the elimination rank perfectly, the scatterplot would look like this:

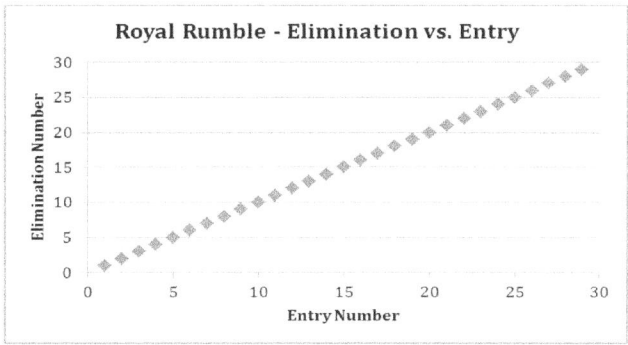

Here, the bivariate data would be perfectly positively correlated—by knowing the wrestler's entry number, you would immediately know his elimination number, too.

Conversely, consider this scatterplot:

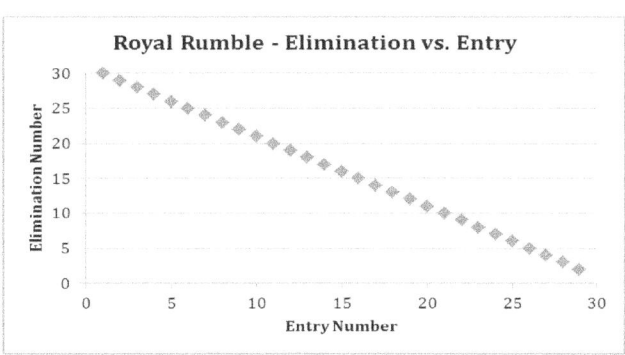

Now we have the completely opposite situation: the wrestler who picks the first spot wins the

---

[*] Son of former professional wrestler Mike Rotunda, a/k/a Irwin R. Schyster (I.R.S.).

Rumble, because the data set is perfectly negatively correlated.

The *strength* of both the perfectly positive scatterplot and the perfectly negative scatterplot is "strong"—meaning the points are tight to a line (the points don't need to be *perfectly* tight to a line to be considered strongly associated, but they do need to hew closely). Look again at the scatterplot of the 2015 *Royal Rumble*. Although positively associated, the data aren't close enough to an imaginary line to consider the data set "strong." Instead, we might term the data "moderate."

If the points hardly form a pattern, either in a positive or negative direction, then we can safely consider the strength of the scatterplot "weak," meaning weakly associated. On the other hand, if the dots are all scattered about, seemingly randomly with no apparent pattern, then that's described as "no association."

Example 4a.-1

Climatologists studying hurricanes wish to determine if there is an association between the average ocean temperature of the Atlantic and the number of tropical storms that develop into hurricanes. They collect two sets of data: the average ocean temperature for a recent ten-year period and the number of named hurricanes in that ten-year period (data courtesy of NOAA).

Year	Average Ocean Temperature (in °C)	Number of Hurricanes
1994	20	6
1995	18	5
1996	19	6
1997	21	9
1998	24	8
1999	22	9
2000	25	10
2001	20	7
2002	21	9
2003	24	11

(a) What are individuals (or observational units, or cases)?

Years.

(b) Which variable is the explanatory variable, and which is the response variable?

Ocean temperature is explanatory; number of hurricanes is response.

(c) Construct a scatterplot of the data.

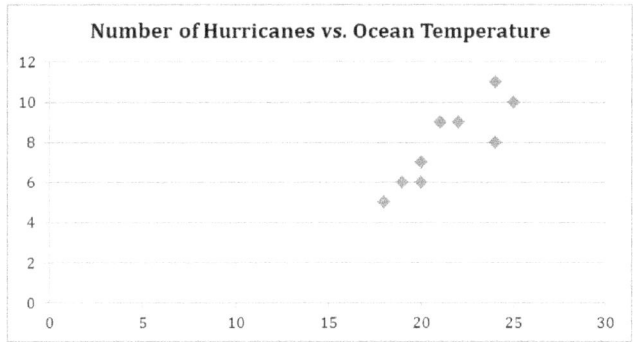

(d) Does the scatterplot reveal any relationship between ocean temperature and hurricanes?

Yes. The higher the ocean temperature, the more hurricanes develop.

(e) Describe the strength, direction, and form of the hurricanes vs. ocean temperature scatterplot. Also, are there any outliers?

The strength is moderate; the direction is positive. The form (or shape) of the plot is linear. There are no outliers.

(f) Predict, based on the scatterplot, how many hurricanes would form if the ocean temperature reached 30°. (This is called *extrapolation*: making predictions outside of the range of the data.)

Approximately 13 hurricanes.

(g) Predict, based on the scatterplot, how many hurricanes would form if the ocean temperature reached 23°. (This is called *interpolation*—making predictions within the range of the data.)

Approximately 9 hurricanes.

Even though the ocean temperature is moderately correlated with the number of hurricanes that formed, we have be careful not to ascribe cause and effect to the proceedings. In other words, from the data gathered here, we do not have enough evidence to claim that more hurricanes form *because* of higher ocean temperatures, *even if* that does indeed describe the mechanism of hurricane formation accurately.

The "*y* = *x*" line is a diagonal line drawn through a scatterplot. It allows you to determine whether one variable *tends* to have higher values than the other variable. Data points that lie on the "*y* = *x*" line have equal values for the variable at that observation.

Example 4a.-2

Consider the following scatterplot of pretest and posttest scores of 23 randomly selected students in the same course. A "$y = x$" line is plotted. (The horizontal and vertical axes have been omitted intentionally.)

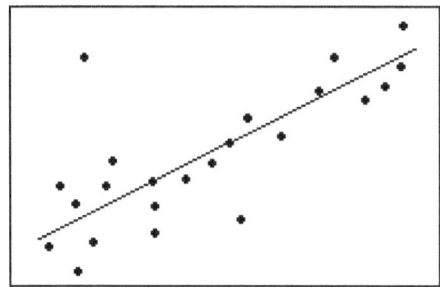

A researcher conducting a study is trying to determine if a student's pretest score can predict his or her posttest score.

(a) Knowing that the explanatory variable must go on the horizontal axis of a scatterplot, and the response variable on the vertical axis, which variable is on the horizontal axis and which variable is on the vertical axis?

Horizontal: pretest; vertical: posttest.

(b) Describe the form of the scatterplot.

Linear, with an outlier.

(c) How many pretest scores are higher than their corresponding posttest scores? Are they above or below the "$y = x$" line?

12; below.

(d) How many posttest scores are higher than their corresponding pretest scores? Are they above or below the "$y = x$" line?

9; above.

(e) How many students received the same score on the pretest as they did on the posttest? How do you know?

2, because these data points lie on the line.

(f) Write a sentence or two describing the outlier.

> The outlier, located at the top left of the scatterplot, is far away from the "$y = x$" line. This student scored much higher on his posttest than his pretest.

Scatterplots can sometimes be deceptive. One of the most famous of this variety of scatterplot was precipitated because of the space shuttle *Challenger*, which was destroyed a little more than a minute after takeoff in January of 1986.[*] The O-rings, which seal the rocket boosters, experienced great thermal distress, largely because of the very low outdoor launch temperature that morning. In fact, the lower the temperature, the more likely that the O-rings wouldn't work. Take a look at data from every shuttle launch prior to the *Challenger* disaster to see.

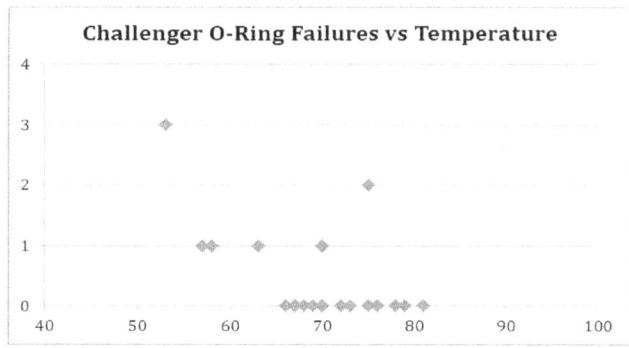

There is a clear, albeit weak, negative association between temperature and O-ring failure. But rather than noting that correlation, NASA dismissed it, based on the following scatterplot, with the set of zero O-ring failures excised.

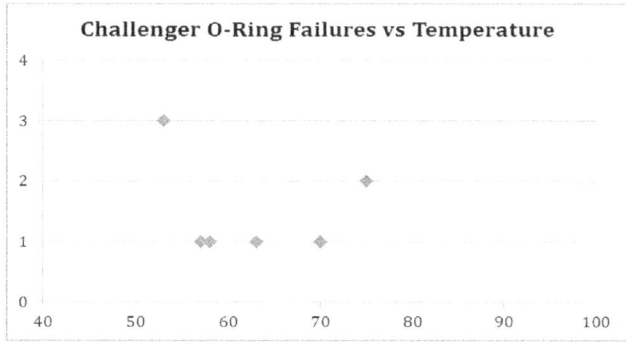

Now there appears to be no substantive linear correlation between O-rings and temperature!

Nobel-prize winner Richard Feynman, one of the most famous physicists of the last century, helped to solve the riddle of the *Challenger*, noting that "I took this stuff that I got out of your

---

[*] What's especially heartbreaking about this disaster is recalling the millions of schoolchildren, including myself, who tuned in at school to watch the launch precisely because the crew included the first-ever school teacher with a ticket to space, Christa McAuliffe.

[O-ring] seal and I put it in ice water, and I discovered that when you put some pressure on it for a while and then undo it, it does not stretch back." Frustrated with the many layers of government bureaucracy, Feynman later wrote,

> It appears that there are enormous differences of opinion as to the probability of a failure with loss of vehicle and of human life. The estimates range from roughly 1 in 100 to 1 in 100,000. The higher figures come from the working engineers, and the very low figures from management. What are the causes and consequences of this lack of agreement? Since 1 part in 100,000 would imply that one could put a Shuttle up each day for 300 years expecting to lose only one, we could properly ask, "What is the cause of management's fantastic faith in the machinery?" It would appear that, for whatever purpose, be it for internal or external consumption, the management of NASA exaggerates the reliability of its product, to the point of fantasy.

Verifying Feynman's computation that a failure probability of 1/100,000 corresponds to an average of one failure every 300 years, assuming a shuttle launch each day, is straightforward: 300 years = 109,500 days. By the way, the engineers were correct: the true failure rate of the space shuttle was 2/135, or about 1.5%, since two shuttles ultimately were destroyed (besides *Challenger*, *Columbia* in 2003).[*]

A single scatterplot can shoehorn in more than just one bivariate data set. In the WWE *Royal Rumble*, just as there was a robust association between entry and elimination rank, another seemingly telling statistic might be the time spent in the ring (in minutes) paired with the entry number. Perhaps those that started earlier tended to remain in the match longer.

Adding a second such bivariate data set to the *labeled scatterplot*, with new icons representing the data points (and a legend to tell the difference), gives us:

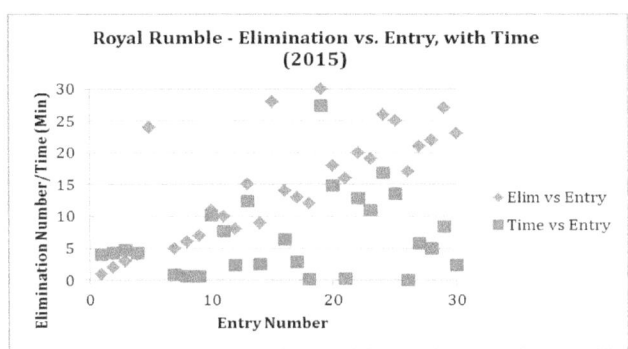

Unlike with comparing the ranks, comparing time with entry doesn't lead us to any obvious conclusions. (The single box-icon observation up on top of the graph represents the winner, Roman Reigns, who entered midway through the *Rumble* and lasted the remainder of the match.)

---

[*] More information about his *Challenger* investigation can be found in Feynman's autobiography, *Surely You're Joking, Mr. Feynman*.

§4b. *Quantifying Relationships*

It's not enough for statisticians to simply wave their hands and pronounce a bivariate data set "strongly positively associated" or "weakly negatively associated"—the discipline of statistics is rooted in methods of quantification. So association (or correlation) unsurprisingly also received the quantification treatment. See the definition below.

Definition 4b.-1

Linear correlation coefficient. A value that measures the strength and direction of the linear correlation between two quantitative variables.

The symbol r is used for the sample linear correlation coefficient, while the Greek letter ρ (pronounced "rho") denotes the population linear correlation coefficient.

The linear correlation coefficient r^*—henceforth, we'll call it simply the correlation coefficient—can be any value between –1 and 1, inclusive. Symbolically, $|r| \leq 1$.

Karl Pearson, who researched natural selection, is generally credited with synthesizing the correlation coefficient we use today (Galton, a man whom Pearson idolized,[†] arrived at the concept of correlation when a flash of insight about the linearity of quantitative data of parents and their children came to him on a train platform in Ramsgate, England), formally, Pearson's version of the calculation is termed the *Pearson product moment correlation coefficient,*[‡] and is found using the tedious formula[§]

$$r = \frac{1}{n-1} \sum_{i=1}^{n} \left(\frac{x_i - \bar{x}}{s_x} \right) \left(\frac{y_i - \bar{y}}{s_y} \right)$$

which can be more simply written with z-scores:

$$r = \frac{1}{n-1} \sum_{i=1}^{n} z_x z_y$$

---

[*] A pirate's favorite letter.

[†] Pearson, like Galton, researched heredity and eugenics—Pearson and Galton together founded the statistics journal *Biometrika*, still in print today, to research heredity—and he even held the Galton Chair in Eugenics at University College, London.

[‡] Pearson founded the discipline mathematical statistics but also, like Galton before him, dabbled in eugenics.

[§] Note the *n*-1 denominator signifying *n* minus one degrees of freedom, à la the variance.

To understand how the formula works, split the first quadrant of the coordinate axis (where all x and y are positive) into four mini-quadrants based on the location of the mean of the explanatory variable (which would be represented as $\bar{x}$) and the mean of the response variable (which would be shown as $\bar{y}$).

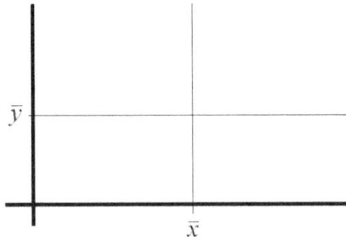

Next, overlay your scatterplot on this arrangement of mini-quadrants. For example,

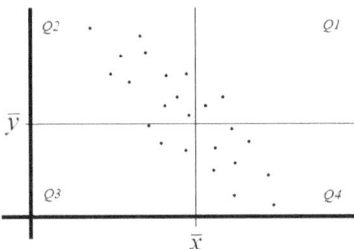

There are four possibilities for each data point:

- In mini-Q1, $z_x > 0$ and $z_y > 0$, so $z_x z_y > 0$.
- In mini-Q2, $z_x < 0$ and $z_y > 0$, so $z_x z_y < 0$.
- In mini-Q3, $z_x < 0$ and $z_y < 0$, so $z_x z_y > 0$.
- In mini-Q4, $z_x > 0$ and $z_y < 0$, so $z_x z_y < 0$.

In the hypothetical negatively associated scatterplot shown above, the majority of data points reside in mini-Q2 and mini-Q4. Look at the correlation coefficient formula again:

$$r = \frac{1}{n-1}\sum_{i=1}^{n} z_x z_y$$

The sum of the z-score products in mini-Q2 and mini-Q4 is negative, so the correlation coefficient ends up negative—which conforms with the direction of the scatterplot (negative).

Of course, leaning on modern technology, we'll never have to calculate the value of r by hand.[*] Although a computer can easily find r (using statistical software, or simply searching for a

---

[*] Unless there's a cataclysmic worldwide electromagnetic or geomagnetic event, akin to the solar storm of 1859 (al-

relevant website), let's see how to use the graphing calculator to obtain the value of r. Input the hurricanes vs. ocean temperature data set, from several pages back, into your calculator as two lists: OCEAN for the ocean temperatures by year, and HURR for the number of hurricanes each year.

Next, make sure that you access your calculator's CATALOG menu. To do this, press 2nd and then 0. Then, scroll down until you find "DiagnosticOn." Press the ENTER button twice. Now, hit the STAT button, go to the CALC menu, and select the option LinReg(a+bx). Before hitting ENTER, select the two lists (with the x-list first), separated by commas, and then hit ENTER.* The value of r should be displayed—r equals 0.8381—along with other measures which we will use later on.

Before we go any further, let's also use the graphing calculator to make a scatterplot. To make a scatterplot of the hurricanes vs. ocean temperature data, press 2nd and Y=, select the first plot, turn it on, and highlight the scatterplot icon. Next, insert the OCEAN and HURR lists in for the x and y lists, respectively, and press ZOOM and then option 9 – ZoomStat. Does the look of the scatterplot, which seems to have a moderate positive correlation, jibe with high value of r, which is 0.8381?

This brings us to our next point: there's only a loose relationship between the "weak," "moderate," "strong" categories and the actual mathematical measurements of r. Nonetheless, what follows is a rule of thumb for translating between the categories and the measurements.

Even though, as we've seen, calculators (and computers) can easily find r, we still need to have an intuitive sense of the value when seeing a scatterplot of bivariate quantitative data.

Example 4b.-1

Make a guess for the value of r for each graph below.

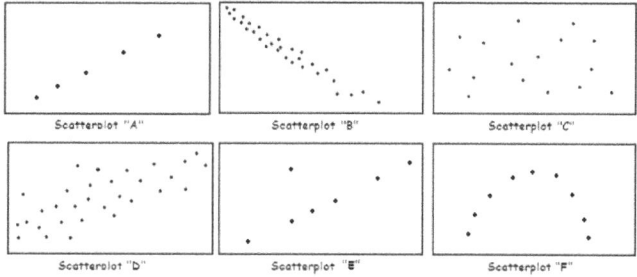

---

so called the Carrington event), which probably would fry most electronics.

---

* You might see a menu that pops up after selecting the LinReg(a+bx) option. If so, be sure to insert the appropriate x and y lists, and then select the Calculate option.

Solution. For A, clearly $r = 1$. For B, though, the relationship is a bit weaker, so perhaps $r = -0.8$. For C, there seems to be no association, so $r = 0$. For D, which looks weak-to-moderate positive, perhaps $r = 0.5$.

It is with Scatterplots E and F that we run into a bit of trouble. E has an outlier; without the outlier, $r = 1$; with it, though, r is dragged down quite a bit, since the correlation coefficient is not resistant and consequently highly sensitive to outliers. Perhaps, then, $r = 0.5$ here. Finally, F shows a clear curvilinear relationship, but r only measures the *linear* relationship between two quantitative variables. Also, the graph is symmetric about a vertical axis that cuts through the peak. Thus, $r = 0$. So even if we know that $r = 0$, we cannot say for sure that the two variables are not associated, only that they are not associated linearly.

Several more properties of r to note. Even if we change the units of measurement for the explanatory variable or response variable—say, from meters to inches—r will not change, since the relationship between the variables has stayed constant. Also, switching the explanatory and response variables in the formulas—in essence, making x equal to y, and y equal to x—won't affect r either.[*] Finally, some linear transformations affect r, while others do not; adding a constant to every x or y value leaves r unchanged, but, let's say, multiplying every x coordinate by -1 flips r's sign.

Correlation alone never implies causation.[†] (And absence of correlation doesn't mean absence of causation, either.) No matter how close $|r|$ is to 1, the measurement of linear correlation in and of itself does not give sufficient evidence to proclaim a relationship causal. Oftentimes the close relationship can be explained plausibly by a *third*, or *lurking*, *variable*.

Definition 4b.-2

Lurking variable. Usually an unmeasured variable that explains the relationship between two others.

The textbook *Workshop Statistics* presents a number of clever, but also at times downright silly, examples: the more ice cream sold, the more drownings occur (lurking variable = summer season), the more fire trucks arrive at a fire, the more damage done by the fire (lurking variable = size of fire), and the more televisions per person in a country, the higher the average life expectancy of that country (lurking variable = wealth). Plenty more such examples exist, but we have to be careful not to confuse *lurking* with *confounding variables*—there's a subtle distinction be-

---

[*] To see why, reexamine the formula for r. Since we're getting a product of the z-scores, and multiplication is commutative, the summation is equivalent.

[†] For more, refer to a semi-famous but totally brilliant *xkcd* comic: two students are discussing whether a statistics class helped to drive home the point that correlation and causation aren't the same—but they can't really be sure the statistics class helped, since, after all, they learned that correlation never implies causation.

tween them. A lurking variable is usually unmeasured and in the background; a confounding variable is "tangled" up with the explanatory variable, with its effects on the response unable to be measured independently.

Sometimes the *direction* of cause and effect also presents confusion. Does increased income cause people to buy more stocks? Is disentangling these two variables even possible? Huff, in *How to Lie with Statistics*, calls this confusion the *post hoc fallacy* (referring to *post hoc ergo propter hoc*: after this, therefore because of this): "In some of these instances cause and effect may change places from time to time or indeed both may be cause and effect at the same time. A correlation between income and ownership of stocks might be of that kind," since the more income earned, the more stocks purchased, and vice versa.[*]

A problem arises when we start discounting *actual* causal connections and attributing *everything* to correlation. (This bears resemblance to the notion of randomania, discussed earlier in the primer; related but even more radical, as John Allen Paulos relays, "…the philosopher David Hume maintained that in principle there is no difference between the two [i.e., correlation and causation].")[†] Take smoking and lung cancer. It's not causation, some experts said; there must be a lurking variable here, such as those who have a genetic predisposition to smoke also have a genetic predisposition to develop lung cancer. Other experts, by the 1940s and '50s, acknowledged the rising trends of cigarette consumption and lung cancer but muddled the issue by also pointing out the concomitant increased consumption of milk. How can we unpack such claims? We'll review some different methods in a subsequent §.

Before leaving this topic, it's important to mention that regardless if causation can be established, mere correlation is sometimes enough to reveal critical information. A *New York Times* article titled "How Companies Learn Your Secrets" (from February 16, 2012) exposed how some corporations are using *analytics*, or data mining to find correlations, to learn about their customers. The *Times* article centers on a Target statistician named Andrew Pole. Over a decade ago, he was posed a question by the marketing department: Could you figure out, based on buying habits alone, if a Target customer was pregnant?

Except in times of upheaval, most people's buying habits remain relatively stable. During pregnancy, however, "old routines fall apart and buying habits are suddenly in flux." It is during these "in flux" times that Target want to, well, target customers—and convince them that *all* of their shopping needs can be met in their stores.

Pole did manage to conjure up an algorithm to predict pregnancy, by carefully looking at patterns of purchases by pregnant women. But his mathematical model worked a little too well, be-

---

[*] And Burton Malkiel presents an even more muddled picture of cause and effect with respect to stocks: "Many of the most important changes that affect the basic prospects for corporate earnings are essentially random, that is, unpredictable."

[†] Marshall McLuhan has a slightly expanded take on Hume's notion, with respect to the "fragmentation" of mechanization: "[A]s David Hume showed in the eighteenth century, there is no principle of causality in a mere sequence. That one thing follows another accounts for nothing. Nothing follows from following, except change." Just because *B* follows *A* doesn't imply that *A* caused *B*.

cause, around a year later, a man in Minneapolis angrily demanded to see the manager. He was holding a wad of coupons that the store had mailed to his daughter.

"My daughter got this in the mail! She's still in high school, and you're sending her coupons for baby clothes and cribs? Are you trying to encourage her to get pregnant?" he asked.

The *New York Times* article describes what happened next:

> The manager didn't have any idea what the man was talking about. He looked at the mailer. Sure enough, it was addressed to the man's daughter and contained advertisements for maternity clothing, nursery furniture and pictures of smiling infants. The manager apologized and then called a few days later to apologize again.

> On the phone, though, the father was somewhat abashed. "I had a talk with my daughter," he said. "It turns out there's been some activities in my house I haven't been completely aware of. She's due in August. I owe you an apology."

In the book *Rise of the Robots: Technology and the Threat of a Jobless Future*, Martin Ford notes that "[s]ome critics fear that this rather creepy story [about Target] is only the beginning and that big data will increasingly be used to generate predictions that potentially violate privacy and perhaps even freedom." The upshot for businesses, though, "where the ultimate measure of success is profitability and efficiency rather than deep understanding, correlation alone can have extraordinary value."

§4c. *Lines of Best Fit*

As described in detail in the comprehensive book *The Lady Tasting Tea*, Francis Galton was fascinated with heredity and inheritance. Galton studied the heights of children of really short couples and of really tall couples. He noticed that these heights tended to what he called "regress," or to lurch backward, toward the mean height of the host population from which they sprang. In other words, short couples didn't necessarily have short children and tall couples didn't necessarily have tall children; if short couples always had shorter children, and tall couples always had taller children, then, over time, we'd have two varieties of people on the earth: really, really tall people (with even taller people yet to be born), and really, really short people (with even shorter people yet to be born).[*] This "regress" Galton formally called *regression toward the mean*, "which explains why pride goeth before a fall and why clouds tend to have silver linings," as Peter L. Bernstein observes.

The phenomenon of regression toward the mean can also be seen in sports. As Jim Albert, author of *Teaching Statistics Using Baseball*, describes, "[T]here is a general tendency for a player's baseball stats from one year to the next to go back, or regress, to the mean. Many of you have heard of 'sophomore slumps' in sports. This happens when someone does well in his/her

---

[*] For the same reason, Einstein's children weren't quite as smart as he was. (Otherwise we might imagine the expression "You're like an Einstein" altered to "You're like one of Einstein's children.")

rookie year and then slumps in the sophomore year…. Here we see that there is a natural tendency for the best performing rookies to slump in their sophomore years."

The behavioral economists Amos Tversky and Daniel Kahneman also studied regression toward the mean, finding it among cadet pilots in the Israeli air force. Frequently, after a good landing, a cadet is praised—and then his next landing isn't as good. And after a bad landing, a cadet is reamed out—and then his next landing is better. Although these results, taken together, would seem to suggest that positive reinforcement isn't effective while negative reinforcement is, rather, these results yet again illustrate regression toward the mean. Here's how Kahneman remembers it:

> This was a joyous moment, in which I understood an important truth about the world: because we tend to reward others when they do well and punish them when they do badly, and because there is regression toward the mean, it is part of the human condition that we are statistically punished for rewarding others and rewarded for punishing them. I immediately arranged a demonstration [with the air force pilots] in which each participant tossed two coins at a target behind his back, without any feedback. We measured the distances from the target and could see that those who had done best the first time had mostly deteriorated on their second try, and vice versa. But I knew that this demonstration would not undo the effects of lifelong exposure to a perverse contingency.[*]

The world of medicine isn't immune, either. Regression toward the mean "can be applied to cardiac surgeons, for whom a few deaths can be statistically ruinous," according to Sandeep Jauhar (from his book *Doctored*). Instead, if we compare cardiac surgeons—or baseball players, or pilots—year-to-year, the less their regression toward the mean, the more attributable skill is to their performance rather than luck (good *or* bad). Consistency through time reduces the probability that luck is the driving force behind their results.

Although a number of individuals are responsible for bringing regression to life, several deserve mention for their contributions. Euler, Roger Cotes, Laplace, and especially Roger Boscovich—who, when figuring out effective ways to measure the shape of the earth, realized that the sum of the absolute values of the errors of a prediction should be minimized—get a share of the credit. Adrien-Marie Legendre, though, mathematically formalized the method of least squares in 1805,[†] which he explained thusly:

> Of all the principles which can be proposed for that purpose, … [the best] consists of rendering the sum of the squares of the errors a minimum. By this means there is established among the errors a sort of equilibrium which, preventing the extremes from exerting an undue influence, is very well fitted to reveal that state of the system which most nearly approaches the truth.[‡]

---

[*] Quote is culled from his modern-day classic *Thinking, Fast and Slow*.

[†] The name comes from the behind-the-scenes mathematical process to obtain the equation: the minimization of the sum of the squares of the residuals. We'll deal with residual values shortly.

[‡] Quoted in *A History of Mathematics (2nd ed.)* by Victor Katz.

Later, Gauss took full credit for Legendre's method, claiming (without evidence of publication) that he had been using least squares before Legendre. Although Gauss was perhaps the greatest mathematician of all time (even the amateur mathematician and Emperor of France Napoleon Bonaparte thought so, sparing a city from attack simply because Gauss lived in it),[*] he also had one of the greatest egos of all time,[†] and this would not be the last time he would "pre-discover" a notable mathematical result: non-Euclidian geometries, geometries in which the parallel postulate is altered, also landed in his sights, supposedly before Lobachevsky, Bolyai, and Riemann's discovery of them. Regardless, Gauss brought a greater rigor to Legendre's method of regression.

Later, as we have seen, Francis Galton introduced the term *regression* (though he initially called the concept "reversion"), and we still make use of the term today, especially with *regression lines*—colloquially known as *lines of best fit*—which neatly package regression toward the mean and the method of least squares into predictive, linear formulas.

Definition 4c.-1

Regression line. An algebraic model of the relationship between two paired quantitative variables; it has a number of aliases: line of best fit, best-fit line, least-squares line, least-squares regression line, least-squares fit, and LSRL, among others.

The equation of the regression line is a line of best fit of the data in the scatterplot. It is given as either $\hat{y} = b_0 + b_1 x$ or $\hat{y} = a + bx$. Let's use the latter equation, since we'll only need two terms: a constant term (represented by a) and a linear term (the coefficient of which is b).[‡] The symbol above the y is called a "hat." The slope of the least-squares line is given by b and the y-intercept is denoted by a. The slope is the rate at which, for every additional unit increase in the explanatory variable (x), the response variable (y) changes.[§] The y-intercept is the value of the response variable when the explanatory variable is equal to zero. The formulas for the slope and y-intercept, respectively, are

---

[*] Newton is another obvious candidate for greatest-of-all-time status, but Euler has his supporters as well (such as mathematician and author William Dunham).

[†] Before he was even six years old, he found errors in his father's financial statements—his father was always deeply annoyed with his son's precociousness—and made mincemeat out of his kindergarten teacher's fatuous arithmetic assignment (a footnote on the latter later).

[‡] The other expressed form, $\hat{y} = b_0 + b_1 x$, generalizes to allow many more terms for nonlinear best-fit equations. For instance, a quartic function of best fit could be expressed in general terms as $\hat{y} = b_0 + b_1 x + b_2 x^2 + b_3 x^3 + b_4 x^4$. In addition, this form is helpful with *multiple regression*—when multiple explanatory variables are used for prediction.

[§] You may also recall the concept of slope from grade school: rise over run, or change in y over change in x.

$$b = r\frac{s_y}{s_x} \text{ and } a = \bar{y} - b\bar{x}$$

Here's a quick derivation of the y-intercept formula. Every regression line, by virtue of its least-squares construction, contains a *point of averages*: $(\bar{x}, \bar{y})$. So, plugging the point of averages into the equation for the regression line, we get $\bar{y} = a + b\bar{x}$ which, when rearranged, becomes $a = \bar{y} - b\bar{x}$.

Example 4c.-1

Let's work through a multi-step regression problem with the hurricanes vs. ocean temperature climatology data we're already familiar with.

(a) Use the 1-Var Stats graphing calculator option, along with LinReg(a+bx), to compute the mean and standard deviation of hurricanes (the response variable, y) and temperature (the explanatory variable, x), as well as the value of the correlation coefficient.

$\bar{x} = 21.4$, $\bar{y} = 8$, $s_x = 2.319$, $s_y = 1.9437$, and $r = 0.8381$.

(b) Find the value of the slope, b.

$$b = 0.8381 \cdot \frac{1.9437}{2.319} = 0.702$$

(c) Find the value of the y-intercept, a.

$$a = 8 - 0.702 \cdot 21.4 = -7.02$$

The fact that the y-intercept and slope look very similar is simply a coincidence.

(d) Write the equation of the line of best fit.

$$\hat{y} = -7.02 + 0.702x$$

(e) Instead of using x and y as variables in our least-squares line, we will sometimes prefer to use *variable names*. Two possible variable names to use are *oceantemp* and *hurricanes*. So let's rewrite the equation for the line below, this time using variable names (make sure to include a "hat").

$$\hat{hurricanes} = -7.02 + 0.702 oceantemp$$

(f) We can also use the calculator to produce the regression line. Use the LinReg(a+bx) option to do so.

You'll end up with a numerically slightly altered version of the regression line (due to rounding).

(g) Recall that the point $(\bar{x}, \bar{y})$ is present on every regression line. Find what $\bar{y}$ equals in terms of $\bar{x}$ for the temperature-hurricanes least-squares line.

$\bar{y} = -7.02 + 0.702\bar{x}$

(h) One of the linear regression lines' major uses is *prediction*—specifically, predicting a *y* value for a given *x* value by simply plugging in that value of *x* into the equation of the least-squares line. Using the regression line you found for ocean temperature and hurricanes, predict how many hurricanes would form if the ocean temperature was 22.5 degrees.

$\hat{y} = -7.02 + 0.702 \cdot 22.5 \approx 9$

(i) Suppose instead you were trying to predict how many hurricanes would form if an extraordinary event caused the temperature of the ocean to hit 55 degrees. (This is extrapolation—predicting beyond the set of values in the data.) What does the least-squares line predict this number of hurricanes to be at 55 degrees? Is this a reliable prediction? Explain.

$\hat{y} = -7.02 + 0.702 \cdot 55 \approx 32$. This is not a reliable prediction because we've extrapolated too far outside the range of the given data.[*] Generally speaking, for problems like this, we should default to $\bar{y}$ as the best possible answer. Think back to our discussion of regression toward the mean to understand why.

(j) Interpret the *y*-intercept of the least-squares line.

If the ocean temperature was zero degrees, −7.02 hurricanes would form. This isn't a meaningful prediction.

(k) Interpret the slope of the least-squares line.

For every one degree increase in temperature, the number of hurricanes increases by

---

[*] Extrapolation is usually safe if the following criteria are met: (1) The scatterplot reveals a consistently linear pattern; (2) The explanatory variable doesn't stray too far away from the largest (or smallest) *x*-value in the data set; and (3) The resulting prediction is reasonable. In the problem above, criterion (2) is clearly not met.

0.702 (the slope amount).

Extrapolation, especially linearly extrapolating, is dangerous. Knowing when it's appropriate to extrapolate isn't easy. A telling *xkcd* comic by Randall Munroe beautifully illustrates this "extrapolation dilemma": a just-married bride is warned that since she had zero husbands last week and one husband this week, she should investigate snagging the "bulk rate on wedding cake" for her dozens of soon-to-be husbands in the weeks to come.

Extrapolation also finds a ready home in the world of sports. At the MLB All-Star Break, the midpoint of the baseball season, homerun and hit totals for the entire season are projected by so-called analysts on-screen—by simply doubling current totals. Rarely do these predictions pan out, because they assume that the second half of the season will have nearly identical conditions as the first half. Perhaps another *xkcd* comic should center on an NFL game like the one Peyton Manning played against the Atlanta Falcons on September 18, 2012, his first Monday Night Football contest quarterbacking the Denver Broncos. On his first series out of the gate, he threw an interception; he followed that up with another interception; and, yet again on his third series, a third interception was thrown. *If this continues*, the on-air commentators breathlessly explained, *Manning will end up with double-digit interceptions tonight*. But surely that wouldn't happen, since—putting aside adjustments in game plans which would be effected—after a handful more picks, Manning surely would have been pulled from play, no matter his star power.

Why do our instincts so often resort to straightforward extrapolation? Paulos notes that "too often people become mesmerized by the technical details of correlation coefficients, regression lines, and curves of best fit and neglect to step back and think about the logic of the situation." And this is doubly true with extrapolation—logic is sometimes is short supply.

Consider the case of Thomas Malthus, for example. He is not a very sympathetic historical figure. A well-known economist, at the end of the 1700s he predicted that the world's population would explode while the production of food and other resources wouldn't be able to keep up, thus leading to worldwide famine and all-around disaster. Malthus obtained his population data from Benjamin Franklin, who incorrectly believed that the world's population would double within 25 years. Malthus said that populations would grow *geometrically* while food resources could only grow *arithmetically*, not keeping pace; this recalls the Red Queen's race in Lewis Carroll's *Through the Looking-Glass*, where running fast is only enough to keep you in place. "If you want to get somewhere else, you must run at least twice as fast as that!" the Red Queen says.

Eventually, Malthus realized the error of his ways, noting that humans would innovate their way out of certain death.* But Malthus is not the only intellectual figure in history to famously

---

* There is something to be said for innovation getting us humans out of trouble. For example, by the late 1800s, there were dire warnings about the increasing volume of manure piling up on city streets as a result of the most popular transportation of the times, horses. Faulty extrapolation would have lead you to believe that, years hence, entire cities would be drowning in excrement. Several decades later, the manure problem was solved: with the advent of horseless carriages. And several decades after that, innovation struck again: the booming iron-lung business took a nosedive when virology advanced enough to permit the emergence of a polio vaccine. Looking to the future, perhaps we can even innovate our way out of the catastrophic environmental consequences of global warming, such as by launching billions of particles in the atmosphere to block some of the sun's rays from penetrating the surface—

bungle forecasts based on the extrapolation of data.

Author Kesten Green, in an article written for the Institute of Public Affairs, lists some other instances of faulty scientific predictions based on misreading data, such as

- *Poisoning by fluoride in drinking water.* Small amounts of fluoride added to drinking water has been shown to help reduce tooth decay; too much, it is true, is harmful to health. As the old adage goes, the dose makes the poison.
- *DDT (and other pesticides) and cancer.* The book that launched the modern environmental movement, *Silent Spring* by Rachel Carson, "forecast that birds would die out and people would be afflicted by cancer due to increasing exposure to the insecticide DDT." But it is very challenging to establish causation when it comes to suspected carcinogens, and there are always unintended consequences when such chemicals are banned. In this case, an increased incidence[*] of malaria worldwide.
- *Famine and exploding population growth.* The Malthusian Paul Ehrlich put forth similar predictions of doom as late as the 1970s.
- *Mad cow disease (bovine spongiform encephalopathy).* A modern-day scare, mad cow attacks the central nervous system using prions, transmitted by eating certain types of meat. Even Oprah Winfrey got into the act on this one, going toe-to-toe against the beef industry. The epidemic never materialized.

What if we do not have the observations of the explanatory and response variables? In other words, what if we don't have the data sets needed to make a scatterplot—can a least-squares line still be made?

Example 4c.-2

Suppose you have the following summary statistics from a study about car weight (in pounds) and miles per gallon (mpg).

$$\bar{x} = 3050,\ \bar{y} = 23,\ s_x = 25,\ s_y = 5,\text{ and } r = -0.730$$

(a) What are the explanatory and response variables of this study?

The explanatory variable is weight; the response variable is mpg.

---

thereby taking a cue from the Mount St. Helens volcano eruption in 1980, which blew enough ash and soot into the sky to lower the earth's overall temperature ever-so-slightly, at least for a time.

[*] Quick note: the *incidence* of a disease is the rate of new cases, whereas the *prevalence* of a disease is the count of those who are both alive and classified as having the disease.

(b) Despite the fact that you cannot view a scatterplot of this data set, you can still describe the strength (weak, moderate, or strong) and direction (positive, negative, or no association) of this data set. So, how would you characterize the strength and direction of this data set?

Based on r, moderately negative. A caution, however: the scatterplot may not look linear.

(c) Calculate the slope.

Simply use the slope formula; it's –0.146.

(d) Calculate the y-intercept.

Use the y-intercept formula; the answer is 468.3.

(e) Write the equation of the least-squares line.

$\hat{y} = 468.3 - 0.146x$, or $\hat{mpg} = 468.3 - 0.146weight$.

(f) Suppose a car weighs 2,750 pounds. What is its predicted mpg?

Plug in 2,750 for x. The predicted mpg is 66.8.

(g) Suppose that Toyota produces a car made of light-weight composite materials that weighs 1000 pounds. Using the least-squares line, predict the mpg of Toyota's car. But does this prediction seem reliable to you? Explain.

322.3 mpg. The prediction isn't reliable, because a linear relationship doesn't model the data well. Extrapolation here is inappropriate.

(h) Interpret the y-intercept of your linear regression line. Is this meaningful?

If a car weighs zero pounds, it will get 468.3 mpg. But cars can't weigh zero pounds!

(i) Interpret the slope of your least-squares line.

For every additional pound of weight, mpg decreases by 0.146. (And for every additional thousand pounds of weight, mpg drops by 146 mpg, clearly not a realistic amount.)

But suppose you were not given all of the summary statistics. Let's work "backward" and calculate descriptive statistics given a least-squares line.

Example 4c.-3

Consider the following summary statistics and regression line from a study about outdoor temperature (in degrees Celsius) and coffee sales (in dollars).

$$\bar{y} = 32,\ s_x = 3,\ s_y = 40,\ \text{and}\ \widehat{coffee} = 300 - 10temp$$

(a) What are the explanatory and response variables?

Explanatory is temperature; response is coffee sold.

(b) Calculate the correlation coefficient.

Plug in as many numbers as you can into the slope equation, and then solve for r:

$$-10 = r\frac{40}{3},\ r = -0.75$$

(c) Calculate the mean temperature.

Plug in as many numbers as you can into the y-intercept equation, and then solve for $\bar{x}$:

$$300 = 32 - (-10)\bar{x},\ \bar{x} = 26.8$$

Example 4c.-4

A study is conducted into foot length and height. Foot length (in inches) is the explanatory variable (the x) and height (in feet) is the response variable (the y). The least-squares line is given as $\widehat{height} = 3.98 + 0.2\,footlength$. You also know that the standard deviation of foot length is 0.9, the standard deviation of height is 0.3, and the mean of foot length is 8.1.

(a) Calculate the correlation coefficient.

$$r = 0.6$$

(b) Calculate the mean height.

$$\bar{y} = 5.6$$

(c) What change in height does the regression line predict for each additional 0.9 inches of foot length?

$r \cdot s_y = 0.6 \cdot 0.3 = 0.18$ feet.

§4d. *Residuals*

Definition 4d.-1

Residual. The difference between an observed and predicted value.

Example 4d.-1

Reconsider the hurricanes vs. ocean temperature data. We've already found the least-squares line to be $\hat{hurricanes} = -7.02 + 0.702oceantemp$. Recall that the slope here is interpreted as for each degree increase in temperature, the number of hurricanes increases by 0.702.

When we constructed a scatterplot of the hurricanes vs. ocean temperature data, it displayed a moderately positive association. This time, here's the scatterplot with the regression line drawn in:

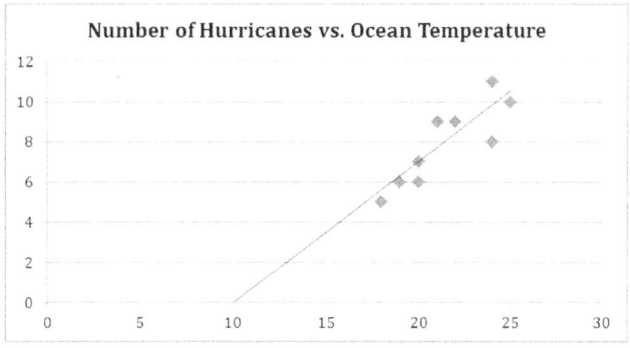

(a) Which observation appears to be right on the line, and what year does the data point represent?

$(20,7)$, which is 2001.

(b) Let's see how good of a prediction the least-squares line makes for that observation. Plug in 20 for the *oceantemp* in the *temperature-hurricanes* regression line. What is the prediction for the number of hurricanes?

7.02

(c) Now, take the actual number of hurricanes that formed in 2001 and subtract the predicted value you calculated in the previous question. This value is called the *residual*.

Residual $= 7 - 7.02 = -0.02$

So the *residual* is the actual (or observed) value minus the predicted value. That is,

$$residual = actual\,(observed) - predicted = y - \hat{y}$$

- The *residual* is equal to zero if the actual value and the predicted values are equal. The least-squares line predicts any observations that have a zero residual perfectly—since these observations are on the least-squares line.
- The *residual* is positive if the actual value is bigger than the predicted value. These observations with positive residuals are above the least-squares line.
- The *residual* is negative if the actual value is smaller than the predicted value. These observations with negative residuals are below the least-squares line.

(d) We know that one observation—from 2001—is predicted almost perfectly by the least-squares line (its residual is nearly zero). Besides the year 2001, according to the scatterplot, how many observations have positive residuals? How many have negative residuals?

Three are positive, and five are negative.

The calculator automatically makes a list of residuals of a set of quantitative data once a least-squares line is made. The list of residuals it creates is called ₗRESID.

If you haven't already done so, enter the hurricane data into the calculator, calling the lists OCEAN and HURR. Then hit STAT, go to the CALC menu, and select the option LinReg(a+bx), and, before pressing ENTER, select the two lists (with the *x*-list first), and hit ENTER. Then, pull up the ₗRESID list on-screen.

(e) Write the first three values of the ₗRESID list below.

−1.02, −0.612, and −0.314.

(f) Create a scatterplot of ocean temperature (as the *x*-list) and residuals (as the *y*-list). How many points are above the *x*-axis? Below the *x*-axis? Compare this to your answer to question (d).

The scatterplot of the residuals is shown on the top of the next page.

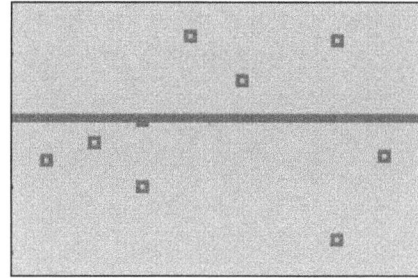

There are three points above and five below the horizontal line (which is the *x*-axis). If a plot of the residuals (also called a *residual plot*) appears *random*, it means that the least-squares line predicts the data well. If a residual plot shows an obvious pattern, the regression equation is not a good model (or "good fit") for the data,[*] and a nonlinear model would be more appropriate.

(g) Do the data in the residual plot seem random? What does this mean?

Yes. The regression line is a good fit for the data set.

We now wish to explore the values of $_L$RESID. We first need to make a copy of the list; store $_L$RESID into a new list called CRSD. Next, run 1-Var Stats on the copied list.

(h) Find the sum of the residuals. Do you suppose this is the sum of any list of residuals?

Zero. Yes.

(i) Find the mean of the residuals. Do you suppose this is the mean of any list of residuals?

Zero. Yes.

(j) Find the median of the residuals. Do you suppose this is the median of any list of residuals?

It's not zero! Remember, the median is the middle number—which could be anything.

We have seen already that the average prediction error—that is, the mean of the residuals—is equal to zero whenever we construct a least-squares regression line. That's because the positive and negative residuals "balance" each other out.

But that doesn't tell us how far off the predictions are on average. Instead, we need to find the

---

[*] This might strike you as a bit strange at first. How can a line of *best* fit not be a *good* fit for a data set? Aren't the two notions mutually exclusive? No. Think of it this way: Suppose you don't believe in a soulmate, but you do otherwise subscribe to the idea of a "person of best fit"—that there's one person out there who's the best-possible fit for you, as compared to all other human beings alive. Now, somewhat depressingly, that "best fit" person might also happen to be a *bad* fit for you, regardless. Sorry.

standard deviation of the residuals, which is given by the following formula; notice that $n-2$ is found in the denominator since we're working with bivariate, not univariate, data:

$$s = \sqrt{\frac{\sum(y - \hat{y})^2}{n-2}}$$

§4e. *The Coefficient of Determination*

There is another quantity that tells us how well the least-squares line predicts values of the response variable *y*. It is called the *coefficient of determination*, or r^2, and is difficult to interpret—essentially, it gives you the percentage of variation in *y* that can accounted for by x^*—but it is an important statistical measure. (In fact, like the regression line, it has many names: "proportion of variability," "proportion of variation," "coefficient of variation," among others.)

Example 4e.-1

Consider the two scatterplots shown.

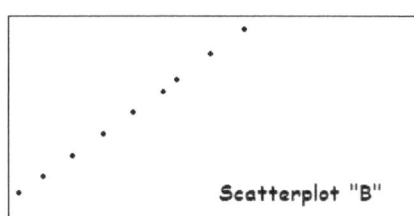

Imagine lines of best fit drawn through the data in each plot. In Scatterplot "A," the points would be relatively close to the line; but in Scatterplot "B," all of the points would be on the line.

(a) The calculated value of *r* for the Scatterplot "A" data is 0.700. What does this mean?

There is a positive, moderate linear association between the variables.

(b) Calculate the coefficient of determination, or r^2, for the Scatterplot "A" data.

$$r^2 = 0.700^2 = 0.490$$

---

[*] Mathematically, the coefficient of determination is the ratio of the squares of the "explained variation" (the deviation of the predicted values of points to the average value of all points) to the "total variation" (the deviation of the actual values of the points to the average value of all points).

(c) The calculated value of r for the Scatterplot "B" data is 1.000. What does this mean?

There is a positive, strong linear association between the variables.

(d) Calculate the coefficient of determination, or r^2, for the Scatterplot "B" data.

$r^2 = 1.000^2 = 1.000$

(e) Which least-squares line does a better job of summarizing the data points in the scatterplot?

"B"

(f) Which scatterplot has a higher value of r^2? So what's the connection between r^2 and the least-squares line?

"B," meaning the points fall closer to the least-squares line than in the other plot. (If $r^2 = 1$, then all of the data points would fall on the line of best fit.)

Example 4e.-2

(a) If r^2 is equal to 1.000, what values could the correlation coefficient be?

−1 or 1.

(b) Suppose that you know that a least-squares line that predicts the reaction time of a certain chemical has a negative slope. You also know that the value of r^2 is 0.810. Find the value of r and explain what it means.

$r = -0.9$, meaning there's a strongly negative association between the variables.

(c) Find the correlation coefficient given that the coefficient of determination is 0.490 and the scatterplot looks as follows.

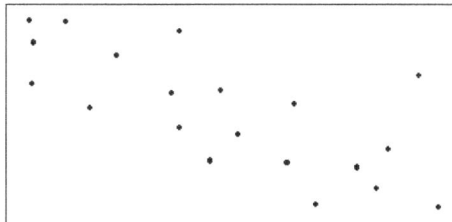

$r = -0.7$

(d) Suppose that you used shoe size to predict foot length and found that r^2 is 0.910. Find the value of r.

0.954

(e) The number of hours that 13 students spent studying for a test and their scores on the test were compared. The coefficient of determination was calculated as 0.851. In this data set, study time can explain how much percentage of the variation in test scores?

85.1%. So most of the variation is accounted for.

If you're still not quite understanding the meaning of the coefficient of determination, then let's take a look at one more example. In the text *Forty Studies that Changed Psychology*, Roger R. Hock summarizes research into the relationship between people's life changes (positive or negative) and onset of illness. Noting that the relationship between "life change units," or LCU (a measure of total life stress factors), and illness is weak, Hock uses r^2 to explain:

> SSRS [The Social Readjustment Rating Scale, which is a listing of the discrete life events constituting the LCU] scores account for only about 10% of the total variation among people who become ill. In other words, if you examine 1,000 people to see who becomes sick over a 6-month period, you will find great variation in the individual factors leading to their illness or lack of illness. If you have them all complete an SSRS, you will find that, considering all the possible reasons for health variation, their LCU scores explain about 10% of it.

§4f. *There's No Substitute for Looking*

The statistician Francis Anscombe (who was John Tukey's brother-in-law), in 1973, constructed four data sets of eleven points each to drive home an important idea: there's no substitute for actually *looking* at graphical displays of the data; glancing at summary measures won't tell the whole story. Here's *Anscombe's quartet*:

Data Set I		Data Set II		Data Set III		Data Set IV	
x1	y1	x2	y2	x3	y3	x4	y4
10	8.04	10	9.14	10	7.46	8	6.58
8	6.95	8	8.14	8	6.77	8	5.76
13	7.58	13	8.74	13	12.74	8	7.71
9	8.81	9	8.77	9	7.11	8	8.84
11	8.33	11	9.26	11	7.81	8	8.47
14	9.96	14	8.1	14	8.84	8	7.04
6	7.24	6	6.13	6	6.08	8	5.25

Data Set I			Data Set II			Data Set III			Data Set IV	
x1	y1		x2	y2		x3	y3		x4	y4
4	4.26		4	3.1		4	5.39		19	12.5
12	10.84		12	9.13		12	8.15		8	5.56
7	4.82		7	7.26		7	6.42		8	7.91
5	5.68		5	4.74		5	5.73		8	6.89

Despite their apparent differences, take a look at the summary information for *all four* of these data sets:

$$n = 11, \ \bar{x} = 9, \ \bar{y} = 7.5, \ \hat{y} = 3 + 0.5x, \ r = 0.82, \ r^2 = 0.67$$

Yet, visually, the four data set look radically different, as shown below; thus, viewing graphical displays of data is paramount before drawing inferences.

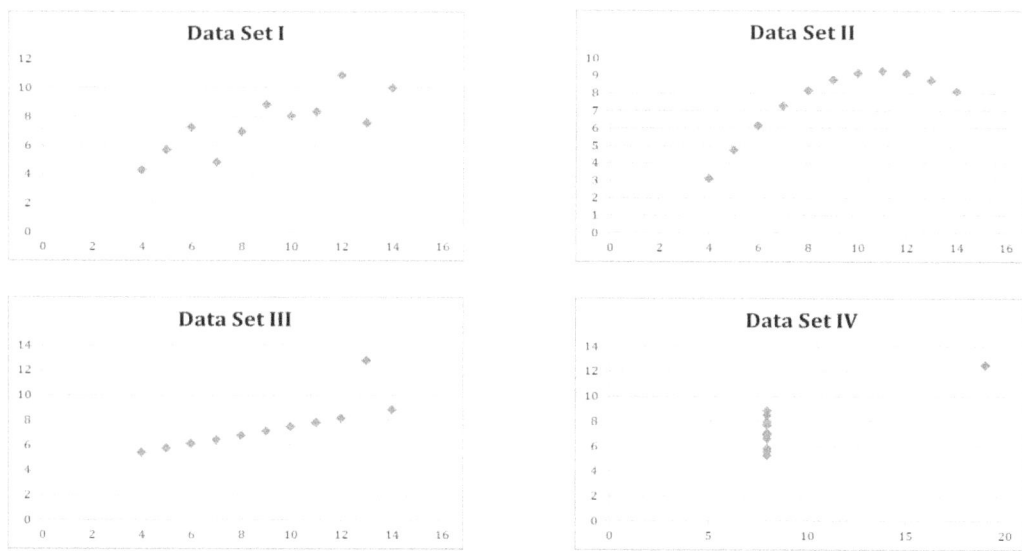

Consider Data Set III above. There's a clear outlier, or observation that doesn't follow the pattern, at $(13, 12.74)$. But let's more formally define an outlier in bivariate quantitative data.

Definition 4f.-1

Outlier (bivariate data). An ordered pair that is a departure from the pattern of data; since r is not resistant, removing the ordered pair affects its value; and the ordered pair has a "large" value for the absolute value of its residual.

The word "large" rests in scare quotes because identifying outliers is oftentimes subject to trial and error: examine the scatterplot with and without the suspected outlier, and compare the summary measures of the data set.

Likewise, *influential observations* are sometimes difficult to spot without using trial and error.

Definition 4f.-2

Influential observation. An ordered pair that usually has a large *x*-value; changing the coordinates of the ordered pair, or outright deleting it, has a substantial effect on the equation of the least-squares line.

Example 4f.-1

Consider the following scatterplot, complete with regression line:

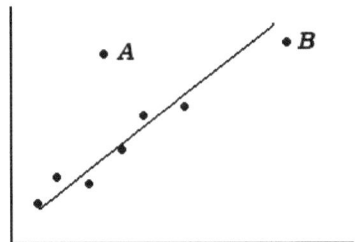

(a) Which point in the scatterplot above is an outlier? Justify your answer.

Point A, since it's outside the pattern of data.

(b) Which point in the scatterplot above is an influential observation? Justify your answer.

Point B, because it has a large *x* coordinate.

§4g. *Computer Output*

Instead of using a calculator, sometimes statisticians use computer software to obtain statistics from data sets; such software produces *computer printouts* of the data analysis. One of the most popular statistical software programs is called Minitab (others include Fathom, SPSS, and R). Computer printouts could contain some relevant information that's labeled in an obvious manner—such as the mean, median, the number of individuals *n*—or in a non-obvious manner—such as the *y*-intercept and slope of a line. Look at the following example.

Example 4g.-1

A random sample of students is collected. Their achievement score on a standardized exam is compared against the average number of hours per week of television they watch. A computer printout of this bivariate data set is shown below.

```
Predictor      Coef      StDev      T          P
Constant       86.14     12.1       0.45       0.101
Hours          -6.12     7.12       2.91       0.005


S = 34.5          R-sq = 84.8%
```

(a) What are the explanatory and response variables?

The explanatory variable is television hours; the response variable is scores.

(b) What is the slope?

$b = -6.120$

(c) Interpret the slope of the regression line in context.

For every additional hour of TV watched, scores on the test decrease by 6.12 points.

(d) What is the y-intercept?

$a = 86.14$

(e) Does the y-intercept make sense?

If a student doesn't watch television, he or she will receive an 86.14%. This seems to make sense.

(f) Write the equation for the least-squares line.

$test\hat{s}core = 86.14 - 6.120 hours$

(g) Predict the test score if a student watches two hours of television.

$test\hat{s}core = 86.14 - 6.120 \cdot 2 = 73.9\%$

(h) Calculate the coefficient of determination.

$r^2 = 0.848$

(i) Interpret the coefficient of determination in context.

84.8% of the variation in test score is accounted for by hours of television viewed.

(j) Calculate and interpret the correlation coefficient.

$r = \pm\sqrt{0.848} = -0.921$. There is a negative, strong association between the variables.

(k) Suppose that Carolyn watched five hours of television and scored a 60 on her test. Find her residual.

$predicted = 55.54$. Residual $= 60 - 55.54 = 4.46$.

§4h. *Linear Transformations of Nonlinear Data*

Take a look at the two residual plots below.

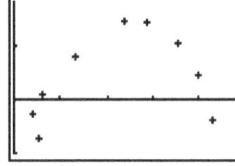
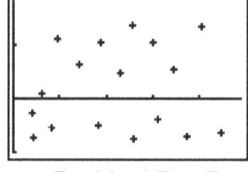

Residual Plot A   Residual Plot B

Residual Plot B has randomly scattered data, meaning the regression line predicts the data well. But Residual Plot A manifests a clear curvilinear (perhaps even parabolic) pattern. How do we proceed?

We usually have two options: either we can attempt to construct a nonlinear equation of best fit, or we can transform the nonlinear data. We'll stick with the latter.

There are a variety of useful data transformations to achieve linearity. Common transformations include taking square roots, and finding logarithms, natural logarithms, or reciprocals. But let's note this upfront: Figuring out which transformation or transformations to use is somewhat of a guess-and-check process. After transforming a variable, or both variables, you should view the new scatterplot—to make sure it now looks linear—and also check the residual plot—to make sure it looks random.

If the scatterplot appears to show an exponential or power relationship, try taking the *logarithm* of the y values, the x values, or both; then, reexamine the scatterplot to see if the data's straightened out.

An example of an exponential relationship might be $y = \left(\dfrac{1}{2}\right)^x$. Graphed on the calculator, this function looks like

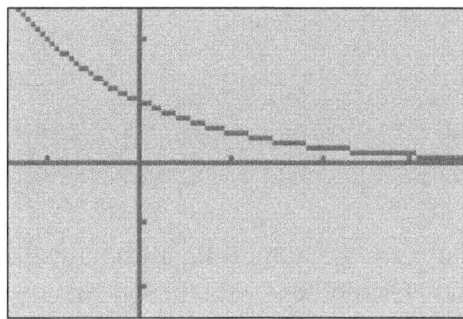

with a y-intercept at 1 and an asymptote at $y = 0$. Functions that have this curved, ever-decreasing shape (decreases slower and slower) are called exponential decay.

Conversely, a functions exhibiting exponential decay, like $y = 2^x$, have this curved, ever-increasing shape (increases faster and faster):

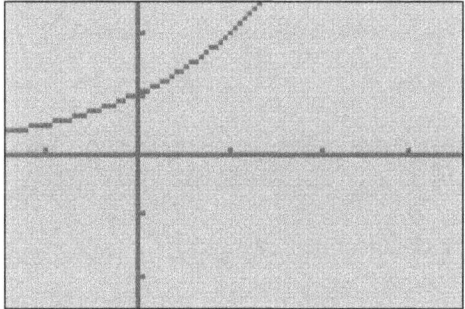

Exponential functions take the form $y = ax^n$, where if $|a|$ is greater than 1, the graph exhibits growth, and if $|a|$ is less than 1, the graph shows decay. To "linearize" the function, take the logarithm of both sides.[*]

$$\log y = \log\left(ab^x\right)$$

$$\log y = \log a + \log b^x$$

$$\log y = \log a + x \log b$$

And if the function is $y = e^x$—the base e being Euler's number (approximated $= 2.718281828$;

---

[*] To understand the next series of steps, you'll need to remember the product and power property of logarithms. The product property: the logarithms of two values multiplied = the sum of the logarithms of the two values; the power property: the logarithm of a value to a power = the logarithm of that value all multiplied by the power. Both properties can be derived by rewriting the logarithms with bases and exponents.

despite the ostensible pattern, the number is irrational)[*]—then the transformation becomes

$$\ln y = x \ln e$$

$$\ln y = x$$

because the inverse function of e is *ln* (the natural logarithm, or the logarithm with "natural" base e), so the functions simply "undo" each other—and disappear.

In addition to exponentials, power relationships, which take the form $y = ax^n$, are also linearly transformed by logarithms. Take $y = 2x^4$, for instance. Graphically, it appears as

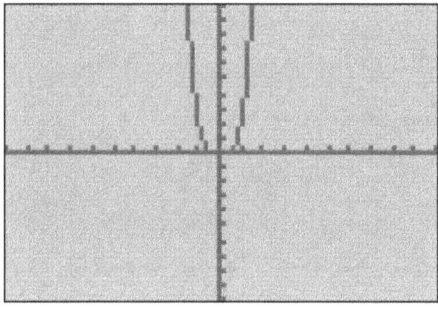

Let's linearly transform this power function. We get

$$\log y = \log 2 + 4 \log x$$

We should also know how to rearrange the equation to "revert" it back to being solved for y.

$$y = e^{\log 2 + 4 \log x}$$

$$y = e^{\log 2} e^{4 \log x}$$

$$y = e^{\log 2} e^{\log x^4}$$

$$y = 2x^4$$

---

[*] Named after the famed eighteenth century mathematician Leonhard Euler (pronounced "Oil-er"), $y = e^x$ doubles as its own derivative (namely, its rate of change with respect to some variable). For convenience, the function is oftentimes written as exp(x). Its value can be approximated by adding up terms of the reciprocal series $1/n!$ where n starts at 0.

Example 4h.-1

The following table compares population (in thousands) with year (since 2000).

year, x	1	2	3	4	5
population, y	20	60	120	190	280

(a) Find the values of the correlation coefficient and the coefficient of determination.

$r = 0.989$ and $r^2 = 0.979$

(b) Comment on both the scatterplot and the residual plot.
The scatterplot is curvilinear, and the residual plot is nonlinear as well.

(c) Is a linear model a good fit?

No. See the answer to (b).

(d) Transform the data set to achieve linearity by taking the logarithm of both variables. Report the transformed equation here, using the variable names *year* and *population*.

The easiest way to do this is to use the graphing calculator to make two new lists—call them logx and logy—that are stored as the logarithms of year and population, respectively. View a scatterplot of these new lists together—the data has indeed straightened out.
After running the LinReg(a+bx) calculator function with the newly transformed lists, the regression equation produced is

$$\log(\widehat{population}) = 1.295 + 1.639 \log(year)$$

(e) Express your equation as a "linear" equation—meaning solve for *population*.

$$\widehat{population} = 10^{1.295 + 1.639 \log(year)}$$

$$\widehat{population} = 10^{1.295} 10^{1.639 \log(year)}$$

$$\widehat{population} = 10^{1.295} 10^{\log(year^{1.639})}$$

$$\widehat{population} = 10^{1.295} year^{1.639}$$

(f) Predict the population for year 6.

$$population = 10^{1.295} \cdot 6^{1.639} = 371.86, \text{ or } 371,860.$$

There are, of course, many more possible transformations besides logarithms. Square roots, cube roots, and reciprocals might also straighten out a data set, permitting the construction of a good-fit linear model. It all depends on the shape of the original data set.

After being drafted and serving in the Vietnam War, a rabid young baseball fan by the name of Bill James was hired at a pork-and-beans factory as a security guard. It was there, amid the boredom of the night shifts, that James went to work on many of the baseball writings that would make him famous. James asked sophisticated quantitative questions about the game that really hadn't been asked before, such as, Which players were most valuable to the team? and Who gives up the most stolen bases?, that required some heretofore new statistical methods to tackle.

Perhaps his best known result, derived after asking another simple question—What is the impact of a single run toward winning a baseball game?—is called the *Pythagorean theorem of baseball*. Not really a theorem but rather a near-equation expressing the relationship between the ratio of wins to losses and runs scored to runs allowed per team, it is stated as[*]

$$\frac{wins}{losses} \approx \left(\frac{runs}{runs\ allowed} \right)^2$$

Let's test out the "theorem" for the 2014 MLB regular season. Taking the logarithm of both sides, we obtain

$$\log\left(\frac{wins}{losses} \right) \approx 2\log\left(\frac{runs}{runs\ allowed} \right)$$

Then, plotting the logarithm of the ratio of wins to losses versus the logarithm of the ratio of runs to runs allowed, the following scatterplot is generated (with a trend line added).

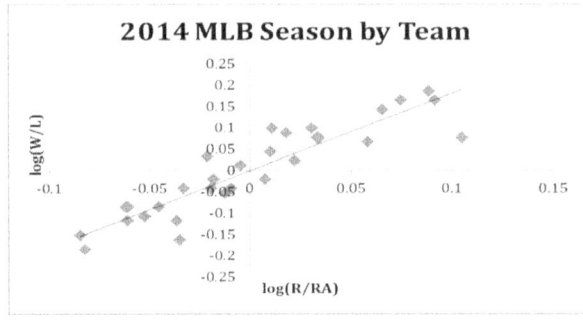

---

[*] See the text *Teaching Statistics Using Baseball* by Jim Alpert for a comprehensive discussion of Bill James' Pythagorean relationship.

Notice the linear relationship between the logarithmic transformations of the ratios. James' "theorem" still works well.

James' original work effectively spawned the field of baseball statistical study known as *sabermetrics*, later popularized by the Michael Lewis book *Moneyball* as well as by the movie of the same name.[*]

It's not only in the study of games that we can learn something from linear transformations. Nate Silver's *The Signal and the Noise* methodically details the nonlinearity of earthquake data. As the magnitude of an earthquake increases, he notes, the number of earthquakes "drops off exponentially"—in other words, quite expectedly, there have been many fewer catastrophic earthquake than small earthquakes. This trend is illustrated in his book by a smooth exponential decay graph, with earthquake magnitude on the *x*-axis and annual frequency on the *y*-axis.

But then Silver linearly transforms the exponential function by turning the *y*-axis into a logarithmic scale. What results is a nearly straight diagonal line travelling down the page. Silver verbally unpacks his linear model:

> This pattern is characteristic of what is known as a power-law distribution. Something that obeys this distribution has a highly useful property: you can forecast the number of large-scale events from the number of small-scale ones, or vice versa. In the case of earthquakes, it turns out that for every increase of one point in magnitude, an earthquake becomes ten times less frequent. So, for example, magnitude 6 earthquakes occur ten times more frequently than magnitude 7's, and one hundred times more often than magnitude 8's.

Although making predictions with the hurricanes-ocean temperature data earlier didn't make much sense, with appropriate massaging of the data and careful attention to detail, extrapolating here with this earthquake data, it seems, doesn't result in quite the shakeup. Nevertheless, Nate Silver, and others making predictions using historical data in general, should heed the warning of Charles Richter, seismologist and creator of the Richter magnitude scale: "Only fools, liars, and charlatans predict earthquakes."

---

[*] For much more on this and other sabermetric ideas, see the texts *Baseball Between the Numbers* and *Teaching Statistics Using Baseball*, from which the latter came the derivation of the Pythagorean theorem of baseball above.

§5. *Sometimes There's No Chance*

 φ

§5a. *Introduction to Probability*

In his *Treatise on Probability*, Keynes notes that "Part of our knowledge we obtain direct; and part by argument. The Theory of Probability is concerned with that part which we obtain by argument, and it treats of the different degrees in which the results so obtained are conclusive or inconclusive." Thus much of probability involves a priori thinking, of deductively reasoning without the benefit of experience—or experiment.

We often need to calculate the likelihood of some "chance process," which may or may not have underpinnings or analogues in lived experience. For instance, you may be asked how likely it is to roll two sixes, or pick a spade and a heart from a single deck of cards. These questions are asking you to calculate the probability of an *event*.

Definition 5a.-1

Event. A collection of the outcomes of a chance process.

Definition 5a.-2

Sample space. An exhaustive enumeration of the outcomes of an event.

Such a listing of the outcomes of dice (called "cases") was completed at least as far back as the Middle Ages, as Gavin Kennedy explains: "…the existence of cases is found in a manuscript that lists the '56 virtues,' one for each of the ways that three dice were believed to be thrown. Apparently, some monks used it to decide which virtue they would practice for the day…," although the amount of cases actually numbers 216 (six to the third power).

The probability of any event, which lies between 0 and 1, can be expressed as a decimal or fraction. (Alternatively, you may instead express the probability as a percentage.) A probability less than 0 is impossible, as is a probability greater than 1. The first formulation of these numerical limits comes to us from Leibniz—co-creator, along with Newton, of the calculus—with respect to property claims: 0 if a property claim has no merit, 1 if the evidence of a claim is overwhelming, or a number between 0 and 1 (exclusive) if the evidence is somewhere in between.

Example 5a.-1

(a) Suppose that in a lottery you must select six numbers, from 1 to 40 each (with no repeats). Determine the lottery combination below you believe is most likely to be drawn, and the lottery combination below you feel is least likely to be drawn.

> a. 5, 10, 15, 20, 25, 30
> b. 2, 11, 13, 27, 31, 39
> c. 4, 6, 10, 18, 34, 40
> d. 1, 2, 3, 4, 5, 6

Although it's difficult to believe at first blush, assuming the lottery game is fair—e.g., there are no weighted balls or no other mechanisms fixing the results—each of the four combinations is equally likely to be drawn. Have you ever seen a ticket with 1, 2, 3, 4, 5, 6 as the winning numbers? As statistician Deborah Rumsey explains, the improbability of 1, 2, 3, 4, 5, 6 being drawn puts into stark relief how unlikely it is to win with *any* particular ticket. The problem with the lottery isn't that "no one ever wins"; rather, the probability that *someone* wins is pretty high—but the probability that *you* win is infinitesimal.[*]

(b) Edgar Allen Poe once claimed that if you roll five 6's in a row with a fair die, your chances of landing a 6 again on the next roll have decreased. Was he correct?

No. Dice don't have brains. This idea has an apropos name: the "memoryless property."[†] (Poe wrote about the die in his short story "The Mystery of Marie Rogêt.")

(c) If the chances of winning a particular scratch-off lottery game are exactly 1 in 100, does that mean that if we buy 100 tickets *exactly* one of them will be a winner?

---

[*] Other problems with the lottery: it's a regressive tax on the poor, because ticket costs constitute a higher proportion of income among people who have less money than those with more money; and those who are poor *tend* to be less educated, less numerate, and more likely to miscalculate how unlikely scoring a jackpot really is—so those who are poor(er) usually buy more tickets than those who are not. In addition, lotteries of all stripes are state-sanctioned gambling—which can lead to or feed into gambling additions, along with other mental health problems.

[†] Also called the gambler's fallacy.

No. A figure like 1 in 100 means that, over time, and with many, many trials, we can expect on average 1 in 100, or 1%, of the tickets to be instant winners.[*]

(d) How many people do we need to gather in a room to *guarantee* that *at least two* of these people have the same birthday?

367. Here's why: suppose we gathered 366 people into a room. Is it *possible* (not necessarily likely, just possible) that these 366 people could each have a different birthday? Sure it is: Person 1 = January 1, Person 2 = January 2, ... , Person 366 = December 31, which covers all potential birthdates through a calendar year, including February 29. Add one more person to the room—the 367th person—and there will be a birthday match.[†]

(e) A couple has conceived five boys so far. A sixth child is coming—do you predict the newborn to be a girl or a boy?

Although if we examine demographic records there is actually a slightly greater chance of a boy being born,[‡] for our purposes in this primer we will treat any given birth equivalent to a coin flip: 50-50.

(f) Suppose a bowl contains two red chips and two white chips. Henry puts a blindfold on and reaches in to the bowl to pick a chip. Putting that chip aside, he takes the blindfold off and reaches in for a second chip. Henry's second chip is white in color. What is the likely color of the first chip?

The first chip was most likely red, since Henry had a 2/3 chance of picking a white chip second (which he did) if he had already picked red, but only a 1/3 chance of picking a white chip second if he had already picked white (since there would have been only one

---

[*] Unlike lotteries, such as the Powerball, which are as close to random as possible, scratch-off tickets are pseudorandomly distributed throughout gas stations, stores, and other venues to ensure that winning tickets don't "clump" together in one particular machine or location. That same sort of pseudorandomness is employed, but for a different reason, with electronic music players, as Jonah Lehrer explains: "When Apple first introduced the shuffle feature on its iPods, the shuffle was truly random; each song was as equally likely to get picked as any other. However, the randomness didn't appear random, since some songs were occasionally repeated, and customers concluded that the feature contained some secret patterns and preferences. As a result, Apple was forced to revise the algorithm."

[†] This is called the *pigeonhole principle*: "the fundamental enumerative principle that if a set of *n* objects is partitioned into fewer than *n* subsets, then at least one subset has at least two members" (from the *Collins Dictionary of Mathematics*). Here, the "*n* objects" are the 367 people, and the "fewer than *n* subsets" are all the possible birthdays, of which there are 366. So there are 367 "pigeons" for 366 "holes"—one subset has at least two members.

[‡] I would be remiss not to mention mathematician John Arbuthnot's explanation, in "An Argument for Divine Providence Taken From the Constant Regularity of the Births of Both Sexes," for more boy than girl births: divine providence. He believed that more boys were born to compensate for men killed in battle.

white chip remaining in the bowl).

We have already seen many methods of descriptive statistics (boxplots, bar graphs, histograms, and the like), but before we can explore inferential statistics, we must delve into probability theory. Probability is a measure of how likely an outcome is to occur. To quote Mario Triola in *Elementary Statistics*, who explains it best,

> Probability is the underlying foundation on which the important methods of inferential statistics are built. As a simple example, suppose that you have developed a gender-selection procedure and you claim that it greatly increases the likelihood of a baby being a girl. Suppose that independent test results from 100 couples show that your procedure results in 98 girls and only 2 boys. Even though there is a chance of getting 98 girls in 100 births with no special treatment, that chance is so incredibly low that it would be rejected as a reasonable explanation. Instead, it would be generally recognized that the results provide strong support for the claim that the gender-selection technique is effective. This is exactly how statisticians think: They reject explanations based on very low probabilities. Statisticians use the *rare event rule for inferential statistics.*[*]

The rare event rule brings to mind a personal anecdote. After having taught a particular class on day, students emptied out as usual, and I had the room free to grade papers, set up for a class later in the day, and complete all of the other minutiae that occupy a teacher's time. Ten minutes elapsed, and a former student of mine waltzed into the empty classroom (empty, that is, except for me) and asked me rather earnestly, "Mr. Lorenzo, do you have a class right now?" I had hardly noticed the abject silliness of the question before thinking how deficient this former student of mine was in the rare-event-rule department. Had I taught him nothing?

Could I have had a class "right now"? Yes, it's possible—I could have forgotten about the class, which would had to have been in a different room than the one I was presently occupying. It's also conceivable that every student from my "class" on that day happened to be absent. But was it *likely* that I had a class at the moment, given that there were no students in the room? No: my former student should have rejected that having-a-class-with-no-one-present explanation out of hand, given its low probability.

As described in the first § of this primer, probability can be classified into three approaches. There's *classical* (or *theoretical*) *probability*, which is probability from an a priori perspective. Classical probability answers such questions as *What are the chances of rolling double sixes?* and *What are the odds that I win the lottery?* A simple formula summarizes classical probability. Assuming there is an equal chance of each sample space outcome, probability is given by

$$\frac{successful\ outcomes}{total\ outcomes}.$$

---

[*] Important to the development of the rare event rule? John Arbuthnot's investigation of boy-girl births. Before collecting data, he assumed the chances were 50-50; not finding that, he fashioned another explanation. (A divine one, until Ronald Fisher, whom we'll meet later, came up with hypothesis testing.)

Next, we have *empirical* (or *experimental*) *probability*, which is probability from an a posteriori perspective. Here, we perform an experiment or observe a procedure and count the number of times a particular event occurs.

Finally, *subjective probability* leans on personal believe. Recall the disagreement with the Sleeping Beauty Problem, which doesn't seem solvable using either of the other two approaches to probability.

It is important to emphasize that, with classical probability, the outcomes of the sample space must be equally likely. A common misconception is to believe that if we know nothing about the probabilities of an event, then the simple events in the sample space are equally likely (recall the economist John Maynard Keynes called this fallacy the Principle of Indifference).

Example 5a.-2

Are the following outcomes "equally likely" or "not equally likely"?

(a) A fair coin landing on heads or tails.

Equally likely.

(b) A fair coin landing on heads or tails, after landing on tails six times in a row.

Still equally likely. Previous flips don't have an effect on future flips.

(c) Life on Titan, Saturn's largest moon.

Not equally likely. This question is best answered (if it can be) using subjective probability. The probability of life on a moon millions of miles away from the earth can't be reduced to a mere coin flip.

(d) A ball landing in one of the thirty-eight slots of a roulette wheel.[*]

Equally likely, but not 50-50. The ball has an equal chance of falling into any one of the thirty-eight slots.

Categorizing events as equally likely and not equally likely brings to mind Gerd Gigerenzer's book *Risk Savvy*. In it, he divides the calculation of probabilities of events into two types: risk versus uncertainty. Risk is quantifiable; uncertainty (unknown risk) is not. Many people today assume that events live in the risk category, when they in fact fall under uncertainty, like stock market projections. With both risk and uncertainty, short-term results are governed by chance;

---

[*] In the U.S., that is; European roulette wheels have thirty-seven slots.

the distinction lies with long-term behavior: risk is somewhat predictable, whereas uncertainty is not. Especially in the financial sector, there is a fetishizing of calculating uncertainty ipso facto; a relative frequency (or frequentist) approach doesn't necessarily conform to the wild movements of the market.

But the bifurcation of risk and uncertainty are not new ideas. Economist Frank Knight, in his *Risk, Uncertainty and Profit* published in 1921, wrote that risk is "a measurable uncertainty," whereas "[u]ncertainty must be taken in a sense radically distinct from the familiar notion of Risk, from which it has never been properly separated." Knight critiqued the relative frequency approach to probability, especially its use in making business decisions, by explaining that any particular "'instance…is so entirely unique that there are no others or not a sufficient number to make it possible to tabulate enough like it to form a basis for any inference of value about any real probability in the case we are interested in." Even though Keynes loathed Knight, he agreed with him that with one-off events, such as war and new discoveries or inventions, "there is no scientific basis on which to form any calculable probability whatever. We simply do not know!"

By the way: the relative frequency approach does not require equally likely outcomes to be present in the sample space. Instead, since we are merely observing a procedure being conducted and recording the results, we care most about the number of trials of the experiment. The more trials, the better—since the more trials, the more accurate our probability estimate is likely to be. This leads us to our next definition, which only applies to the relative frequency approach to probability.

Definition 5a.-3

The law of large numbers. As a process is conducted more and more times, the empirical probability converges to the true probability.

Jacob Bernoulli was the first to prove the law of large numbers, which turns on the distinction between a priori and a posteriori reasoning: casting aside a priori, Bernoulli leaned on past results to make probable (but not necessarily perfect) predictions about the future. But how? As he explained, "[We assume that] under similar conditions, the occurrence (or non-occurrence) of an event in the future will follow the same pattern as was observed in the past."

Essentially, empirical probability results from conducting an experiment, making observations, or performing a simulation. It's a probability calculated from real life. For example, if you toss a fair coin 50 times and obtain 30 heads, the experimental probability of obtaining heads would be 30/50, or 0.6. But toss a coin 50 times again, and this probability may change.

Experimental probability differs from theoretical probability in that theoretical probability represents what you would "expect" from the theory or the description of the situation. The classical probability of a coin flip is 1/2, so tossing a coin 50 times would "theoretically" give you a ratio of 25/50 heads, but not always—because in real life, remember, there is variability.

As you increase the number of times a probability experiment is repeated, the empirical probability of an event approaches the theoretical probability of the event. This is Bernoulli's powerful statistical concept known as the law of large numbers: that an estimate from a simulation generally gets closer to the actual probability as the number of repetitions increases. As the number of trials increases, the variation between the experimental results and the true probability decreases. In other words, repeat a process enough times and, *over time*, the truth emerges.[*] Expressed differently: a posteriori → (converges to) a priori. Many trials lend a "moral certainty" (Bernoulli's words) to the probability.[†]

For instance, as discussed in a prior §, a total of 135 space shuttle launches resulted in two that catastrophically failed: *Challenger* in 1986 and *Columbia* in 2003. According to physicist Richard Feynman, engineers believed that the true probability (or, as you may wish to think of it, the classical probability, as if we're finding the probability of rolling double sixes) of catastrophic failure was 1 in 100. The 2 in 135 figure—or roughly 1.5%—conforms closely to the engineers' predictions and, more or less, illustrates the law of large numbers in action: conducting a process (of launching shuttles) many times to capture a true probability (of catastrophic failure).

Example 5a.-3

Consider an experiment in short-run and long-run behavior. Examine the plot below, which shows the relative frequencies of rolling a "6" on a fair, six-sided die, and answer the questions.

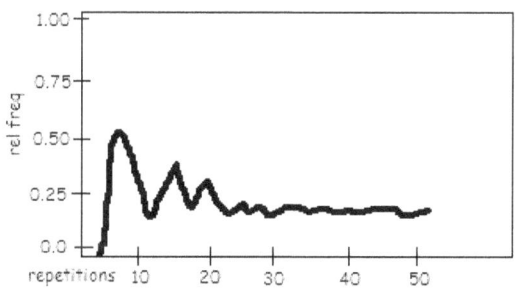

(a) According to the results of the experiment, what is the empirical probability of rolling a "6" after about 15 rolls? How about after about 40 rolls?

Approximately 0.3; approximately 0.2.

---

[*] There is also a *weak law of large numbers*, known as Bernoulli's theorem, that statistician George Cobb explains the genesis of as follows: What Bernoulli wished to find was, "'How many trials do you have to observe in order to be able to estimate *p* with a given precision?' He wasn't able to solve this problem, and had to settle for the limit theorem that bears his name. Why couldn't he solve it? Because he couldn't compute the tail probabilities," which were a summation of many terms of the binomial distribution (we'll encounter this important distribution shortly) that would await easy calculation only by the end of the twentieth century—courtesy of computer processing power. Bernoulli substituted raw calculation for approximation via sums of geometric series.

[†] Bernoulli anticipates the *significance level* of hypothesis testing when suggesting that there be some numerical limit for moral certainty: "It would be useful if the magistrates set up fixed limits for moral certainty." More on hypothesis testing later.

(b) What do you notice about the shape of the plot? In other words, what happens as the number of die rolls is increased? How does this illustrate the law of large numbers?

The shape "stabilizes." The empirical probability approaches the true probability (1/6) after many repetitions of the experiment.

Even Aristotle noted relative frequency concepts with dice—"…to repeat the same throw ten thousand times with the dice would be impossible, whereas to make it once or twice is comparatively easy"—the Greeks' bias against a posteriori reasoning didn't permit him (as we explored earlier in the "Foundations" §) or his ancient contemporaries to flesh out any serious theory of probability.

Example 5a.-4

Consider a fair coin.

(a) When flipping the fair coin, do you think it's more likely to obtain about 5 tails out of 10 flips or about 5,000 tails out of 10,000 flips? Justify your answer.

It's more likely to obtain roughly 5,000 tails out of 10,000 because of the law of large numbers.

(b) Again, consider flipping the fair coin. Do you think it is more likely to obtain exactly 5 tails out of 10 flips or exactly 5,000 tails out of 10,000 flips?

Unlike the previous question, here it's more likely to obtain 5 tails out of 10 flips. The distinction lies with the word *exactly*—there are many fewer possibilities when flipping a coin 10 times rather than 10,000. Although flipping 10,000 times will give you a roughly 50-50 split of heads and tails, landing exactly 5,000 tails is very unlikely.

Now consider a gold coin that you think might be weighted (biased). You decide to flip the coin repeatedly to see.

(c) Suppose you flipped the gold coin 1000 times and it came up heads 501 times. Would you be convinced the coin is weighted?

Probably not; we'd expect a fair coin to land approximately 50% of the time on heads.

(d) Suppose instead the coin came up heads 550 times. Would you now be convinced the coin's weighted?

Perhaps. According to the rare event rule, if our assumption going into this experiment is that the coin's fair, any significant divergence from 50-50 should lead us to reject our assumption.

(e) Suppose the coin actually came up heads 600 times. Would you be convinced now?

Probably.

(f) Out of 1000 flips, how many times x would the coin have to land on heads for you to be completely convinced that the coin is weighted?

This sort of question asks us to create a threshold, a concept that Ronald Fisher, one of the fathers of inferential statistics, formalized. We'll get to Fisher's take later on.

Imagine a coin flipped ten times. Better yet, don't imagine; let's use the graphing calculator to simulate the ten flips. A "0" will represent heads, while a "1" will represent tails. Press the MATH key, scroll over to the PRB menu, and scroll down to the randInt function. The function takes three arguments: first, the lower bound (here, 0); then, the upper bound (here, 1); and finally, the number of repetitions (here, 10). Enter the arguments, separated by commas, and press the ENTER button. Scroll right to see all of the numbers.

When I ran the simulation, I obtained the following set of coin flips: 1, 0, 0, 0, 1, 1, 1, 0, 0, 0. Perhaps you got something similar; you probably won't get the same set because of the calculator's random selection of the integers.[*] Are you surprised that there are "streaks" or "runs" in the string of numbers? That's to be expected; such streaks are more common than we might think. (In fact, one of the ways data forensic analysts can spot human-generated attempts at randomness is by noting the presence or absence of such streaks.) Although it is true that in the long run 0's and 1's (or heads and tails) will appear half the time, what is a myth is there is some probabilistic "rule" that corrects any such imbalance trial to trial in the outcomes.[†] Coins don't have brains, so they can't remember what side they landed on previously. This is called the "memoryless" property, which leads to the gambler's fallacy and *the law of averages*—the belief in the myth of a perpetual balance.[‡] And, again, it's something that Edgar Allen Poe couldn't accept, since he ar-

---

[*] Actually, calculators and computers do not select numbers randomly but *pseudorandomly*, or according to some deterministic process—usually a mathematical function or algorithm that simulates randomness.

[†] A belief in the "hot hand" theory among sports players, coaches, and commentators is widespread, but lacks evidence. Simply because a player has made his last few shots (or goals, or whatever) doesn't necessarily mean he (or she) is more likely to make the next shot. Masses of basketball data of Philadelphia 76ers players were studied, but no correlations to indicate a hot hand amongst players were found. Instead, the usual presence of streaks in random data more than likely explains the hot hand phenomenon.

[‡] Even just last night, on a sports-commentary show, a color commentator was weaving tales of the virility of an athlete, noting that he "goes to a bar each night and asks girls out until someone finally goes home with him. That increases his odds. It's the law of averages at work." No! It's the law of large numbers—the sample size grows so that there is a convergence to the truth of the proposition—although, as in so much of life, how closely a mathematical law models the vagaries of such complex real-world events is suspect.

gued that if you roll five 2's in a row, your chances of getting a 2 on the next roll are less than one-sixth, which is certainly not true—dice don't have brains either.

§5b. *Simulation*

Using the graphing calculator to represent heads and tails brings to mind *simulation*.

Definition 5b.-1

Simulation. A correspondence or matching between real-world processes and artificial modeling, usually on a calculator, computer, random number table, or other physical mixing device.

Simulations are not new. Perhaps you've heard of the fifty-year-old Strat-O-Matic game, which, in its most popular incarnation, released baseball cards containing a variety of statistics; these statistics would be used with dice and decoding tables to permit the player a "managerial" decision-making role in a simulated baseball game.

The role-playing game *Dungeons & Dragons* spurred the imaginations of preteens and teens in the 1970s and '80s; like the Strat-O-Matic, it also made use of dice (of many sides) and statistics to code simulations.

Example 5b.-1

Danica crashes in 8% of her races. Simulate 10 races; report the proportion of crashes.

Solution. There are several ways we can complete this simulation. First, though, we need to set a correspondence between crashes and numbers.

- Let the digits 01 to 08 represent a crash;
- Let the digits 09 to 99, and 00, represent no crash.

Using the Random Number Table (Table B in the back of this primer), pick a random column and random row (you can close your eyes and point), read off digits, in pairs, until 10 two-digit numbers have been gathered. Then report results.

For instance, suppose we picked row 22, column 1:

Row 22	88861	56842	57442	40841	80247	45291	03655	54639	91122	52990

Starting from the leftmost digit, we find the following:

88 = no crash; 86 = no crash; 15 = no crash; 68 = no crash; 42 = no crash;
57 = no crash; 44 = no crash; 24 = no crash; *08 = crash*; 41 = no crash.

According to the results of the simulation of ten races, Danica crashed one time, or at a rate of 0.1 (10%).

Alternatively, we could use the graphing calculator's randInt feature to simulate races. The function would be input as randInt(0,99,10).

We could even use a computer to run hypothetical races. In Excel, for example, you could utilize the RANDBETWEEN(*lower bound, upper bound*) function.[*]

Regardless of how you create a correspondence between potential real-world and simulated outcomes, the process must cohere and be *realistic* enough so that the hypothetical results of the simulation can be trusted.

Example 5b.-2

A well-known theory of stock market analysis is the random walk hypothesis, as popularized in *A Random Walk Down Wall Street* by Professor Burton Malkiel. The term *random walk*, applied to stock prices, implies that short-term changes in stock prices cannot be predicted but rather are completely random; as Malkiel describes the theory, "Taken to its logical extreme, it means that a blindfolded monkey throwing darts at the stock listings could select a portfolio that would do just as well as one selected by the experts," because the next move in the price of a stock (up or down) is unpredictable on the basis of what prices changes happened before. "The efficient-market hypothesis explains why the random walk is possible," he writes, since "[i]t holds that the stock market is so good at adjusting to new information that no one can predict its future course in a superior manner." We can summarize with the following statement:

> *Even though the market quickly adjusts to new information (= the efficient-market hypothesis), this new information arrives at unpredictable, random times (= the random walk theory).*

Malkiel tested his theory by creating a hypothetical stock worth $50, whose close each day was determined by a coin flip: pressed higher by a half-point if heads was obtained, and pushed lower by a half-point with tails. Without revealing the hypothetical nature of the financial results, Malkiel then presented the data to experts—who immediately urged the professor to buy the pseudo-stock.

Let's use a simplified model to simulate the stock prices changes of a hypothetical company, RBG, Inc., over a 15-day period of trading.

(a) The initial stock price of RBG is $50. Use the random number table to select 15 digits. Record them below.

---

[*] This is how the primer's random number table (Table B) was constructed.

Suppose we randomly selected row 22 again:

Day	1	2	3	4	5	6	7	8	9	10	11	12	13	14	15
Digit	8	8	8	6	1	5	6	8	4	2	5	7	4	4	2

(b) Next, we need to use each digit to determine whether the stock price increases or decreases on that particular trading day. How do you suppose we could do this?

We set up a correspondence of digits to stock movements. Perhaps 0 to 4 represents "increase," while 5 to 9 represents "decrease." Or an even digit signifies "increase," with an odd digit signifying "decrease."

(c) If the price increases on a trading day, assume it increases by $1. If it decreases on that day, assume it decreases by $1. With this in mind, fill in the cells in the table below to get a final price of RBG after 15 days of trading.

Using even-odd correspondences,

Day	Initial	1	2	3	4	5	6	7	8	9	10	11	12	13	14	15
Price ($)	50	51	52	53	54	53	52	53	54	55	56	55	54	55	56	57

(d) Now you have an initial ($50) and a final price (the price of the stock at day 15). Take the final price and subtract it from the initial price. For example, if your final price was $47, take $50 – $47 = $3. Do this calculation below.

We earned $7.

(e) Create a scatterplot of the previous days' prices compared to the previous days'. In addition, find the correlation coefficient, and interpret your results.

PreviousDay	50	51	52	53	54	53	52	53	54	55	56	55	54	55	56
NextDay	51	52	53	54	53	52	53	54	55	56	55	54	55	56	57

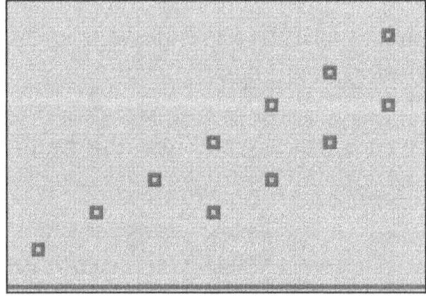

And $r = 0.861$. Although this seems to be good news about using past results to make future predictions about the stock market—after all, the correlation coefficient shows a positive, strong correlation, and the scatterplot has a nice, clear linear upward trend—remember two things: (1) The price data were generated randomly, and (2) The prices could only fall or rise a single unit per day. Unfortunately, taken together, these issues confound any predictions, making them as meaningless as Malkiel's experts'. ▨

We could also make use of simulation when the calculation of probabilities by brute-force methods is either too difficult or too time consuming, such as with more complex lotteries and other games.

§5c. *Probability in Theory, Probability in Practice*

Time to sharpen our a priori tools again to help solve probability problems—without much recourse to experiment or simulation. First, though, a word on notation.

- We usually use P to represent probability. We've already encountered this with normal distribution problems.
- Also, capital letters usually denote events in the sample space.

In any probability experiment, it is very important to identify, exhaustively, all the possible outcomes. Again, a list of all of the different outcomes in a probability experiment is called a sample space. Consider set S, a sample space, that is composed of a finite number of events, each of which is equally likely to occur. The sum of the probabilities of all of the elements in this sample space is equal to one (also called "unity"); the probability of any particular event can be no less than zero and no greater than one.

Example 5c.-1

(a) List the elements of the sample space S of flipping a coin twice.

HH, HT, TH, TT.

(b) List the elements of the sample space S of families with three children.

BBB, BBG, GBB, BGB, GGG, GBG, BGG, GGB.[*]

(c) If a family has had three children, what's the probability that all three are boys? All three

---

[*] There are eight elements here since there are two possibilities for the first child (boy or girl), two possibilities for the second, and two for the third—hence, two multiplied by two twice over gives us eight.

are girls? All three are the same gender? All three aren't the same gender? Exactly two are the same gender?

Count elements of the sample space in part (b) for these answers: 1/8; 1/8; 1/4; 3/4; 3/4.

(d) Suppose a set of students in a class is queried about how many pencils they own. Let x denote the number of pencils. Define the sample space.

$$S = \{x : x \geq 0\}$$

(e) Suppose the sample space is $S = \{1,2,3,...,50\}$. Find the probability of selecting a factor of 30.

There are 7 factors of 30: 1, 2, 3, 5, 6, 10, 15. Thus, the probability is 7/50.

Example 5c.-2

A game in which you sum the values of rolling two, six-sided dice is played.

(a) What is the sample space of the sums?

	1	2	3	4	5	6
1	2	3	4	5	6	7
2	3	4	5	6	7	8
3	4	5	6	7	8	9
4	5	6	7	8	9	10
5	6	7	8	9	10	11
6	7	8	9	10	11	12

(b) Suppose event A is "sum is 7." Find $P(A)$.

6/36 = 1/6

(c) Suppose event B is "sum isn't 7." Find $P(B)$.

30/36 = 5/6

(d) Suppose event C is "sum is 4." Find $P(C)$.

3/36 = 1/12

(e) Find $P(A) + P(B)$.

Unity, since all elements in the sample space are accounted for.

(f) Find $P(A) + P(C)$.

$9/36 = 1/4$

We can also think of the sample space S as a finite (enclosed) area and event E as a segment of that area, as shown below.

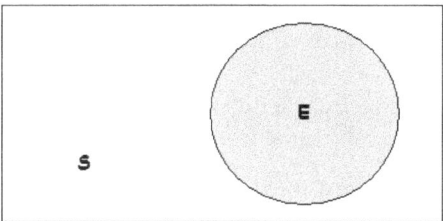

Example 5c.-3

A dart is thrown at the board below (not drawn to scale; units are in centimeters).

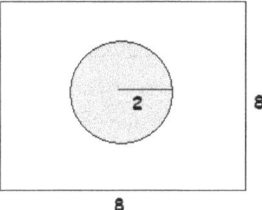

(a) What's the probability that the dart will hit the circular area?

$$P(hit) = \frac{\pi \cdot 2^2}{8^2} \approx 0.2$$

(b) What's the probability that the dart won't hit the circular area?

$$P(hit') = 1 - \frac{\pi \cdot 2^2}{8^2} \approx 0.8$$

We all like to receive compliments, such as *you look nice today* or *that tie is fantastic*. Complements, like *that tie really complements your shirt*, are a little different; a complement *completes*.

In probability theory, a complement completes a sample space, bringing it to unity. If A is an event in S, then the event consisting of all outcomes in S not in A is the *complement* of A, written as A' (read "A prime"). You could also see the complement of A written as A^C.

Mathematically, we can express the relationship of an event to its complement as $P(A') = 1 - P(A)$. To show why, we need to quickly introduce two new symbols here, which we will spend more time on shortly: $\cup$, which means "union" and conjoins two events together, accounting for everything in both of them; and $\cap$, which means "intersection" and only gathers what's in common between the events. So, $P(A') = 1 - P(A)$ because

$P(S) = 1$	The probability of the entire sample space is 1;
$P(S) = P(A \cup A')$	The probability of the sample space is also A <u>or</u> not A;
$P(A \cap A') = 0$	The probability of A *and* not A is zero: they are mutually exclusive;
$P(S) = P(A) + P(A')$	Since events A and not A are mutually exclusive;
$1 = P(A) + P(A')$	By substitution from the first line;
$P(A') = 1 - P(A)$	Which is what we wanted to show.

Recall that when events are termed *mutually exclusive*, they cannot happen together and have no set elements in common.[*] A synonym for mutually exclusive is *disjoint*.

When events are *exhaustive*, the set of those events constitute the entire sample space. So, if $S = A \cup B \cup C$, then A, B, C are the exhaustive events of sample space S.

The null (or empty) set has a probability of zero; this is expressed symbolically as $P(\varnothing) = 0$. Here's why: Suppose $A = \varnothing$; then A', the complement of A, must be the entire sample space: $A' = S$. Thus, $P(\varnothing) = 1 - P(A') = 1 - P(S) = 1 - 1 = 0$.

Example 5c.-4

(a) Consider a fair, six-sided die. Suppose event A is "sum is 10." Find $P(A^c)$.

$33/36 = 11/12$

---

[*] Real-world example: in African society, the Kiswahili divide all people, living or dead, into three, mutually exclusive sets: those who are still alive; the *sasha*, or living-dead, those who have died but whose time on earth overlapped with people who are still alive; and the *zamani*, or the dead, in which no one currently alive was alive when they were corporeal. Mathematician John Allen Paulos offers up an example of mutual exclusivity too, but with respect to sexual history. There are three mutually exclusive (and thus non-overlapping) groups of people: virgins, those who've had a single partner, and those who've had more than one sexual partner.

(b) Consider a standard deck of 52 playing cards. Suppose event A is the probability of "picking an Ace." And suppose event B is "picking a King." Find $P(A \cup B)$. Are these two events disjoint?

Yes, these are mutually exclusive events because they cannot happen together: you cannot pick a "KingAce" card. The probability of A or B is the sum of each: $4/52 + 4/52 = 2/13$.

Example 5c.-5

Bonnie is a sales associate at a large Chevrolet auto dealership. She believes in Fermi's Law: "If you don't fail a good part of the time you're not doing your job." So she motivates herself by using probability estimates of her sales. For a sunny Saturday in April, she estimates her car sales as follows:

Cars sold:	0	1	2	3
Probability:	0.3	0.4	0.2	0.1

(a) Is this a legitimate probability distribution?

Yes, since the probabilities add up to one and each probability is between zero and one.

(b) Find the probability of Bonnie selling more than one car.

$0.2 + 0.1 = 0.3$

(c) Find the probability that Bonnie will *not* sell three cars.

$1 - 0.1 = 0.9$

When we're trying to find probabilities involving two events, a *two-way table* (also called a *contingency table*) can display the sample space in a way that makes calculations easier.

Example 5c.-6

A poll was conducted among high school students asking: Do you want more study halls? One hundred and twenty students were asked, and what follows is a two-way table of the results (split by gender).

	Yes	No	Unsure
Male	10	53	7
Female	14	25	11

(a) If you choose a student surveyed at random, what is the probability that the student wants more study halls?

24/120 = 0.200

(b) If you choose a student at random who answered "No" to the survey question, what is the probability that the student is a female?

25/78 = 0.321

(c) If one person is randomly selected, find P("Unsure").

18/120 = 0.150

(d) If one person is randomly selected, find the probability of getting someone who said "Yes" or "No."

1 – 0.150 = 0.850

(e) If one person is randomly selected, find the probability of getting someone who is a male or who said "No."

There are overlapping events—they are not mutually exclusive, since they can happen together—so we have to be careful not to double count anyone (i.e., count people twice).

70/120 + 78/120 – 53/120 = 0.792

(f) If one person is randomly selected, find P("Unsure" *or* Female).

18/120 + 50/120 – 11/120 = 0.475

§5d. *Using Venn Diagrams*

Venn diagrams are named after mathematician John Venn, although the mathematician Euler was the first to use such a graphic. This is just one of many examples of Stigler's law of eponymy (or misonomy), which states that discoveries in science and mathematics aren't properly named for their creators. (Ironically, Stigler's law is itself a member of this set, since statistics professor Stephen Stigler didn't originate the law.)

Venn diagrams can be used to illustrate the relationship between events in a sample space. A *union* of two events, such as A and B, written as $A \cup B$, consists of all outcomes in A or B (or both). Note that, by default, the union is an *inclusive or*—both events are implicitly included, even without the "or both" qualifier appended—not exclusive (denoted *xor*).

The *intersection* of two events, such as A and B, written as $A \cap B$, consists of *all* outcomes in A and B; think of an intersection as what's in common between events A and B.

Take a look at these Venn diagrams:

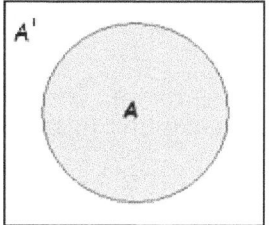 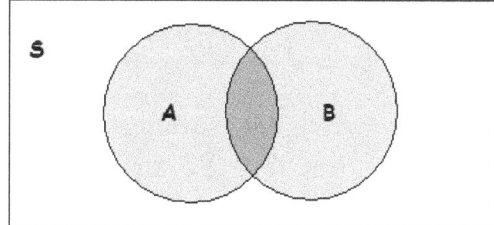

The diagram on the left shows the complement of A—all of the outcomes that aren't in A. On the right, the shaded area represents $A \cap B$, and the total of the shaded area—that is, everything in the crescents A and B, along with the common region in the middle—represents $A \cup B$.

With that explained, we can now introduce the *addition rule* (sometimes called the *general addition rule*) to account for the overlap (and the double counting).

$$A \cup B = A + B - A \cap B$$

Slapping probability notation onto the addition rule is simple:

$$P(A \cup B) = P(A) + P(B) - P(A \cap B)$$

Note that in a set of mutually exclusive (also called disjoint) events, there is no intersection, since mutually exclusive events cannot both happen together—they have no overlapping events:

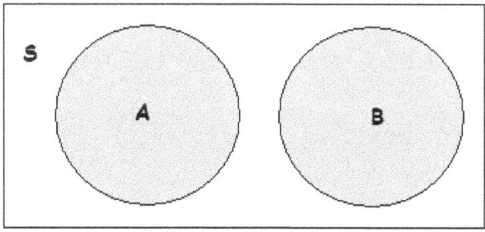

For instance, whereas the set of Aces and the set of Kings—four cards in each set in a standard deck of 52 cards—are mutually exclusive, the set of Clubs and the set of Kings aren't disjoint, since there is a card common to both sets: the King of Clubs.

Before proceeding any further, though, we need to take a closer look at the addition rule. First, consider the Venn diagram of $A' \cap B$ below, which we will make use of to derive the addition rule:

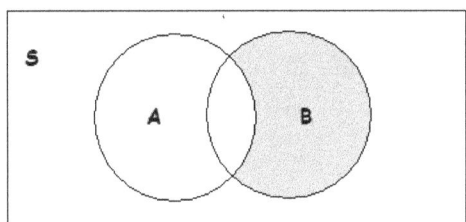

Note that $A \cup B = A \cup (A' \cap B)$, since event A "shades in" the remainder of the union between A and B (as shown above). But this can also be rewritten strictly as a probability statement: $P(A \cup B) = P(A \cup (A' \cap B))$ or, better yet, $P(A \cup B) = P(A) + P(A' \cap B)$, because of the mutual exclusively (i.e., non-overlapping nature) of those two pieces.

We might be better served to think of event B this way, as a composite between two pieces of the diagram, as shown below:

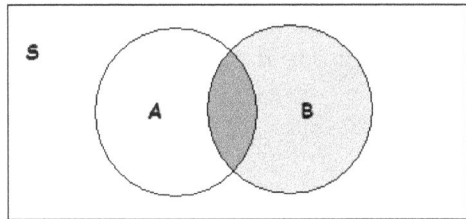

Symbolically, this works out to be $B = (A \cap B) \cup (A' \cap B)$ which, because of the mutual exclusively of the two "shaded" pieces, is equivalent to $P(B) = P(A \cap B) + P(A' \cap B)$, giving us, via substitution, the addition rule:

$$P(B) = P(A \cap B) + P(A' \cap B)$$

$$P(A' \cap B) = P(B) - P(A \cap B)$$

Recall, from above, $P(A \cup B) = P(A) + P(A' \cap B)$, so

$$P(A \cup B) = P(A) + P(B) - P(A \cap B)$$

We don't have to restrict our attention to examining just two events in a sample space at a time. If there are three events, A, B, and C, the addition rule still applies. (See the top of the next page for the Venn diagram illustrating three overlapping events.)

When subtracting out all three double-overlaps (the second-darkest filled shade), we have subtracted a bit too much—and have to add back in the intersection of all three (the darkest shade). Hence, the formula is

$$P(A \cup B \cup C) = P(A) + P(B) + P(C) - P(A \cap B) - P(A \cap C) - P(B \cap C) + P(A \cap B \cap C)$$

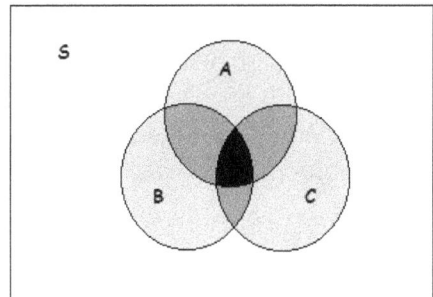

Example 5d.-1

(a) Consider the set of all cars, C. Now consider V, the set of all VW Beetles. Find $C \cap V$ and $C \cup V$.

Since V is a subset of C, we know that $C \cap V = V$. And, by the addition rule,

$$C \cup V = C + V - C \cap V = C + V - V = C$$

(b) Find the probability of selecting a heart or a face card.

$$P(H \cup F) = P(H) + P(F) - P(H \cap F)$$

$$P(H \cup F) = 13/52 + 12/52 - 3/52 = 0.423$$

(c) Find the probability of selecting an Ace from a deck of cards or rolling a "6" with a die (or both).

$$P(A \cup 6) = P(A) + P(6) - P(A \cap 6)$$

$$P(A \cup 6) = 4/52 + 1/6 - 0 = 0.243$$

(d) Suppose the sample space is given as $S = \{1,2,3,4,5,6,7,8\}$. What is the probability that the number you select is a prime number or a factor of 6?

There are 4 factors of 6: 1, 2, 3, and 6. There are 4 prime numbers: 2, 3, 5, and 7 (the number 1 is not considered prime, because 1 could repeat as a factor of any number an arbitrary number of times—unlike any other prime factor). Thus, we have

$$P(PRIME \cup FACT\,6) = P(PRIME) + P(FACT\,6) - P(PRIME \cap FACT\,6)$$

$$P(PRIME \cup FACT\,6) = 4/8 + 4/8 - 2/8 = 0.750$$

(e) Back in the mid-to-late 1990s, viewers were presented with a choice of top-tier wrestling cable television programs to watch on Monday nights: WWF Monday Night Raw or WCW Monday Nitro. Suppose that, on a particular Monday night, 16% of the cable-viewing population watched WWF, 9% tuned into WCW, and 3% switched between both. What percentage of the cable viewers watched neither show that night?

Making a Venn diagram is helpful here.

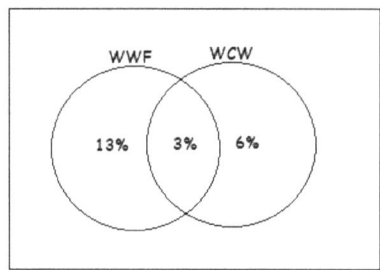

The reason why 13% lies inside the WWF crescent and not the earlier stated 16% is because the 16% figure *includes those who watch both, which is 3%*. Same idea with the 6% inside the WCW crescent. The crescents, then, are *exclusive*: the percentage of viewers of *only* WWF is 13%, for example, and the percentage of people who watch only a single wrestling program is 13% + 6% = 19%.

Adding together all three percentages in the Venn diagram gives us the percentage of viewers of either WWF or WCW (or both): 22%. (So, just to be clear, $P(WWF \cup WCW) = 0.22$.)

To obtain the percentage of people who watched neither, simply take that number from 100%. The answer is 88%.

(f) If $P(A) = x$, $P(B) = y$, and $x > y$, what is the range of probabilities for $P(A \cap B)$?

The sum of x and y cannot exceed 1. So, the minimum that this sum could be is $1 - (x + y)$, which is also, coincidentally, the minimum value of $P(A \cap B)$.

Since $x > y$, $P(A \cap B)$ would reach its maximum only if B were a subset—i.e., entirely contained within—A. If that were to happen, $P(A \cap B) = b$.

§5e. *Conditional Probability*

L et's return to the poll, conducted among high school students, asking: Do you want more study halls? One hundred and twenty students were asked, and here is a two-way table of the results (split by gender).

	Yes	No	Unsure
Male	10	53	7
Female	14	25	11

If we know that a randomly selected student answered "Yes," then the probability that the student is male is 10/24. If we know that the student is male, then the probability that the student said "Yes" is 10/70. Although these two probability statements sound alike, they are asking very different things. When information is "given"—as in the probability statements here—the question narrows focus from everything in the sample space to a portion of the sample space, funneling the answer within a carefully circumscribed subset. Probabilities that restrict the sample space in such a way are called *conditional*.

Definition 5e.-1

Conditional probability. A probability that has been explicitly circumscribed to a subset of the sample space courtesy of a "given" statement; the vertical bar | denotes "given."

The formula for conditional probability is $P(A \mid B) = \dfrac{P(A \cap B)}{P(B)}$, as long as $P(B) \neq 0$.

Example 5e.-1

Back to the study hall data again.

   (a) If we know that a randomly selected said "Yes," then what is the probability that the student is male?

$$P(Male \mid "Yes") = \frac{P(Male \cap "Yes")}{P("Yes")} = \frac{10}{24} = 0.417$$

   (b) If we know that a randomly selected student is male, then what is the probability that the student said "Yes"?

$$P("Yes" \mid Male) = \frac{P("Yes" \cap Male)}{P(Male)} = \frac{10}{70} = 0.143$$

It's also important to recognize the close link between conditional property and *independence*, which is defined below.

Definition 5e.-2

Independence. When the occurrence of one event has no effect on the occurrence of any other event.

Example 5e.-2

Determine if the following events are independent or dependent (i.e., not independent).

(a) Flipping a fair coin repeatedly.

Independent.

(b) Gender of offspring.

Independent.

(c) Selecting an Ace then a heart out of a single deck of cards.

Dependent.

(d) Selecting an Ace then a heard out of different decks.

Independent.

To mathematically determine if two events are independent,[*] both of these formulas need to be utilized:

$$P(A \mid B) = P(A) \text{ and } P(B \mid A) = P(B)$$

Both formulas, translated into English, state the following: the additional information, the "givens," make no difference—the probability of A or B is unaffected.

Example 5e.-3

Recall that we asked 21 people about Misery Loves Company (MLC) towels—whether they like the towels or not. The survey data were organized into the following contingency table:

---

[*] Independence can extend to more than just two events: three events, for instance, are *mutually independent* if the events are *pairwise independent*, meaning that every pair of events is independent.

	Like	Dislike	Total
Male	2	12	14
Female	1	6	7
Total	3	18	21

Are the events "Male" and "Like" independent?

Solution. To confirm independence, find each of the conditional probabilities and compare them to the probabilities without the "given" statements.

First, check $P(M \mid L) = P(M)$:

$$P(M \mid L) = \frac{P(M \cap L)}{P(L)} = \frac{2}{3}, \text{ which is the same as } P(M) = \frac{14}{21} = \frac{2}{3}.$$

Then, check $P(L \mid M) = P(L)$:

$$P(L \mid M) = \frac{P(L \cap M)}{P(M)} = \frac{1}{7}, \text{ which is the same as } P(L) = \frac{1}{7}.$$

Since both check out, the events "Male" and "Like" are independent.

If two events are independent and we want to calculate the probability of them both happening together, we can use the *general multiplication rule*.

Definition 5e.-3

The general multiplication rule. Used to find the probability of two (or more) events occurring together.

The general multiplication rule formula accounts for independence (or not), as shown below.

$P(A \cap B) = P(A) \cdot P(B \mid A)$, but if A and B are independent, then the "given" portion is irrelevant to the calculation and the formula becomes simply $P(A \cap B) = P(A) \cdot P(B)$.

Usually, the word *and* clues you in to having to use the general multiplication rule (as opposed to *or*, which usually means you'll have to utilize the general addition rule).

Example 5e.-4

(a) What is the probability of flipping a coin twice and getting heads both times? First, solve this problem by listing out the sample space. Then, use the idea of independent events to calculate the probability.

There are four events in the sample space: HH, HT, TH, TT. Only one event is HH, so the probability is 1/4.

Using the idea of independent events (along with the general multiplication rule), we find the probability of getting successive heads is $\dfrac{1}{2} \cdot \dfrac{1}{2} = \dfrac{1}{4}$, matching our answer above.

(b) What's the probability that a couple has 5 boys (the couple has 5 children)?

$$\left(\dfrac{1}{2}\right)^5 = \dfrac{1}{32}$$

(c) What's the probability a couple doesn't have 5 boys (the couple has 5 children)?

$$1 - \left(\dfrac{1}{2}\right)^5 = \dfrac{31}{32}$$

(d) A test has 10 true/false questions. What's the probability of guessing on every question and getting a 100% on the test?

$$\left(\dfrac{1}{2}\right)^{10} = 0.000977$$

(e) Natasha has 3 rock, 4 country, and 2 punk CDs in her car. One day, before she starts driving, she pulls 2 CDs from her CD carrier without looking. What's the probability that both CDs are rock?

$\dfrac{3}{9} \cdot \dfrac{2}{8} = \dfrac{1}{12}$. Notice that these two events are not independent.

(f) There are 20 toy cars, 10 tops, and 5 yo-yos all mixed up in a big vat. You reach in without looking and select one toy. Then you throw the toy back into the vat and reach back in, again without looking, and select a second toy. This is called making selections *with replacement*. Find P("first toy is a car" *and* "second toy is a car"). Also, are these independent events?

$\left(\dfrac{20}{35}\right)^2 = 0.327$. These events are independent.

(g) Same situation as part (f), except after you reach in for your first toy, you put it aside before snagging the second. Find *P*("first toy is a car" *and* "second toy is a car"). And are these independent events?

$\dfrac{20}{35} \cdot \dfrac{19}{34} = 0.319$. These events are not independent.

(h) Twenty-three people are chosen at random. What is the probability that at least two of these people have the same birthday?

This problem, usually called the Birthday Problem, relies on a complement. If we selected only one person, the chances that he would have the same birthday as himself is, well, 100%:

$\dfrac{366}{366}$

Now let's select a second person. The chance that he *wouldn't* have the same birthday as the first person is

$\dfrac{366}{366} \cdot \dfrac{365}{366}$

Now, the chances that those two people *would* have the same birthday is calculated as

$1 - \dfrac{366}{366} \cdot \dfrac{365}{366}$

To find the probability that twenty-three people don't have the same birthday, simply extend that logic:

$\dfrac{366}{366} \cdot \dfrac{365}{366} \cdot \dfrac{364}{366} \cdots \dfrac{344}{366}$

So, finally, the complement of twenty-three people not having the same birthday is the same as saying that at least two of those twenty-three people have the same birthday (i.e., *anything but all people with the same birthday = two or more birthdays are identical*). It is calculated as

$1 - \dfrac{366}{366} \cdot \dfrac{365}{366} \cdot \dfrac{364}{366} \cdots \dfrac{344}{366} \approx 0.5$

Let's consider only independent events in our next example. Before proceeding, first recall that the probability of the complement of A, or A', is given as $P(A') = 1 - P(A)$. Other than seeing the word "none" in a problem, a clue that you'll have to make use of a complement is the presence of the words "at least one."

Example 5e.-5

Jamie is playing basketball. The probability that she misses a free throw is 0.6. What's the probability that

(a) she misses two free throws in a row?

$0.6^2 = 0.36$

(b) she misses six free throws in a row?

$0.6^6 = 0.046$

(c) she makes three free throws in a row?

$0.4^3 = 0.064$

(d) out of three free throws she takes, she makes at least one?

Making "at least one" is the opposite of missing them all. So,

$1 - 0.6^3 = 0.784$

(e) out of ten free throws she takes, she misses at least one?

$1 - 0.4^{10} = 0.999$

Example 5e.-6

Six out of every 1,000 bags checked in at an airport are lost. You check in your bag at the airport.

(a) What's the probability that your bag will be lost?

0.006

(b) What's the probability that your bag won't be lost?

0.994

(c) Suppose 100 bags are on your plane. What's the probability that none of these bags will be lost?

$0.994^{100} = 0.548$, assuming that independence holds in this case.

(d) Again, suppose that 100 bags are on your plane. What's the probability that at least one of these bags will be lost?

$1 - 0.994^{100} = 0.452$, again assuming that independence holds. Realize that 0.006^{100}, which effectively equals zero, means that *all* bags were lost. Which is unlikely—unless the plane is lost, too.

Example 5e.-7

(a) The $P(A) = 0.50$, $P(B) = 0.40$, and we know that events A and B are mutually exclusive. Find $P(A \cup B)$.

Because the events cannot happen together, $P(A \cap B) = 0$. Thus,

$P(A \cup B) = 0.50 + 0.40 - 0 = 0.90$

(b) The $P(A) = 0.50$, $P(B) = 0.40$, and we know that events A and B are independent. Find $P(A \cup B)$.

Because the events are independent, $P(A \cap B) = P(A) \cdot P(B) = 0.50 \cdot 0.40 = 0.20$. Thus,

$P(A \cup B) = 0.50 + 0.40 - 0.20 = 0.70$

(c) Can two events that are mutually exclusive also be independent? Explain.

If two events are mutually exclusive, then $P(A \cap B) = 0$, and if two events are independent, then $P(A \cap B) = P(A) \cdot P(B)$. So, setting these two equations equal, we see that $P(A) \cdot P(B) = 0$, if two events have the properties of both mutual exclusivity and independence.

According to the zero product property of real numbers, if $ab = 0$, then either $a = 0$, $b = 0$, or both a and b are zero. Likewise, for events to be both disjoint and independent, either $P(A) = 0$, $P(B) = 0$, or both events have probabilities equal to zero.

Assuming independence where it isn't true can lead to all sorts of errors, some of which may have real-life consequences. For instance, as statistician Nate Silver describes in *The Signal and the Noise*, faulty assumptions about independence in the real estate market may have led to the financial crisis of 2008-09.

Silver asks us to imagine five mortgages, any one of which has a 5% chance of defaulting. There are a number of mortgage "pools," essentially bets on the possibility of default, created, including an Alpha Pool, which doesn't pay out only if all five mortgages default, and an Epsilon Pool, which doesn't pay if *any one of the five* mortgages defaults. Clearly, the Epsilon Pool is much, much riskier than the Alpha Pool; hence, it pays out much more.

It should be clear how to calculate the probabilities of not being paid for each pool. For the Alpha Pool, it is simply $0.05^5 = 0.000000313$, whereas for the Epsilon Pool, it is $1 - 0.05^5 \approx 0.999$.

All fine and good, if mortgage defaults are independent of one another. But is that really the case?

Insurance companies try to avoid insuring mortgages to homeowners on the same block, for instance, because if some sort of event strikes one house—such as flood, fire, earthquake, etc.—other houses on that block might also be affected. Thus, here, the mortgages wouldn't be independent of each other. Likewise with insurance policyholders from the same family: a financial wipeout of one family member might spread to others like wildfire.

Thinking on a larger scale, if the economy is in relatively good shape, the assumption of independence seems like a good bet. But if there's some sort of macro-level economic event that "ties the fate of these homeowners together," as Silver says, then "[n]ow you've got trouble: if one borrower defaults, the rest might also. The risk of you losing your bet has increased dramatically. [This] scenario was what came into being in the United States beginning in 2007, and largely explains why there was a financial crisis in 2008."

Sometimes, jurors also have to understand the notions of probability, such as independence. And it's often not as easy as it first appears.

A common mathematical misunderstanding in the courtroom is called the *prosecutor's fallacy*. Statistician Mario Triola defines the prosecutor's fallacy as "a misunderstanding or confusion of two different conditional probabilities: (1) the probability that a defendant is innocent, given that the forensic evidence shows a match; (2) the probability that forensics shows a match, given that a person is innocent." Falling into the trap of the prosecutor's fallacy also often hinges on whether or not the assumption of independence is warranted.

There are a number of legal cases, such as the infamous Sally Clark and Lucia de Berk convictions, that hinged on a faulty categorization of events as independent. However, let's work through an example involving the prosecutor's fallacy modeled after a 1964 court case that took place in California.

Example 5e.-8

In late April, Shiri parks her 1963 Volkswagen Beetle in downtown Philadelphia. Suddenly, as she's stepping out of the vehicle, her keys are snatched, she's thrown to the pavement, and the thief races off in her Bug, leaving a trail of burnt rubber and exhaust behind. Shiri and several other eyewitnesses are able to describe the thief as a brown-haired man, around six feet tall, between 20 and 30 years old, with a small beetle tattoo on his arm (that at least explains his affinity for VWs).

Later on that day, the police pick up a man fitting the description. At the trial, the prosecutor relays the following probabilities (all obtained with the help of an expert statistician using demographic data from the 1.5 million people in Philadelphia):

- Being male: 1 in 2 chance;
- Having brown hair: 1 in 3 chance;
- Being about six feet tall: 1 in 12 chance;
- Being between the ages of 20 and 30: 1 in 4 chance;
- Having a beetle tattoo on arm: approximately a 1 in 200 chance.

The prosecutor says, "Because all of these characteristics are independent of each other, we can multiply them all together to obtain the probability of one person having these characteristics." The prosecutor continues: "Since the chances that any one person would have all of these characteristics is so incredibly small, and the accused has them all, he is most certainly guilty of the Beetle-jacking!"

(a) Find the value the prosecutor claims is the probability of one person having all of the characteristics.

$$\left(\frac{1}{2}\right)\cdot\left(\frac{1}{3}\right)\cdot\left(\frac{1}{12}\right)\cdot\left(\frac{1}{4}\right)\cdot\left(\frac{1}{200}\right) = 0.000017361$$

(b) Now calculate the correct probability that the accused is the real culprit. Does this value surprise you?

We would estimate there to be $0.000017361 \times 1,500,000 \approx 26$ people in Philadelphia fitting the description. Therefore, the chance we have actually napped the real culprit is 1/26, or 3.85%—certainly less than a reasonable doubt.

(c) Why is the prosecutor's assumption of independence in this case not correct?

For instance, being male and six feet tall are not independent of each other.

Definition 5e.-4

Bernoulli trial. A trial in a probability experiment in which there are two possible outcomes: success and failure; the probabilities of success and failure must stay the same throughout the experiment.

When sampling from a population, as long as the Bernoulli trials are independent, calculations are easily completed. Look again at this setup:

> *There are 20 toy cars, 10 tops, and 5 yo-yos all mixed up in a big vat. You reach in without looking and select one toy.*

If we sample two toys *with replacement*—selecting one and then throwing it back into the vat before selecting another—independence is satisfied. In this case, as we'll see shortly, the *binomial distribution* will allow us to calculate a wide variety of such probabilities.

But if we sample the two toys without replacement, the condition of independence is not met. Instead of the binominal distribution, we'd have to use the *hypergeometric distribution*. The mean and variance formulas of the binomial and hypergeometric distributions are different.

Moving away from this artificial setting, most sampling in real life—with polls and surveys—is done without replacement. In other words, if you are randomly selected by a pollster to be called and asked about an upcoming election, your phone number is not put back into the population to be conceivably randomly selected again. Sampling without replacement here means that the assumption of independence with Bernoulli trials[*] is violated.

As Peck, Olsen, and Devore helpfully explain in their *Introduction to Statistics and Data Analysis,*

> You might think that the differences between sampling without replacement and sampling with replacement should be a source of concern to the practicing statistician. If the inferential methods we study assume sampling with replacement and if, in practice, sampling is done without replacement [like in opinion polls and surveys], won't this undermine the credibility of the results?
>
> It turns out that the answer is no. The size of the sample is typically small compared to the size of the population. Whether we are sampling cards, people, light bulbs, or concentrations of metals in rivers and streams, the theory of sampling with replacement can coexist with the practice of sampling without replacement because of the [10% Condition].

Definition 5e.-5

The 10% Condition. As long as the sample size n of the random sample selected is no larger than 10% of the host population size N, the sampling can be treated *as if* the selections are with

---

[*] Bernoulli trials = you are either selected to be in the sample (= success) or not (= failure) with an initial probability of n/N which changes as people are sampled but not replaced.

replacement.

The upshot is that calculations assuming independence can be utilized even when the sampling is done without replacement, as in surveys and opinion polls, as long as the sample size is small relative to the size of the population (i.e., no larger than 10% of the population).

A multiplicative *finite population correction factor* could be used, if you wish, to correct for the sampling without replacement. The factor is

$$\frac{(N-n)}{N-1}$$

However, as N grows to be much bigger than n, the correction factor converges to 1. For instance, if the sample size is 10 and the population size is 1,000, then

$$\frac{(N-n)}{N-1} = \frac{(1000-10)}{1000-1} = \frac{990}{999} = 0.991$$

making the need for such a multiplicative correction factor effectively moot.

§5f. *Tree Diagrams*

When there aren't too many things going on in a probability problem, we can use tree diagrams to quite literally "map out" all of the possibilities. Tree diagrams can be made from independent or dependent events, so the diagrams are quite versatile, and they can be especially helpful understanding conditional probabilities.

A tree diagram starts at a single *node*, and branches out rightward to include all of the possibilities of the sequence of events. In probability problems, each separate probability should be written on each branch, and the probabilities are multiplied together. All branches' probabilities emanating from any particular node must sum to one.

Example 5f.-1

Consider this data set of students at a large college party watching one of two wrestling shows, WWF or WCW, on a Monday night in the mid-1990s.

	WWF	WCW	total
male	52	39	91
female	3	6	9
total	55	45	100

Suppose we choose two students at random. Draw a tree diagram to represent the sample space, and find the probability that both students are female.

Solution. The tree diagram must take into account sampling without replacement.

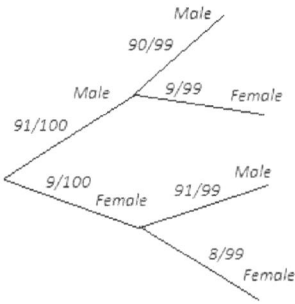

To find probability of selecting two females, look for the branches that connect two females—which are the bottommost. Then multiply those probabilities together:

$$\frac{9}{100} \cdot \frac{8}{99} = \frac{2}{275} = 0.0073$$

Now recall the formula for dependent events (a/k/a the general multiplication rule):

$$P(A \cap B) = P(A) \cdot P(B \mid A)$$

Let's rearrange the formula (and the order of events) to solve for the probability of an event given that some preceding event has occurred.

$$P(A \mid B) = \frac{P(A \cap B)}{P(B)} \text{, as long as } P(B) \neq 0.$$

Thus, the above shows where the *conditional probability formula* comes from.

However, we can do better: let's note that $P(A \cap B)$, or the probability of events A and B occurring together, can be found by taking $P(B \mid A) \cdot P(A)$, regardless of whether the two events are independent or dependent. Taking this into account, we can rewrite the conditional probability formula:

$$P(A \mid B) = \frac{P(B \mid A) \cdot P(A)}{P(B)} \text{, as long as } P(B) \neq 0.$$

This revised formula is called *Bayes Theorem*, named for the Rev. Thomas Bayes (of which this theorem and its consequences were published posthumously). As new evidence comes in, Bayes tells us, we need to revise our prior beliefs, which are called *prior probabilities*; the revised beliefs form what are called *posterior probabilities*.

Example 5f.-2

Every high school students wants to get out of having to go to school, especially when the weather's brutally cold. Suppose that the chances of it being "brutally cold" (below 20 degrees Fahrenheit outside) on any given winter day are 30%. Furthermore, assume that if it's brutally cold, the electrical power in a high school has a 15% chance of going out; if it's not brutally cold, the probability decreases to about 1%. (If the power's out, all students are sent home.) Given that the power's out, what are the chances that it's brutally cold outside?

Solution. A tree diagram can effectively organize this sort of question, replete with conditional probabilities:

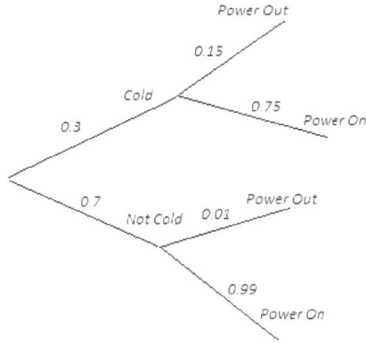

We update our prior probabilities—of the power going out or not—based upon new information—of whether it's brutally cold outside or not. Here,

$$P(Cold \mid PowerOut) = \frac{P(PowerOut \mid Cold) \cdot P(Cold)}{P(PowerOut)} = \frac{0.15 \cdot 0.3}{0.15 \cdot 0.3 + 0.99 \cdot 0.7} = 0.061$$

Knowing that the power's out, the chances that it's also brutally cold outside are 6.1%.

Example 5f.-3

A blood test for diabetes is 98% accurate for those who have the disease, and 95% accurate for those who don't. The prevalence of diabetes among Americans is roughly 5.9%. Suppose you take a blood test for diabetes and you obtain a positive result. What are the chances you have diabetes?[*]

---

[*] This question is asking for the *positive predictive value*, or the probability that you have the disease, given that you obtained a positive result on a diagnostic test. The *sensitivity* is the conditional probability that the diagnostic test

Solution. At first you might think that the chances that you have the disease are 98%, since the test is "98% accurate for those who have the disease." If so, you are in good company. Psychologist Gerd Gigerenzer sampled many medical doctors who thought the same thing.

But it's not that simple. The chances of a false positive result here are 5%; the more dangerous number, the false negative, is 2%. Those values need to be taken into account. Also important to consider: the incidence of the disease. The smaller the incidence, the less chance that a positive result means you're disease-ridden.

Let's make use of Bayes' Theorem:

$$P(D \mid +) = \frac{P(+ \mid D) \cdot P(D)}{P(+)} = \frac{0.98 \cdot 0.059}{0.98 \cdot 0.059 + 0.05 \cdot 0.941} = 0.56$$

So the chances that you have diabetes given you obtained a positive results on the blood test are only 56%—effectively a coin flip—which shouldn't be that surprising since not that many people have diabetes in the population to begin with.

Your first order of business? Take another test. If that second test shows positive as well, your chances of not having the disease have decreased dramatically, since your prior probability has changed—the probability of having diabetes, obtained from the result of the first test, is $P(D) = 0.56$:

$$P(D \mid +) = \frac{P(+ \mid D) \cdot P(D)}{P(+)} = \frac{0.98 \cdot 0.56}{0.98 \cdot 0.56 + 0.05 \cdot 0.44} = 0.96$$

Obsessively taking the test a third time, given the positive results of the other two, brings the probability of you having diabetes close to 100% (if the third test also relays a positive result).

Both Gerd Gigerenzer and Nate Silver rail against the misuse and misinterpretation of all medical diagnostic tests,[*] especially when it comes to screening low-risk, youngish, apparently healthy women for breast cancer.

---

identifies the disease, given that the person has the disease in question.

[*] The misrepresentation of diagnostic tests can hinge on a misunderstanding of conditional probability called "confusion of the inverse." For instance, in the diabetes example, instead of finding the probability that you have diabetes *given* a positive result, you instead calculate the probability you've obtained a positive result *given* that you have diabetes.

§5g. *Random Variables*

Consider a four-sided die, specifically in the shape of a *tetrahedron*, one of the Platonic solids. The polyhedron die has four triangular faces of the same size. On each face, let's affix a number: 1, 2, 3, and 4. The chance of this fair die landing on any particular number is, of course, 1/4. The chances of the die landing on the same face twice in a row? One-fourth multiplied by itself, giving 1/16.[*] Exploring more properties of this tetrahedral die leads us to the central idea of this §, *random variables*.

Example 5g.-1

Consider rolling two fair tetrahedral dice.

(a) Construct the sample space of the differences of the faces—specifically, the smaller face subtracted from the larger face.

	1	2	3	4
1	0	1	2	3
2	1	0	1	2
3	2	1	0	1
4	3	2	1	0

(b) Construct the probability distribution.[†]

X (Difference)	0	1	2	3
P(X)	4/16	6/16	4/16	2/16

(c) Draw a histogram below representing the probability distribution for X = difference of rolling two fair tetrahedral dice.

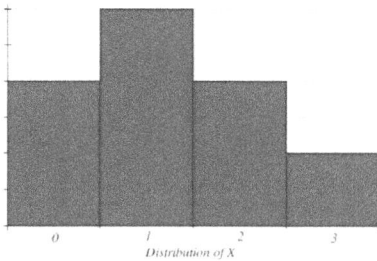

Distribution of X

---

[*] The events are independent here, so the first roll doesn't affect the second.

[†] Recall that in order for a probability distribution to be legitimate, each probability must be between 0 and 1, and the sum of all probabilities in the sample space must equal 1.

(d) What is the probability that when we roll two fair tetrahedral dice, we get a sum of at least 2? Symbolically written, find $P(X \geq 2)$.

$$P(X \geq 2) = \frac{4+2}{16} = \frac{6}{16}$$

Definition 5g.-1

Random variable. A variable whose values represent outcomes from a probability experiment; usually we denote random variables with capital letters, such as X.

Definition 5g.-2

Discrete random variable. A random variable that can only assume a finite, or countable, number of possible values.

Our tetrahedral die example above, with random variable X, is discrete, since X can only assume a countable number of values: 0, 1, 2, or 3. Usually, if you can affix the word "fewer" with the outcomes—and are correct grammatically—the variable is discrete. (To wit: "When rolling our four-sided die twice, we can get two differences *fewer* than 2, namely, 0 and 1.") As a general rule, if a probability distribution exhaustively listing the values of X can be constructed, then X is discrete.

Definition 5g.-3

Continuous random variable. A random variable that can assume an infinite number of possible values, or all possible values over some interval (a, b).

Random variables that are normally distributed are continuous, since there are not a fixed or preset number of outcomes the variable could assume. Typically, if you can use the word "less" correctly grammatically when discussing the values, the random variable is discrete.[*]

Example 5g.-2

Determine if the following random variables are discrete or continuous:

(a) The volume of liquid in a soda can?

---

[*] That pesky "15 Items or Less" sign hanging in front of the express lines at many supermarkets should read "15 Items or *Fewer*," since the items, by their very nature, would have to be countable for you to determine if you even qualify to get in the express line!

Continuous. "He has *less* liquid than she does."

(b) The number of touchdowns thrown by a quarterback in a football game.

Discrete. "There were two *fewer* touchdowns thrown this Sunday than last."

(c) The amount of snow that fell on Philadelphia in a winter.

Continuous. There was *less* snow in Philly this year than last.

(d) The length of time needed to complete a quiz.

Continuous. Victoria needed *less* time to finish the quiz than most of her classmates.

(e) The number of electronic devices each student has in a certain class.

Discrete. Emma, owning only a calculator, had fewer electronic devices than nearly everyone else, including her teacher.

Example 5g.-3

Does $P(X) = \dfrac{x}{4}$, where $X = 1, 2$, and 3, determine a probability distribution?

Solution. Construct a table using the function.

X	1	2	3
P(X)	1/4	2/4	3/4

Since the sum of the probabilities exceeds unity, the function doesn't determine a probability distribution.

Example 5g.-4

Suppose $f(x) = \dfrac{c}{2}x^3$ for $x = 1, 2, 3$, and 4. Determine what the constant c must equal so that $f(x)$ can be considered a probability mass function.

Solution. We know that $\dfrac{c}{2}\left(1^3 + 2^3 + 3^3 + 4^3\right) = 1$, since the sum of the probabilities from any probability distribution must equal one. So we need to solve for c:

$$\frac{c}{2}(100) = 1$$

$$50 \cdot c = 1$$

$$c = \frac{1}{50}$$

Definition 5g.-4

0^+. Stands for "effectively zero"—meaning the outcome can occur, but there's virtually no chance of it happening.

Mario Triola, in *Elementary Statistics*, introduces this useful notation (which has nothing to do with calculus limits, such as "approaches from the right"). When thinking of events with an effectively zero probability, you may wish to remind yourself of a funny interaction from the film *Dumb and Dumber*, in which Lloyd Christmas (Jim Carrey) attempts to ask Mary Swanson (Lauren Holly) out.

LC: What do you think the chances are of a guy like you and a girl like me ending up together?

MS: Well, Lloyd, that's difficult to say. I mean, we don't really...

LC: Hit me with it! Just give it to me straight! I came a long way just to see you, Mary. The least you can do is level with me. What are my chances?

MS: Not good.

LC: You mean, not good like one out of a hundred?

MS: I'd say more like one out of a million.

[beat]

LC: So you're telling me there's a chance. *YEAH!!!*

Perhaps Mary directly stating "effectively zero" or "0^+" would have even gotten a more positive reaction from Lloyd—after all, that's what she meant.

§5h. *Working with Discrete Random Variables*

Recall the idea of *weighted averages* (also called *weighted means*): When a mean can't be calculated simply as an arithmetic average because the values all don't have the same weights, finding a weighted average is an alternative.

Definition 5h.-1

Expected value. Synonymous with the weighted average and also called *mathematical expectation*, it can be thought of as: (1) The long-run average of a process; or (2) The average or ending value we would expect to obtain if we could conduct trials of a process forever.

The expected value for a random variable X is denoted as $E(X)$; however, since the expected value is equivalent to the weighted average, we can say:

$$E(X) = \mu_X = mean$$

The weighted average is found using a "multiply-add algorithm": multiply a value by its weight, and add that to the next product (found the same way), until all values have been accounted for.

Example 5h.-1

Suppose we are playing a game with a fair, four-sided die. Each face reveals a different point total: if 1 is rolled, you receive 1 point; if 2 is rolled, 2 points; if 3 is rolled, 3 points; but if 4 is rolled, you lose 6 points. Find the expected value of this game; also determine if this is a *fair game*.

Solution. The random variable is defined here as X = The points won or lost based on a roll of the four-sided die. The probability distribution is

X	1	2	3	-6
P(X)	1/4	1/4	1/4	1/4

To find the expected value, we do the following:

$$E(X) = 1 \cdot \tfrac{1}{4} + 2 \cdot \tfrac{1}{4} + 3 \cdot \tfrac{1}{4} + (-6) \cdot \tfrac{1}{4} = 0$$

A fair game has an expected value of zero, so this is a fair game.

Example 5h.-2

(a) Bonnie is a sales associate at a large Chevrolet auto dealership. She motivates herself by using probability estimates of her sales. For a sunny Saturday in April, she estimates her car sales as follows (notice that her selling more than three cars is extremely unlikely):

Cars sold:	0	1	2	3	> 3
Probability:	0.3	0.4	0.2	0.1	0^+

Find the expected value of cars sold.

$E(X) = 1.1$ cars. (Note that the concept of a "fair game" doesn't make sense here in context.)

(b) Suppose a spinner from a board game has five colors: red, blue, green, yellow, and orange. The spinner is twice as likely to land on red than on any other color; the spinner is equally likely to land on any other color besides red. If the spinner lands on red, the player loses 10 points; if it lands on any other color, the player gains 2 points. Is this game fair?

	Red	Blue	Green	Yellow	Orange
Probability (x)	2x	x	x	x	x
Probability	1/3	1/6	1/6	1/6	1/6
Points	-10	2	2	2	2

$E(X) = -2$, so the game is not fair.

(c) Suppose the random variable X is the difference of the faces from two four-sided dice. Construct a probability distribution to find the mean, and interpret your results.

X (Difference)	0	1	2	3
P(X)	4/16	6/16	4/16	2/16

$E(X) = 1.25$, meaning you'd expect, on average, a difference of a little above 1.

(d) Suppose the chances of a thirty-year-old nonsmoker dying during the course of a calendar year are 0.001. That thirty-year-old nonsmoker wishes to purchase a one million dollar life insurance policy. What is the minimum that the policy could be priced at?

$E(X) = 0.001 \cdot 1{,}000{,}000 + 0.999 \cdot 0 = \$1{,}000$ for the year, but that expected value calculation assumes zero profit for the insurance company. More realistically, a loading fee of

10% to 15% will be imposed. So, to make a reasonable profit, the insurance company should charge between \$1,100 and \$1,150 for this policy.

(e) The following is a question posed to interviewees at Goldman Sacks (taken from the book *Young Money* by Kevin Roose): Imagine yourself playing a coin flip game. If you flip heads, you have to pay \$1, and the game's over; flip tails, and you are permitted to flip for a second time. If you flip heads on this second trial, you'll pay \$2; otherwise, you'll be allowed to flip again. The game continues, increasing the amounts to \$4, \$8, etc. How much should you be paid to play the game?[*]

The question is one "solved" using expected value, although there is no right answer—the interviewers simply want to gauge the risk-adverseness of the interviewees. For any number of trials, the calculation of expected value is

$$E(X) = \frac{1}{2} \cdot (-2)^0 + \frac{1}{2^2} \cdot (-2)^1 + \frac{1}{2^3} \cdot (-2)^2 + \cdots$$

No casino game is a fair game; instead, all casino games have a slightly negative expected value (for the player) so that, over time and many games played (especially among the pathologically addicted),[†] casinos can rake in millions of dollars. If a particular casino game had a mathematical expectation that was *too* negative, though, that game's frequency of play would decrease.

(f) A roulette wheel has 18 black, 18 red, and 2 green slots. Suppose you wager a paltry \$1 on black. What is your expected profit?

	Black	Red	Green
X (\$)	1	-1	-1
P(X)	18/38	18/38	2/38

$E(X) = -0.05$, meaning you can be expected to lose about 5 cents for every dollar bet under these conditions.

The expected value and the mean μ_X of a discrete random variable X are the same because both describe the long-run average outcome. For example, in the roulette example above, $E(X) = -0.05$, meaning that you lose about 5 cents for every 10 dollars you bet—over time, not on a single spin of the wheel.

---

[*] This game, known as the Petersburg Paradox, originated with Nicolaus Bernoulli hundreds of years ago.

[†] See the chapter "Autism as a Design Principle: Gambling" in Matthew Crawford's *The World Beyond Your Head* for a multifarious study of gambling—minus the mathematics.

Definition 5h.-2

Standard deviation of a random variable. The average deviation around the mean of a random variable.

Remember that the variance is the standard deviation squared. The formula for the variance of a random variable is

$$\sigma_X^2 = \sum (x_i - \mu_X)^2 p_i$$

But there's an easier-to-use, equivalent "shortcut" formula for the variance:

$$\sigma_X^2 = \sum x_i^2 p_i - \mu_X^2$$

The "shortcut" formula is derived as follows. First, note that the variance of X is given by the expected value of the square of the difference of x and the mean—symbolically, that translates to $E\left[(x - \mu)^2\right]$. Expanding out the binomial in the parentheses, we obtain $E\left[(x^2 - 2x\mu + \mu^2)\right]$. Although this is jumping the gun a bit, we'll soon see that the mean of the sum of independent variables is equal to the sum of their means, so the expected value expression can be broken up into three parts: $E(x^2) - E(2x\mu) + E(\mu^2)$, which is the same as $E(x^2) - 2\mu \cdot E(x) + \mu^2$. But recall that $E(X) = \mu = mean$, so we can rewrite the expression as $E(x^2) - 2\mu \cdot \mu + \mu^2$, which equals $E(x^2) - 2\mu^2 + \mu^2$ —and immediately simplifies to $E(x^2) - \mu^2$. Thus we have our "shortcut" formula.

The standard deviation σ_X, of course, is the square root of σ_X^2. Let's try some problems.

Example 5h.-3

Back to Bonnie, the sales associate at a large Chevrolet auto dealership. She motivates herself by using probability estimates of her sales. For a sunny Saturday in April, she estimates her car sales as follows.

Cars sold:	0	1	2	3	> 3
Probability:	0.3	0.4	0.2	0.1	0^+

(a) Find the mean, variance, and standard deviation of this probability distribution.

$$E(X) = 1.1$$

$$\sigma_X^2 = 0^2 \cdot 0.3 + 1^2 \cdot 0.4 + 2^2 \cdot 0.2 + 3^2 \cdot 0.1 + (>3)^2 \cdot 0 - 1.1^2 = 0.889249$$

$$\sigma_X = \sqrt{0.889249} = 0.943$$

(b) Find the minimum and maximum "usual" values.

$$1.1 \pm 2 \cdot 0.943 = (-0.786, 2.986)$$

(c) Would Bonnie selling 3 cars on a sunny Saturday in April be considered "unusual"? Explain.

Yes, since $3 > 2.986$.

Using the 1-Var Stats function, the graphing calculator can automatically find the mean and standard deviation of a random variable.

Example 5h.–4

The number of yearly motor accidents per student at a local college is shown below.

Accidents	0	1	2	3	4	5
Students	260	500	425	305	175	45

Construct a probability distribution; then use the calculator's 1-Var Stats function to find the mean, variance, and standard deviation of the data and interpret the results.

Solution. Construct a relative frequency distribution table to obtain a probability distribution:

Accidents	0	1	2	3	4	5
Students	260	500	425	305	175	45
Relative Freq.	0.152	0.292	0.249	0.178	0.102	0.026

To find the mean and standard deviation using the calculator, type in the top row—call the list $_L$ACC—and the bottom row—call the list $_L$STU. Then, input 1-Var Stats $_L$ACC, $_L$STU (or, if a menu pops up, make $_L$STU the "FreqList") and press $\boxed{\text{ENTER}}$. The results are

$$\mu_X = 1.865, \ \sigma_X = 1.307, \text{ and } \sigma_x^2 = 1.307^2 = 1.705.$$

The mean number of accidents is a bit less than two per year, and most of the values differ from the mean by about 1.3 accidents.

Recall that, in an earlier §, we examined what happens to the measures of center and spread when a constant is added or multiplied to each value of a data set. What happens, though, when similar linear transformations are applied to discrete random variables?

Example 5h.–5

Bonnie, the sales associate at a large Chevrolet auto dealership, collects a small cash bonus of $200 for each car she sells. Her new probability distribution is

Cash bonus:	$0	$200	$400	$600	> $600
Probability:	0.3	0.4	0.2	0.1	0^+

(a) Find the mean amount of her cash bonus.

$E(X) = 0 \cdot 0.3 + 200 \cdot 0.4 + 400 \cdot 0.2 + 600 \cdot 0.1 = \$220$

(b) Demonstrate how to calculate the mean of the cash bonus in a different way.

The expected value of Bonnie's cars sold is 1.1. Taking 1.1 and multiplying it by $200 also gives us $220.

(c) Find the standard deviation of the cash bonus.

The standard deviation of the cash bonus is $188.60.

(d) Demonstrate how to calculate the standard deviation of the cash bonus in a different way.

The standard deviation of the number of cars sold is 0.943. Taking 0.943 and multiplying it by $200 also gives us $188.60.

§5i. *Working with Continuous Random Variables*

Acontinuous random variable generally doesn't allow for its probabilities to be exhaustively summarized in a probability distribution table. Instead, probabilities must be calculated by finding areas underneath its *probability distribution function* between given intervals.

Consider the *uniform distribution* first. The uniform distribution graphs as a rectangle, modeling some process that has an equal probability of occurring at any value over an interval. Calculating areas—and, thus, probabilities—with a uniform distribution is especially easy, since the area of a rectangle is just *width x height*. The mean is also easy to calculate: it's *(lower bound + upper bound)/2*. In other words, the mean's located at the middle of the rectangle (as is the median).

The variance of a uniform distribution, though, makes use of a more complicated (and not readily intuitive) formula: $\sigma_x^2 = (b-a)^2/12$, where a and b are the lower and upper bounds, respectively.

Example 5i.-1

Suppose the probability of the length of time a library patron at the Huntingdon Valley Library actually stays in the library is given by the following uniform probability distribution.

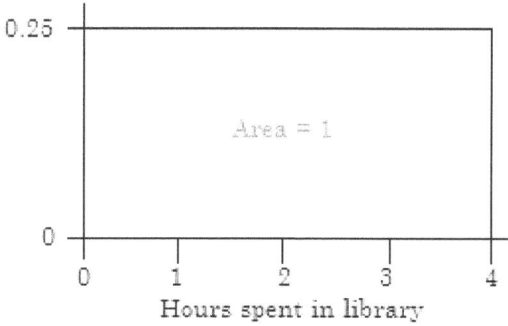

(a) Find the probability that a randomly selected library patron spends less than 2 hours in the Huntingdon Valley Library. Mathematically stated, $P(X<2)$.

$P(X<2) = 2 \cdot 0.25 = 0.5$

(b) Find the probability that a randomly selected library patron spends more than 3 hours in the Huntingdon Valley Library. Mathematically stated, $P(X>3)$.

$P(X<2) = 1 \cdot 0.25 = 0.25$

(c) Find the probability that a randomly selected library patron spends between 1 and 4 hours in the Huntingdon Valley Library. Mathematically stated, $P(1<X<4)$ or $P(1 \le X \le 4)$.[*]

$P(1<X<4) = P(1 \le X \le 4) = 3 \cdot 0.25 = 0.75$

---

[*] These two notation variants symbolically express the same thing, since all continuous probability models assign a probability of zero to every individual outcome. At any particular value for X, the probability is zero—because the vertical line above the X value has no thickness, thus the area under the curve at that spot is zero as well.

(d) Find the mean, variance, and standard deviation of the Huntingdon Valley Library probability distribution.

$$\mu_X = \frac{0+4}{2} = 2$$

$$\sigma_X^2 = \frac{(4-0)^2}{12} = 1.33$$

$$\sigma_X = \sqrt{1.33} = 1.15$$

More important than uniform distributions, however, are normal distributions. We've already calculated z-scores and looked them up on a table to find probabilities. Let's refresh our memory on normal distribution calculations by answering several questions.

Example 5i.-2

For IQ scores, $N(100,15^2)$.

(a) Find $P(X > 109)$.

$$z = \frac{109-100}{15} = 0.60 , \; 1-0.7257 = 0.2743$$

(b) Find $P(109 < X < 113)$.

$$z = \frac{113-100}{15} = 0.87 , \; 0.8078 - 0.7257 = 0.0821$$

Example 5i.-3

Suppose that X is a continuous random variable with a probability density function (pdf) of

$$f(x) = \frac{x^2}{3}$$

over the interval $0 < x < c$. What is the value of the c that makes $f(x)$ a valid pdf?

Solution. Like the area underneath a uniform distribution and a normal distribution, the area underneath $f(x)$ must be equal to one. Thus, we have a calculus integration problem:

$$\int_0^c \frac{x^2}{3} dx = 1$$

Integrating over the limits of 0 and c,[*] we find that the left side of the equation is

$$\frac{1}{3} \cdot \left[\frac{x^3}{3} \right]_{x=0}^{x=c} = \frac{1}{3} \cdot \left[\frac{c^3}{3} - \frac{0^3}{3} \right] = \frac{1}{3} \cdot \left[\frac{c^3}{3} \right] = \frac{c^3}{9}$$

Finally, then, we can set the simplified quotient equal to one and solve for c:

$$\frac{c^3}{9} = 1, \ c^3 = 9, \text{ so } c = \sqrt[3]{9}.$$

Example 5i.-4

Suppose that X is a continuous random variable with a probability density function of

$$f(x) = \frac{x^2}{3}$$

over the interval $0 < x < \sqrt[3]{9}$. Find $P(X < 1)$.

Solution. To find $P(X < 1)$, we must perform integration:

$$\int_0^1 \frac{x^2}{3} dx = \frac{1}{3} \cdot \left[\frac{x^3}{3} \right]_{x=0}^{x=1} = \frac{1}{3} \cdot \left[\frac{1^3}{3} - \frac{0^3}{3} \right] = \frac{1}{3} \cdot \left[\frac{1}{3} \right] = \frac{1}{9}$$

Let's now take a look at what happens when simple linear transformations are applied to continuous random variables.

Example 5i.-5

A standardized test has a mean of 500, a standard deviation of 100, and is approximately normally distributed. The testing company decides to radically scale the scores, by multiplying the raw scores by 2 and adding 20.

(a) Define the random variable X to be the raw scores, and Z to be the "radically" scaled

---

[*] Such integrals can be evaluated rapidly on the *WolframAlpha* website using the "integrate" function.

scores. What does Z equal?

$$Z = 2X + 20$$

(b) Find the mean of Z.

The mean is affected by both the multiplicative linear transformation as well as the additive linear transformation. Therefore, the new mean is

$$\mu_Z = 2 \cdot 500 + 20 = 1020$$

(c) Find the standard deviation of Z.

The standard deviation is not affected by the additive portion of the linear transformation, but it is affected by the multiplicative piece.

$$\sigma_Z = 2 \cdot 100 = 200$$

(d) What is the probability of a student obtaining a "radically" scaled score of 1100 or greater?

$$P(Z > 1100) => z = \frac{1100 - 1020}{200} = 0.40, \ 1 - 0.6554 = 0.3446$$

§5j. *Random Variable Combinations*

When combining random variables together, one of the first things we need to know is if the variables are independent of each other—independence here meaning that the occurrence of any event from one variable has no effect on the occurrence of any event from the other.

When finding means (or expected values), independence makes no difference—we can simply add together the variables. But we are not allowed to calculate variance or standard deviation unless independence between the variables holds.[*]

In general, the mean of the sum of independent variables is equal to the sum of their means; likewise with their differences. Stated symbolically,

$$E(X + Y) = E(X) + E(Y) \ \text{and} \ E(X - Y) = E(X) - E(Y)$$

---

[*] For instance, the math and verbal scores of an SAT test are not independent, so the variance and standard deviation between these two variables (math and verbal scores) does not make sense to calculate.

or, expressed differently,

$$\mu_{X+Y} = \mu_X + \mu_Y \text{ and } \mu_{X-Y} = \mu_X - \mu_Y$$

Also, if and only if the random variables are independent of each other, then the sum *or* the differences of their variances is equivalent to the sum of their variances. Stated symbolically,

$$\text{var}(X+Y) = \text{var}(X-Y) = \text{var}(X) + \text{var}(Y)$$

or, expressed differently,

$$\sigma^2_{X+Y} = \sigma^2_{X-Y} = \sigma^2_X + \sigma^2_Y$$

Example 5j.-1

Suppose *A*, *B*, *C*, and *D* are independent random variables with $\mu_A = 5$, $\mu_B = 10$, $\mu_C = 12$, $\sigma_A = 3$, $\sigma_B = 4$, and $\sigma_C = 5$, and $D = A + B + C$. Find the mean, variance, and standard deviation of random variable *D*.

Solution. The mean of the random variables is equal to the sum of their means:

$$\mu_D = 5 + 10 + 12 = 27$$

The variance of the random variables is equal to the sum of their variances—only because the independence condition holds here (otherwise, we couldn't calculate the variance):

$$\sigma^2_D = 3^2 + 4^2 + 5^2 = 50$$

And the standard deviation of the random variables is given by the square root of their variance:

$$\sigma_D = \sqrt{50} = 7.07$$

Note that you cannot simply add together the standard deviations of the random variables to obtain the standard deviation of their sum:

$$\sigma_D = 7.07 \neq 3 + 4 + 5$$

Instead, the variance must always be obtained first; then, the standard deviation can be backed into by square rooting.

Example 5j.-2

Here is a joint probability distribution[*] of discrete random variables X and Y.

			X	
			1	**2**
	3		0.20	0.60
Y	**4**		0.10	0.10

(a) Find the probability distributions of X and Y.

To construct the probability distribution of X, sum the probabilities downward; for Y, sum them across:

X	1	2
P(X)	0.3	0.7

Y	3	4
P(Y)	0.8	0.2

(b) Find $E(X)$ and $E(Y)$.

Using the probability distributions found in part (a), $E(X) = 1.7$ and $E(Y) = 3.2$.

(c) Find $E(X+Y)$.

$E(X+Y) = 1.7 + 3.2 = 4.9$

(d) Find $E(X-Y)$.

$E(X-Y) = 1.7 - 3.2 = -1.5$

(e) Find $E(2Y)$.

---

[*] A "regular" probability distribution table for a discrete random variable would merely have the probabilities of each value that the discrete random variable could assume. With the joint probability distribution table, you get to see the probabilities of two discrete random variables simultaneously.

$$E(2Y) = 2 \cdot 3.2 = 6.4$$

(f) Find μ_{2X+2Y}.

$$\mu_{2X+2Y} = 2 \cdot 1.7 + 2 \cdot 3.2 = 9.8$$

(g) Find $E(X+6)$.

$$E(X+6) = E(X) + E(6) = 1.7 + 6 = 7.7$$

(h) Find $\text{var}(X)$.

$$\sigma_X^2 = 1^2 \cdot 0.3 + 2^2 \cdot 0.7 - 1.7^2 = 0.21$$

(i) Find $\text{var}(Y)$.

$$\sigma_Y^2 = 3^2 \cdot 0.8 + 4^2 \cdot 0.2 - 3.2^2 = 0.16$$

(j) Find $\text{var}(X+Y)$.

If random variables X and Y are independent, $\sigma_{X+Y}^2 = \sigma_X^2 + \sigma_Y^2 = 0.21 + 0.16 = 0.37$.

(k) Find $\text{var}(X-Y)$.

Same answer as part (j).

(l) Find $\text{var}(2Y)$.

When dealing with a constant inside a variance calculation, we need to make use of the formula $\text{var}(a \cdot x) = a^2 \cdot \text{var}(x) = a^2 \cdot \sigma^2$.

$$\sigma_{2Y}^2 = 2^2 \cdot 0.16 = 0.64$$

(m) Find σ_{x+2Y}.

If random variables X and Y are independent, $\sigma_{X+2Y} = \sqrt{0.21 + 0.64} = 0.922$

(n) Are random variables X and Y independent? Explain.

No. Recall the general multiplication rule for independent events:

$$P(A \cap B) = P(A) \cdot P(B)$$

Take the events $X = 1$ and $X = 3$, for instance. If these were independent events, then

$$P(X = 1 \cap Y = 3) = P(X = 1) \cdot P(Y = 3)$$

But that's not the case, since

$$P(X = 1 \cap Y = 3) = 0.2 \neq P(X = 1) \cdot P(Y = 3) = 0.3 \cdot 0.8 = 0.24$$

Finding even one counterexample to the general multiplication rule for independent events disqualifies the two random variables from being independent of one other.[*] Thus, parts (j), (k), and (m) above cannot be answered using the information provided.

§5k. *Combinatorics*

*C*ombinatorics is the "branch of mathematics concerned with the theory of enumeration, combinations, and permutations in order to solve problems about the possibility of constructing arrangements of objects to satisfy specified conditions."[†] There is a tight connection between combinatorics and *graph theory*, a discipline of mathematics devoted to the study of graphs. So let's begin with graph theory.

Example 5k.-1

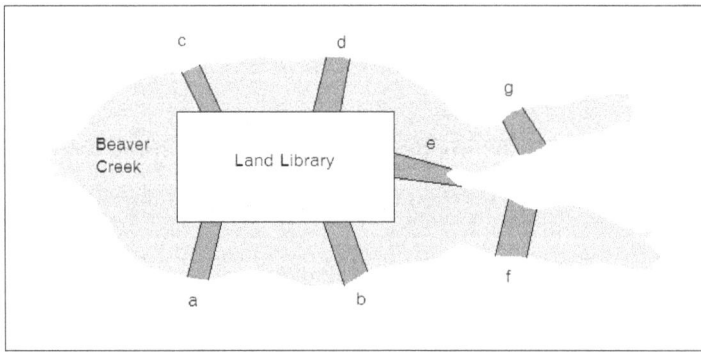

---

[*] Another way to test for independence between several discrete random variables: when calculating the variance of X and Y, along with the variance of $X+Y$, if the sum of the individual variances of X and Y equals the variance of $X+Y$, then the random variables X and Y are independent.

[†] From the *Collins Dictionary of Mathematics*.

When it was just in the planning stages, some powerful university administrators believed that the new Land Library would greatly benefit by having a peaceful creek dug around it. "Beaver Creek," as it came to be known, would have seven bridges constructed over it to allow students and faculty access to the library and the surrounding areas, as shown in the diagram on the previous page. (The bridges are labeled with letters.)

However, when students on campus got wind of the plan, they weren't pleased. A protest was organized to prevent the creek's creation. But the students told the administrators that they would not protest the digging of Beaver Creek around the Land Library if the following condition was met:

> *We, the students of the University, must be able to traverse these seven bridges in one continuous walk without re-crossing any bridge.*

The students were convinced that it couldn't be done—and, hence, that they would not have to deal with a new body of water in their midst. But the administrators believed that all seven bridges could be traversed in one continuous walk without re-crossing any of the bridges, and thus were ready to go ahead with the plan. No one has yet figured out who has the correct argument. Can you?

Solution. Before attempting to answer the question, let's make a simpler diagram of Beaver Creek and the seven bridges (the bridges are labeled with the same letters as above). The dots at which bridges meet on the graph below are called *vertices.*[*]

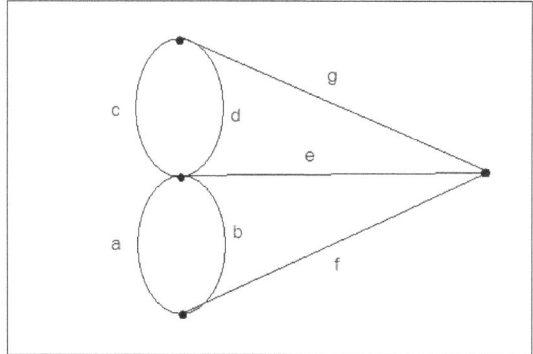

(a) Start at any vertex, and try to trace your pencil around all of the bridges on the graph above without lifting your pencil (this is akin to walking on the bridges on Beaver Creek). Can you draw over every bridge only once without lifting your pencil? Or do you have to retrace your path to cover every bridge?

(b) Consider a different setup of bridges. Assume that there were only the four bridges, as is shown on the graph below. Beginning at any vertex, can we cross every bridge without

---

[*] We will wait to reveal the answers of the parts of this example until its end.

re-crossing any? Use your pencil again to trace the paths.

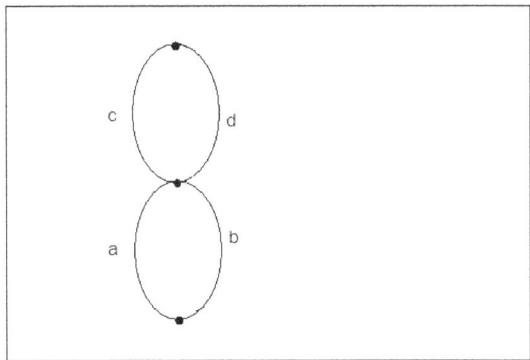

(c) Fill in the tables associated with each of the following graphs of bridges. (The vertices have been numbered.) Then see, using your pencil, if paths can be traced through all of the bridges without re-crossing any bridges. Consider why it is important to know if the number of bridges coming from a vertex is odd or even. The first one has been done for you.

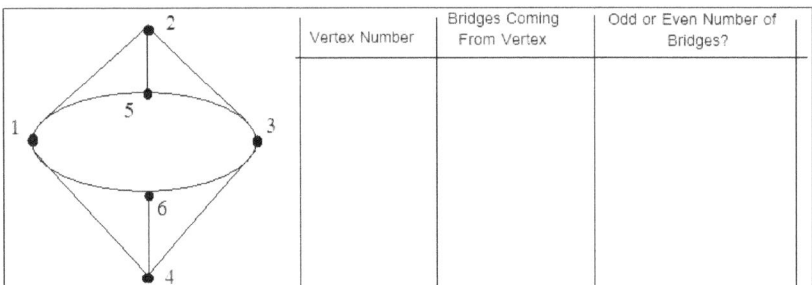

Vertex Number	Bridges Coming From Vertex	Odd or Even Number of Bridges?
1	3	Odd
2	3	Odd
3	2	Even

Vertex Number	Bridges Coming From Vertex	Odd or Even Number of Bridges?

Vertex Number	Bridges Coming From Vertex	Odd or Even Number of Bridges?

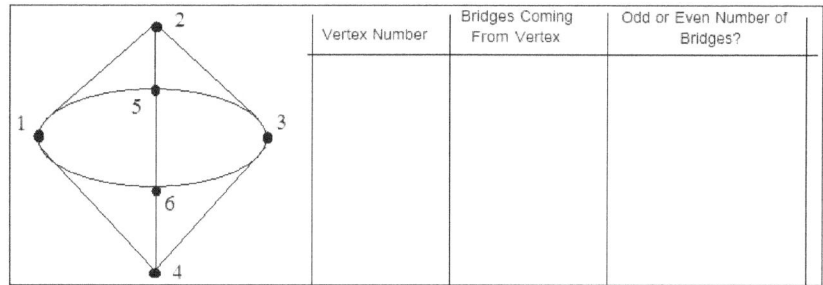

Vertex Number	Bridges Coming From Vertex	Odd or Even Number of Bridges?

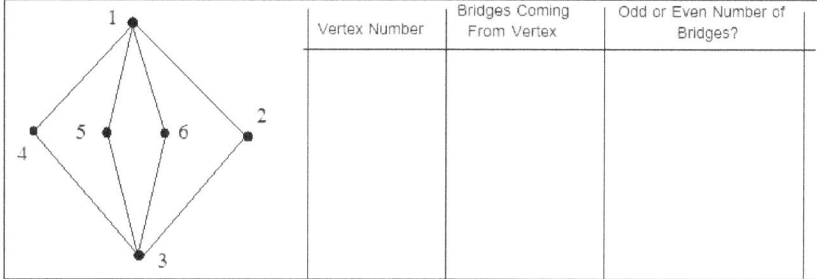

Vertex Number	Bridges Coming From Vertex	Odd or Even Number of Bridges?

And, finally, take another look at the Beaver Creek graph, and complete the table:

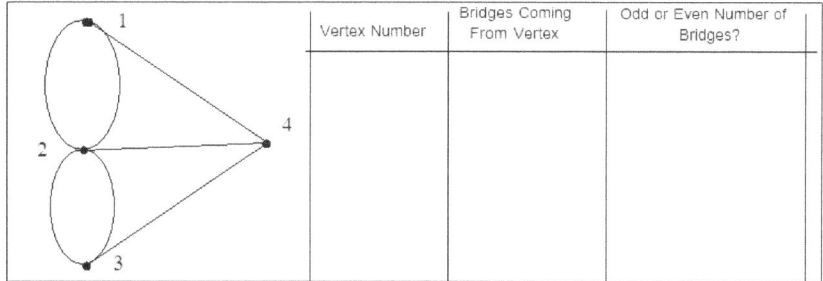

Vertex Number	Bridges Coming From Vertex	Odd or Even Number of Bridges?

(d) Explain why (or why not) it is (or is not) possible to traverse the seven bridges of Beaver Creek in one continuous walk without re-crossing any of the bridges. (Consider the importance of the number of bridges coming from each vertex in your answer.) Also attempt to generalize: How can we know, given any graph, if it is possible to cross all of the graph's bridges without re-crossing any?[*]

The Beaver Creek problem is really a famous, and old, mathematics problem in disguise called The Seven Bridges of Königsberg. As Keith Devlin elaborates in *The Language of Mathematics*,

> Euler solved the problem in 1735. He realized that the exact layout of the islands and the bridges is irrelevant. What is important is the way in which the bridges connect—that is to say, the network formed by the bridges.... Using terms we'll define precisely, in Eu-

---

[*] Draw your own graphs if you wish, but remember: just because we have completed many graph examples in an attempt to reveal some latent rules of graph theory, we haven't *proven* anything—we've only postulated.

ler's network, the bridges are represented by edges, and the two banks and the two islands are represented by vertices. In terms of the network, the problem asks whether there is a path that follows each edge exactly once.

Euler argued as follows: Consider the vertices of the network. Any vertex that is not a starting or a finishing point of such a path must have an even number of edges meeting there, since those edges can be pared off into path-in-path-out pairs. But, in the bridges network [i.e., the Seven Bridges of Königsberg], all four vertices have an odd number of edges that meet there. Hence there can be no such path. In consequence, there can be no tour of the bridges of Konigsberg that crosses each bridge exactly once.

Author Italo Calvino's novel *Invisible Cities* describes a metropolis with a complication perhaps needing a graph theorist's help to untangle:

In Ersilia, to establish the relationships that sustain the city's life, the inhabitants stretch strings from the corners of the houses, white or black or gray or black-and-white according to whether they mark a relationship of blood, of trade, authority, agency.

We can no doubt consider Ersilia as a urban muse, a readymade fount, to math-problem writers everywhere.

So, with that in mind, let's look at a graph theory problem termed NP-complete—a special class of *nondeterministic polynomial time* questions, meaning that although solutions to them can be verified in *polynomial time* (relatively quickly, especially by a computer), finding the solutions themselves may take longer than the time of the universe's existence (thus far).[*]

Example 5k.-2

The people who have just moved into the newly built houses on University Road (shown in the schematic below) are furious: there's no electricity, cable television, or telephones operating!

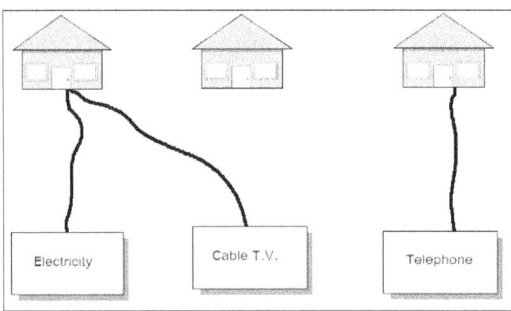

---

[*] One of the most famous NP-complete problem is called the Travelling Salesman Problem, which goes like this: A salesman has a list of many cities in which to peddle his wares. What is the shortest possible route he can take between these cities, making sure to visit every city on his list? (But thank your lucky stars for NP-complete problems, otherwise credit card information you've posted online would be stolen in a flash—by intrepid prime factorization algorithm coders.)

The contractors, though, realized that there's a problem: when making the power-line connections, no power lines can intersect, otherwise, no house will have electricity, cable, or telephone access. They have already connected a couple of power lines—as you can see in the schematic—but they're too afraid to continue without a mathematician's help. Think through the following questions.[*]

(a) Draw lines from the power boxes to the houses without intersecting any lines (several have already been done for you). Can all three houses be connected to electricity, cable, and telephone access simultaneously without any power lines crossing? Or will the people in a house have to sacrifice something to ensure that the whole power grid doesn't go out?

(b) What if there were only two houses? What about four houses? Explain how many, if any, power lines would intersect. Relay any generalizations that you encounter.

(c) How does this problem relate to the Beaver Creek problem?

Like Beaver Creek, this telecommunication lines problem can be reimagined in graph theory terms using vertices and edges; it's also similar to the four-color problem,[†] proved by mathematicians Haken and Appel with a brute-force computer algorithm in 1976. The salient point to be taken from the Beaver Creek and telecommunication lines problems is this: that shifting problems between *seemingly different types of mathematical disciplines* may lead to insights, so long as isomorphism is ensured.[‡]

Graph theory—and its close cousin, and what this § is primarily focused on, combinatorics—involves new ways to count or enumerate objects. Here is another famous, albeit easier than what we've thus far dealt with, graph theory/combinatorics problem:

Example 5k.-3

Suppose there are 1,500 students in a high school. If everyone shakes everyone else's hand exactly once, how many total handshakes will there be?

Solution. Unsurprisingly called the handshake problem, it's best solved using mathematical induction: With two students, there is one handshake; with three, there are three handshakes; with

---

[*] Again, answers—but only some—will be revealed at the end of the example.

[†] That four is the minimum number of colors on a two-dimensional map such that no contiguous boundaries share a color.

[‡] Think of René Descartes, who bridged algebra and geometry—now appropriately called algebraic geometry—with his coordinate plane.

four, we're up to six, and with five students, there are ten handshakes. Eventually, a pattern emerges:

$$\frac{n(n-1)}{2} \text{, where } n \text{ represents the total number of students}$$

Thus, for our high school here, the answer is 1,124,250 possible handshakes.[*]

Definition 5k.-1

The basic (or Fundamental) counting principle. If one event has m outcomes, and another event has n outcomes, the number of outcomes of both events occurring together is given by $m \cdot n$; in addition, this formula easily generalizes to more than two outcomes.

We have, in part, already worked with the basic counting principle: when calculating the probabilities of independent events using the general multiplication rule.

Example 5k.-4

(a) How many different 5-letter "*LINGO*" words are there?

   If we permit 5-letter strings like *tqfer* to be words, then there are

   $$26 \cdot 26 \cdot 26 \cdot 26 \cdot 26 = 26^5 = 11,881,376$$

(b) How many 5-letter "words" are there that begin with A or B and end with Y or Z?

   $$2 \cdot 26 \cdot 26 \cdot 26 \cdot 2 = 4 \cdot 26^3 = 70,304$$

(c) How many 5-letter "words" are there that begin with A or B and have no letters which repeat?

   $$2 \cdot 25 \cdot 24 \cdot 23 \cdot 22 = 607,200$$

---

[*] There is a connection between the handshake problem and the mathematician Carl Friedrich Gauss who, when he was only in kindergarten, derived a formula to answer a tedious problem assigned by his sadistic teacher: sum the numbers from 0 to 100. Gauss noticed that if you write out the sum in ascending order—0+1+ … +99+100—and, right underneath, lay out the sum term by term in reverse order—100+99+ … +1+0—and then add, column-wise, downward, you'll have one hundred and one sums of 100 each, so all that's needed is to divide by two, eliminating the double-counting, to obtain the final result. His generalized formula for the sums: $n(n+1)/2$, a hair's breath away from the handshake problem formula.

(d) A MLC doll can be dressed with five different shirts, two pairs of pants, and three hats. How many different outfits for the doll are possible?

$5 \cdot 2 \cdot 3 = 30$

(e) Refer to the previous problem. What is the probability that, without looking, you and I dress our individual MLC dolls in the same way?

Probability is successful outcomes over total outcomes. There is precisely one successful outcome here—the "matched" way of dressing the dolls. Therefore, the probability is 1/30.

(f) Refer again to part (c). What are the *odds* that, without looking, you and I dress our individual MLC dolls in the same way?

Odds are found by calculating the $P(success)/P(failure)$. Thus, the odds are 1/29, or $O(Success) = 1/29$, since there's one way to be successful here—i.e., matching the clothing arrangements of the dolls—and 29 paths to failure.

The basic counting principle can be utilized to determine the number of ways that n objects can be arranged in an order. Such orderings of objects are called *permutations*. If asked to find a permutation, it's important to realize that the *order of arrangement matters*.

A handy, compact notation for navigating permutation problems is the *factorial*, represented simply by the exclamation point (!). For instance, $5! = 5 \cdot 4 \cdot 3 \cdot 2 \cdot 1 = 120$.[*] Note that, by definition, $0! = 1$.

Example 5k.-5

(a) How many ways can five people line up in a row?

$5! = 5 \cdot 4 \cdot 3 \cdot 2 \cdot 1 = 120$

(b) How many ways can zero people be lined up in a row?

$0! = 1$, because there's only one way to line up no one: by not doing anything at all.

(c) Write all of the permutations (orderings) of A, B, and C. How many are there?

---

[*] The factorial symbol is read as "factorial." So, when reading "5!" out loud, don't shout "FIVE!"—say, "Five factorial." Also, to bring up the factorial button on the graphing calculator, press the MATH button, and scroll over to the PRB menu.

There are two ways to do this. First, let's list them out. Then, let's use a factorial.

ABC, ACB, BCA, BAC, CAB, CBA.

$3 \cdot 2 \cdot 1 = 3! = 6$

(d) How many different permutations of A, B, C, D, E, and F are possible?

$6! = 720$

(e) How many different ways can we arrange the letters in the word LAMP?

$4! = 24$

(f) How many different ways can we arrange the letters in the word HERBIE?

Here's where we can get into trouble. Unlike LAMP, which has no repeating letters, HERBIE has two E's and thus can produce what are called *indistinguishable permutations*—permutations you cannot tell apart. For example, here's a set: EHRBIE and EHRBIE. Look the same? Of course! But the E's were switched between them. Perhaps if the E's had subscripts these two indistinguishable permutations would be more obvious: E_1HRBIE_2 and E_2HRBIE_1. See the difference between them now? Since the letters don't permit subscripts, however, we need to take into account this double counting of permutations. Thus, the number of ways to arrange the letters HERBIE is

$$\frac{6!}{2} = 360$$

(g) How many ways are there to arrange the letters in the word MISSISSIPPI?

Now there are multiple repeated letters: four S's, four I's, and two P's. All possible arrangements of these multiple letters must be divided out when calculating the permutations:

$$\frac{11!}{4!4!2!} = 34,650$$

(h) How many different numbers can be formed by various arrangements of the six digits 1, 1, 1, 1, 2, and 3?

The numbers here can be thought of as letters. Thus, there are four 1's which must be accounted for:

$$\frac{6!}{4!} = 30$$

(i) Permutations also arise when people sit around a round table. These are called, appropriately enough, *circular permutations*. For example, try to calculate how many ways 5 people can sit around a round table. (But careful: when "lining up" the people around the table, ask yourself: Who's first in line?)

When arranging five people around a perfectly round table, there is no "first person" in line—any one of the five people sitting at the table could be regarded as being "first"; hence, being "first" is indistinguishable from any other position at the table. Therefore, the number of ways to arrange five people around a round table is

$$\frac{5!}{5} = 24$$

(j) Charms on bracelets can also be thought of as circular permutations, but since bracelets can be turned over, many permutations are indistinguishable. Suppose a bracelet has 10 charms. Find this bracelet's number of possible arrangements.

$$\frac{10!}{2} = 1,814,400$$

(k) Your graphing calculator's screen is made up of pixels. Each pixel can be on or off. If it is on, it displays a dot; if it's off, it doesn't display a dot. The screen is 40 pixels high and 80 pixels wide. How many different images can your calculator display?

If the calculator had one pixel, there would be two images: on or off. If there were two pixels, then there would be four images; three pixels, eight images; and so forth. Since there $40 \cdot 80 = 3200$ pixels, the number of possible images is 2^{3200}.

(l) In a famous passage from his book *Unweaving the Rainbow*, Richard Dawkins writes:

> We are going to die, and that makes us the lucky ones. Most people are never going to die because they are never going to be born. The potential people who could have been here in my place but who will in fact never see the light of day outnumber the sand grains of Arabia. Certainly those unborn ghosts include greater poets than Keats, scientists greater than Newton. We know this because the set of possible people allowed by our DNA so massively exceeds the set of actual people. In the teeth of these stupefying odds it is you and I, in our ordinariness, that are here. We privileged few, who won the lottery of birth against all odds, how dare we whine at our inevitable return to that prior state from which the vast majority have never stirred?

So, genetically, how many "potential people" are there?

This a difficult question to answer—there is much disagreement—but the arrival at a solution makes use of the multiplication rule. On the website *Stanford at The Tech Understanding Genetics*,[*] a graduate student asks effectively the same question—and it is addressed by Dr. Rama Balakrishnan of Stanford this way:

> Genes can be of almost any length and can have many, many combinations of the four bases [A, G, C, T]. This fact puts the number of possible genes pretty close to infinite.
>
> Let's start with a gene that has 2 bases (no such gene exists). If we allow any base at each one of the 2 positions, then we have 4x4 (42) or 16 possible combinations of sequences. Similarly, for a sequence length of 4, the total # of combinations are 4x4x4x4 (44) or 256.
>
> How about a sequence that is 300 bases long? How many different combinations of the 4 bases are possible in these 300 bases? The total number of possibilities will be 4300 which to me is something like zillions.
>
> You see how many differences can occur in just one gene? Now expand that to 25,000 and you see what we're up against!

Balakrishnan goes on to note that the calculations aren't quite that simple, either, since many changes in DNA have no discernable effect, while others would be instantly lethal. So both of those sets (the no effect and the lethal) would need to be excluded from the $\left(4^{300}\right)^{25,000}$ possible permutations.

Scientific discovery has not yet arrived at the point where it is known precisely *which* permutations to exclude, or how to determine them, therefore we can't give an exact answer to Dawkins' implied question in *Unweaving the Rainbow*.

For some counting problems, the order of arrangements doesn't matter. Choosing *m* objects from a set of *n* objects is called forming a *combination of n objects taken m at a time*.

For instance, if you wish to find out how many ways there are to get an Ace and two Kings from a single deck of cards, the order in which you pull them from the desk makes no difference.

The combinations' formula, which takes into account repetitions of subsets by dividing them out, for the number of combinations of *n* elements taken *m* at a time is:

$$\binom{n}{m} = \frac{n!}{(n-m)!\, m!}$$

---

[*] Located online at http://genetics.thetech.org/

The *binomial coefficient* $\begin{pmatrix} n \\ m \end{pmatrix}$ is read as "*n* choose *m*," so $\begin{pmatrix} 3 \\ 2 \end{pmatrix}$ would be read "3 choose 2." $\begin{pmatrix} 3 \\ 2 \end{pmatrix}$

is calculated as follows: $\begin{pmatrix} 3 \\ 2 \end{pmatrix} = \dfrac{3!}{(3-2)!2!} = \dfrac{3 \cdot 2 \cdot 1}{1 \cdot 2 \cdot 1} = 3$, meaning that choosing 2 elements out of a

possible 3 can be completed in 3 ways if order doesn't matter. Your graphing calculator can also evaluate combinations. To find "3 choose 2," first type 3. Then press the MATH button, go to the PRB menu, select the nCr function, type 2, and press ENTER.

Example 5k.-6

(a) In how many ways can 3 letters be chosen at random from A, B, C, D, and E if order *doesn't* matter? List out the possibilities.

ABC, ABD, ABE, BCD, BCE, BDE, CDE, ACD, ACE, ADE. There are 10 ways.

(b) Recalculate the previous problem using a binomial coefficient.

$\begin{pmatrix} 5 \\ 3 \end{pmatrix} = 10$

What oftentimes makes combination problems especially difficult is that you're not always told explicitly that order doesn't matter—you might have to make that determination yourself. Remember, if order did matter, you'd be solving a permutation problem, not a combination one.

Example 5k.-7

(a) Nine pizza toppings are available; in how many ways can you choose three of them?

The order in which you select toppings makes no difference. Therefore, the answer is given by the combination $\begin{pmatrix} 9 \\ 3 \end{pmatrix} = 84$.

(b) How many committees of 5 students can be selected from a class of 25 students?

Usually, when we spot the word "committee," order doesn't matter. Here, it doesn't matter in what order the students were selected; we're only concerned with the total number of sets of 5 students among a group of 25. The answer is $\begin{pmatrix} 25 \\ 5 \end{pmatrix} = 53{,}130$.

(c) A box contains 5 white and 6 black marbles. In how many ways can any 4 marbles be chosen?

$$\binom{11}{4} = 330$$

(d) How many different ways are there to select 5 cards from an ordinary deck of 52 playing cards?

$$\binom{52}{5} = 2,598,960 \text{, or the total number of five-card poker hands.}$$

(e) What is the chance of your five-card hand being a royal flush (Ace, King, Queen, Jack, and Ten all of the same suit)?

There are four royal flushes, since there are four suits (spades, diamonds, hearts, and clubs). So, the probability is 4/2,598,960.

(f) How many ways are there to divide up seven cans of sodas among three people (and no one shares)?

We will use the *stars and bars method*, originated by the mathematician William Feller, to solve this problem. Essentially, treat each of the sodas as a "star":

★★★★★★★

Next, place a "bar" (or divider) to represent grouping the cans of soda among three people:

★ | ★★ | ★★★★

In the stars and bars scenario above, the first person receives one soda, the second two sodas, and the third four sodas.

Since two items are bars and seven items are stars, there a total of nine items; from these, there are subsets of two bars that form the arrangements. So, the total number of ways to divide up seven soda cans among three people is $\binom{7}{2} = 21$.

There is a connection between Pascal's triangle and combinations. The top row, or 0th row, of the triangle has but one number: 1. Note that the binomial coefficient $\binom{0}{0} = 1$, since there is exactly one way to "choose" nothing amongst nothing: by not making a choice at all. The 1st row of the triangle contains two numbers: 1 and 1. Increment the combinations slightly, and we obtain: $\binom{1}{0} = 1$ and $\binom{1}{1} = 1$; observe that no matter how many objects we have, there is only one way to "choose" none of them: by not making a choice at all.

Continuing down Pascal's triangle, the 2nd row is 1 2 1—and, combination-wise, we have binomial coefficients $\binom{2}{0} = 1$, $\binom{2}{1} = 2$, and $\binom{2}{2} = 2$, matching the triangle again. The pattern continues, ad infinitum, since Pascal's triangle is constructed iteratively: the previous row's entries are summed in pairs to create the next row:

$$
\begin{array}{c}
1 \\
1\ 1 \\
1\ 2\ 1 \\
1\ 3\ 3\ 1 \\
1\ 4\ 6\ 4\ 1 \\
1\ 5\ 10\ 10\ 5\ 1
\end{array}
$$

There are many patterns that emerge from the triangle.[*] Here's one of the more obvious: each row sums to that row's power of 2. For instance, row 3's sum is 8, and $2^3 = 8$. Many more such patterns can be found in the intriguing book *The Number Devil: A Mathematical Adventure*.

The close connection between Pascal's triangle and combinations is helpful when calculating the binomial coefficients of the *binomial theorem*, which can be used to systematically expand a binomial of any whole number power, such as $(x+y)^{10}$. Using summation notation, the binomial theorem is given as

$$
(x+y)^n = \sum_{i=0}^{n} \binom{n}{i} x^{n-i} y^i
$$

Here's a rule of thumb relating to all probability problems (which you should be aware of by now):

- If you see the word *and*, it usually means to multiply;
- If you see the word *or*, it usually means to add.

---

[*] By the way, Pascal's triangle is another example of Stigler's law of eponymy.

Example 5k.-8

(a) A box contains 12 black and 8 green marbles. In how many ways can 3 black *and* 2 green marbles be chosen?

$$\binom{12}{3}\binom{8}{2} = 6{,}160$$

(b) A box contains 12 black and 8 green marbles. In how many ways can 3 black *or* 2 green marbles be chosen?

$$\binom{12}{3}+\binom{8}{2} = 248$$

(c) An urn contains 8 white, 6 blue, and 9 red chips. How many ways can 6 chips be selected if

    a. All chips are red.

$$\binom{9}{6} = 84$$

    b. Three are blue, 2 are white, and 1 is red.[*]

$$\binom{6}{3}\binom{8}{2}\binom{9}{1} = 5{,}040 \, .$$ Note that "9 choose 1" is equal to 9, since there are nine ways to choose one object among nine: by choosing each object one at a time.

    c. Three are blue or 2 are white.

$$\binom{6}{3}+\binom{8}{2} = 48$$

    d. Exactly 4 are white.

$$\binom{8}{4}\binom{15}{2} = 7{,}350 \, ,$$ since you need two chips of any other color besides white.

---

[*] Realize that all events separated by commas here have implied *and*s.

(d) In gym class, a group of 15 students is asked to pick a captain; that captain is then asked to pick a group of 4 students. How many ways can this occur?

$$\binom{15}{1}\binom{14}{4} = 20{,}475$$

(e) An urn contains 4 white, 6 blue, and 3 green chips. How many ways can 5 chips be selected if 2 are white or 3 are green?

$$\binom{4}{2}\binom{9}{3} + \binom{3}{2}\binom{10}{3} = 864$$

(f) According to their website, "Powerball® [is a] combined large jackpot game and cash game. Every Wednesday and Saturday night at 10:59 p.m. Eastern Time, we draw five white balls out of a drum with 69 balls and one red ball out of a drum with 26 red balls." Find the probability of winning the jackpot.

$$\frac{1}{\binom{69}{5}\binom{26}{1}} = \frac{1}{292{,}201{,}338}$$

Suppose we can divide a big set of items into two smaller mutually exclusive sets. The first small set has n items, and the second small set has m items. We choose N items from the big set, hoping that i items among the N chosen come from the first small set (which has n items). (All selections are made without replacement.) The probability of selecting i of these items from the first small set is

$$\frac{\binom{n}{i}\binom{m}{N-i}}{\binom{n+m}{N}}$$

This *probability mass function* (pmf) is called the *hypergeometric distribution*. Several examples will make the math clearer.

Example 5k.-9

(a) There are 10 white balls and 5 red balls in an urn. What is the probability of selecting 3 white balls and 2 red balls?

$$\frac{\binom{10}{3}\binom{5}{2}}{\binom{10}{5}} = 0.399$$. The numerator takes care of the selection of colors; the denominator

enumerates all possible sets of five balls selected among the ten.

(b) In a standard deck of 52 cards, what is the probability of selecting

a. Three clubs and 3 diamonds?

$$\frac{\binom{13}{3}\binom{13}{3}}{\binom{52}{6}} = 0.004$$

b. Three clubs or 3 diamonds?

$$\frac{\binom{13}{3}+\binom{13}{3}}{\binom{52}{3}} = 0.026$$. (This setup does *not* conform to a hypergeometric model.)

c. Three kings or 2 jacks?

$$\frac{\binom{4}{3}}{\binom{52}{3}} + \frac{\binom{4}{2}}{\binom{52}{2}} = 0.004$$. (And neither does this.)

For more combinatorics problems in the vein of this §, refer to the Alan Tucker text *Applied Combinatorics* (6th ed.)—you might want to jump directly to the fifth chapter, on general counting methods, for some educational and entertaining problems: Tucker's deep abiding love of *Lord of the Rings* shines through in the examples he pens.

§51. *Binomial and Geometric Random Variables*

Definition 51.-1

Binomial random variable. A random variable used to model a probability experiment that meets the following criteria: (1) There are two options, commonly called "success" and "failure"; (2) Each trial is independent; (3) The probability of "success" and "failure" stay the same trial to trial; and (4) There are a fixed number of trials, set in advance of the experiment.[*]

Example 51.-1

Binomial situations often arise in real life. Determine whether each of the following situations is binomial or not. Justify your answer.

(a) Flipping a fair coin ten times.

This is binomial, since there are two options (heads or tails), each trial is independent, the probabilities stay constant (50-50), and there are a fixed number of trials (ten).

(b) Flipping a fair coin until heads appears.

This is not binomial because there is not a fixed number of trials. There could be one trial, or ten trials, or an infinite number (although even ten trials before heads is obtained is exceedingly unlikely).

(c) Picking three Kings out of a single deck without replacement.

Not binomial, since the trials are not independent of each other and the probabilities of "King" and "No King" don't stay the same trial to trial.

(d) Picking three Kings out of a single deck with replacement.

Yes, this is binomial. The issues from part (c) have been addressed.

Example 51.-2

Assume independence for all of the following problems.

(a) Suppose the probability that a light bulb is defective is 0.1. What is the probability that

---

[*] Another way to note that we repeat the same experiment over and over again, unchangingly, is by using the term *stationarity*.

three light bulbs are all defective?

$0.1^3 = 0.001$

(b) Again, suppose the probability that a light bulb is defective is 0.1. What is the probability that exactly two out of three light bulbs are defective?

There are three possible ways this situation can occur:

$DDN : 0.1 \cdot 0.1 \cdot 0.9 = 0.009$

$DND : 0.1 \cdot 0.9 \cdot 0.1 = 0.009$

$NDD : 0.9 \cdot 0.1 \cdot 0.1 = 0.009$

Adding the probabilities, we get 0.027.

(c) Look again at the previous question. How many ways are there to pick two light bulbs out of a possible three? (Your answer will be called a *binomial coefficient*.)

$$\binom{3}{2} = 3$$

Let's use the answers from parts (b) and (c) above to help us develop a formula for calculating binomial probabilities outright. Each binomial experiment has two options: success, with a probability of p, and failure, with a probability of $1 - p$. The experiment is repeated n times—with all trials independent of each other. The probability of k successes is

$$P(X = k) = \binom{n}{k} p^k (1 - p)^{n-k}$$

And, to repeat, the "n choose k" coefficient is called a binomial coefficient.

So, if X is a binomial random variable with n independent observations and a probability of success equal to p, we can represent the random variable symbolically as $X \sim B(n, p)$.

Example 51.–3

Assume independence for all of the following problems.

(a) Suppose the probability that a light bulb is defective is 0.1. What is the probability that exactly three out of eight light bulbs are defective?

$$\binom{8}{3}(0.1)^3(0.9)^5 = 0.033$$

(b) Suppose the probability of having to attend a certain meeting at work is 0.125. What is the probability that exactly one person out of six employees will have to attend the meeting?

$$\binom{6}{1}(0.125)^1(0.875)^5 = 0.385$$

(c) A regular, fair die is rolled three times. Find the probability of rolling exactly one 1.

$$\binom{3}{1}\left(\tfrac{1}{6}\right)^1\left(\tfrac{5}{6}\right)^2 = 0.347$$

(d) The MLC (Misery Loves Company) manufacturers towels. About 1 in 100 towels is not threaded to specifications. You purchase 35 towels. What is the probability that four of them are not threaded to spec?

$$\binom{35}{4}\left(\tfrac{1}{100}\right)^4\left(\tfrac{99}{100}\right)^{31} = 0.000383$$

(e) A fair, regular die is rolled five times. What is the probability of rolling *at least* three 1's?

Seeing *at least* three 1's implies we could see three 1's, four 1's, or five 1's out of five rolls of the die. We need to add these probabilities together (since they're linked with the word *or*).

$$\binom{5}{3}\left(\tfrac{1}{6}\right)^3\left(\tfrac{5}{6}\right)^2 + \binom{5}{4}\left(\tfrac{1}{6}\right)^4\left(\tfrac{5}{6}\right)^1 + \binom{5}{5}\left(\tfrac{1}{6}\right)^5\left(\tfrac{5}{6}\right)^0 = 0.035$$

(f) The MLC also manufacturers snow globes, which are quite fragile. Ten of these snow globes are lined up at the top of an unstable shelf on the vibrating assembly line in the MLC factory. The probability of any one snow globe accidently falling (thereby shattering) is 0.95. What is the probability that *fewer than eight* of the snow globes fall off the shelf?

"Fewer than eight" is the same as saying "zero to seven." A complement here might be easier to use:

$$1-\left[\binom{10}{8}(0.95)^8(0.05)^2+\binom{10}{9}(0.95)^9(0.05)^1+0.95^{10}\right]$$

(g) Barry Bonds, perhaps the greatest MLB baseball player of all time (well, except for Babe Ruth), had a career batting average of .298. Ignoring walks, what was the probability that he would get at least one hit in four times at bat?

Notice the presence of "at least one" in the problem description. We can thereby bypass binomial coefficients and the like and instead proceed straight to using a complement:

$$1-0.702^4 = 0.757$$

Although we can always write out all of the terms to find answers to binomial questions, it is more efficient to have the graphing calculator to find binomial probabilities. We will use two functions: binompdf(and binomcdf(, both of which are available in the DISTR menu (you can get there by typing 2nd and VARS). The function takes three arguments: n,p,x, in that order, where n is the number of trials, p is the probability of success, x is number of successes when using the pdf function and *no more than x* successes when using the cdf function.[*]

Example 51.-4

According to a recent survey, about 7% of people in the U.S. don't discard their mattress until it starts smelling—badly. Suppose that we randomly select 45 people. Let x = the number of people, out of the group of 45, who don't throw out their mattress until it starts smelling badly.

(a) Use your graphing calculator to find the probability that none of the 45 people throw their mattress out only when it starts smelling badly.

binompdf(45,0.07,0) = 0.038

(b) Next, let's find the probability that no more than three of the people don't throw their mattress out until it starts smelling badly.

binomcdf(45,0.07,3) = 0.613

(c) Find the probability that at least two of the people don't throw their mattress out until it starts smelling badly.

1 – binomcdf(45,0.07,1) = 0.167

---

[*] The "cdf" stands for "cumulative density function," and finds the cumulative (sum of) probabilities over a given interval.

(d) Find the probability that between four and eight of the people don't throw their mattress out until it starts smelling badly.

$$\text{binomcdf}(45,0.07,8) - \text{binomcdf}(45,0.07,3) = 0.384$$

While we can make use of the usual formulas for the mean and standard deviation of a binomial random variable, there is a simpler set of formulas we can use.

A generalized distribution for a binomial variable can be written as

X	0	1
P(X)	1-p	p

since there are two options—success ($X = 1$) and failure ($X = 0$).

Now, the mean is of this random variable easy to calculate.

$$\mu_X = \sum 0 \cdot (1-p) + 1 \cdot p = p$$

The variance of the random variable is a little more complicated.

$$\sigma_X^2 = 0^2 \cdot (1-p) + 1^2 \cdot p - p^2$$

$$= p - p^2$$

$$= p(1-p)$$

But this doesn't complete the process—for either the mean or variance. We need to consider counting the number of successes in n independent trials. Each trial has a probability of success of p. Let's call this sum of random variables a new name: random variable Y.

$$\mu_Y = \mu_{X_1} + \mu_{X_2} + \cdots + \mu_{X_n} = p + p + \cdots + p = np$$

$$\sigma_Y^2 = \sigma_X^2 + \sigma_X^2 + \cdots + \sigma_X^2$$

$$= p(1-p) + p(1-p) + \cdots + p(1-p)$$

$$= np(1-p)$$

So the standard deviation of a binomial random variable is

$$\sigma_Y = \sqrt{np(1-p)}$$

Example 51.-5

(a) Suppose you flip an unfair coin ten times. The coin has a 70% chance of landing on heads. Find the mean and standard deviation for the number of flips, out of ten, likely to land on heads, and interpret the results.

The mean (or expected value) is very intuitive: if a coin has a 70% chance of landing on heads, and the coin is flipped ten times, you'd expect the coin to land on heads seven times. Formally, the calculation is $\mu = np = 10 \cdot 0.7 = 7$.

The standard deviation is $\sigma = \sqrt{np(1-p)} = \sqrt{10 \cdot 0.7(1-0.7)} = 1.45$, the interpretation of which is as follows: the number of heads obtained out of ten flips of this unfair coin differs from the mean number of heads by about 1.45.

(b) According to a recent survey, about 7% of people in the U.S. don't discard their mattress until it starts smelling badly. Suppose that we randomly select 45 people. Find the mean and standard deviation of this binomial situation.

$$\mu = 45 \cdot 0.07 = 3.15$$

$$\sigma = \sqrt{45 \cdot 0.07(1 - 0.07)} = 1.71$$

Under certain conditions, the binomial distribution can be used to *approximate* the normal distribution. First, let's discuss why we would ever want to do this. Then, we'll explore those "certain conditions," in addition to looking at some examples.

Suppose we were asked to solve a problem like this:

Suppose a fair coin is flipped 100 times. Find the probability of obtaining at least 55 heads.

Clearly, this is a binomial situation: there is independence, the probabilities of success (heads) and failure (tails) stay the same with each trials, and there is a fixed number of trials (one hundred). But—without a calculator—the calculation itself is incredibly tedious:

$$\binom{100}{55}(0.5)^{55}(0.5)^{45} + \binom{100}{56}(0.5)^{56}(0.5)^{44} + \cdots + \binom{100}{100}(0.5)^{100}(0.5)^{0}$$

That's quite a few terms to evaluate.[*]

---

[*] Such sums are easily calculated with computer software, however; even the graphing calculator can find the probability. Approximations of the binomial using the normal were very useful well before our era of powerful electronic computation; as George Cobb writes, "[T]he normal distribution and the Central Limit Theorem arose *as a by-product of a 30-year struggle with a computing impasse*" (his italics).

And there's another point to consider. When selecting a random sample from a population, we may need to obtain the probability of success p of members of that sample having some characteristic. But most of the time, sampling is done without replacement and is thus not independent—leading to calculations that may set new records in tedium.

Luckily, we can instead use the normal distribution to approximate the binomial distribution as long as the sample size in question is no larger than 10% of the population. (Notice that, in the example above of one hundred coin flips, the sample—which is the one hundred flips—is much less than 10% of the population—which is an infinite number of flips of the coin.)

It's not only n we have to worry about, though. The shape of the binomial distribution is affected by n, the number of trials, and also by p, the probability of success of each trial. The closer p is to zero, the more skewed right the binomial distribution is—leading to a poor normal approximation for the distribution, since the normal distribution is symmetrical. Conversely, the closer p is to one, the more skewed left the binomial distribution is, likewise leading to an uneasy matching with the normal approximation.

Counteracting large or small p, though, is n. As long as n is large enough—and our rule of thumb is that n can't be more than 10% of the population size, if we're sampling; with "infinite" processes, such as coin flips, there are no restrictions on the size of n—the normal distribution will approximate the binomial distribution sufficiently well.

There are some subtle disagreements in the literature on precisely how to perform the approximation, however. Although all sources agree that the mean and standard deviation of the normal approximation to the binomial distribution are

$$\mu_X = E(X) = np \text{ and } \sigma_X = \sqrt{np(1-p)}$$

The sources also mostly concur on the rule of thumb—also called the "technical conditions," or "assumptions," or "requirements"—to check and see if n is large enough:

$$np \geq 10 \text{ and } n(1-p) \geq 10, \text{ provided also that } n \leq \tfrac{1}{10}N,$$
if a random sample has been selected (without replacement)

That is, there must be at least ten expected successes and ten expected failures,[*] and the sample size must be no larger than 10% of the population.

But here's where the crux of the disagreement in the literature lies: on using a *continuity correction*, a factor designed to smooth the transition from using a continuous distribution (in this case, the normal distribution) to approximate a discrete one (the binomial distribution). As seen in the representative diagram below, the binomial distribution overlaid atop the normal distribution is jagged-edged; as n increases, though, the width of the boxes decreases.

---

[*] Some statisticians and textbook authors, which of course are not mutually exclusive sets of individuals, prefer the less conservative value of five: that is, five expected successes and five expected failures.

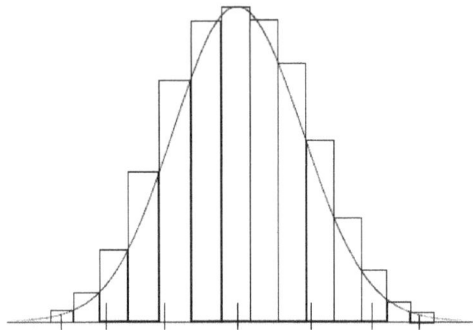

The continuity correction, used in a z-score, is designed to counteract the jaggedness, and is usually set at a value of 0.5 to "split the difference" of the value of interest. However, not every statistician uses a c.c. (although with big enough sample sizes using a c.c. hardly matters anyway—and needlessly complicates things), so we won't in this primer, either.

Example 51.-6

Suppose a fair coin is flipped 100 times. Using the normal approximation of the binomial distribution, find the probability of obtaining at least 55 heads.

Solution. First, let's check to make sure the technical conditions are satisfied here.

$$np \geq 10$$
$$100 \cdot 0.5 = 50 \geq 10$$

$$n(1-p) \geq 10$$
$$100 \cdot (1-0.5) = 50 \geq 10$$

Both conditions are satisfied. So let's now obtain the mean and standard deviation:

$$\mu_X = np = 100 \cdot 0.5 = 50$$

$$\sigma_X = \sqrt{np(1-p)} = \sqrt{100 \cdot 0.5(1-0.5)} = 5$$

Finally, making use of a z-score, we see that the probability of at least 55 heads is

$$z = \frac{55-50}{5} = 1, \; 1-0.8413 = 0.1587$$

Another distribution, called the *Poisson distribution*, also serves to approximate the binomial distribution, but under very different circumstances than the normal.[*]

Definition 51.-2

Poisson distribution. A distribution that emerges from the binomial distribution as the number of trials increases dramatically, while the probability of success *p* decreases dramatically as well.

If the Poisson distribution is to model a probability experiment, the experiment must have three properties: (1) Two options, success or failure; (2) Independence between trials; and (3) The probability staying the same trial to trial.

Example 51.-7

At a dangerous intersection, there are, on average, two traffic accidents per day. Find the probability that in the next day, there will be three traffic accidents.

Solution. First, let's subdivide the interval of one day into *n* tiny intervals, all of equal duration:

$$0 \qquad\qquad time \qquad\qquad 1$$

Now, assume that in any tiny subinterval, there is either zero or one traffic accident. In addition, assume that each tiny subinterval is independent of the others (although this might be a bit unrealistic when it comes to traffic accidents—one accident may increase the probability of another one—we'll stick with the simplifying assumption here). And, finally, assume that the probability *p* of a traffic accident occurring remains constant, subinterval to subinterval (trial to trial).

The number of traffic accidents, then, would have a binomial distribution:

$$P(k \text{ traffic accidents}) = \binom{n}{k} p^k (1-p)^{n-k}, \text{ where } n \text{ is very big—since there are many subintervals}$$

As $n \to \infty$, np, the mean of a binomial random variable, remains constant (for instance, as *n* doubles, *p* is divided by 2). We signify this *np* constant using the Greek letter λ (pronounced "lambda"). Then we have

$$P(k \text{ traffic accidents}) \Rightarrow \lim_{n\to\infty} \binom{n}{k} p^k (1-p)^{n-k} = \frac{e^{-\lambda} \lambda^k}{k!}$$

---

[*] The word *poisson* is French for fish, by the way.

And, finally, to be consistent with other textbooks, we replace k in the formula with r. Therefore, we have the Poisson distribution showing the probability of r successes over an interval given as

$$P(X = r) = \frac{e^{-\lambda}\lambda^r}{r!}$$

λ is the *mean* number of successes over time, volume, or area. So, the population mean and standard deviation of a Poisson distribution are

$$\mu = \lambda \text{ and } \sigma = \sqrt{\lambda}$$

And the probability of three traffic accidents can be found by using the probability mass function:

$$P(3) = \frac{e^{-2}2^3}{3!} = 0.180$$

Generally, the Poisson distribution can be utilized any time many independent intervals centered on observing an event that is very rare can be constructed. We can use the Poisson to approximate the binomial if the number of trials n is very large and the probability of success p is very small. Specifically, the Poisson approximates the binomial well iff[*]

$$n \geq 100 \text{ and } \lambda = np < 10$$

Before we move on from the binomial distribution, note that the binomial is a special case of the *multinomial distribution*: a discrete distribution that can account for n trials of an exhaustive set of k categories of probabilities.

Example 51.-8

There are many toy cars, tops, and yo-yos all mixed up in a huge vat. The probabilities of pulling a toy car are 0.4, a top 0.5, and a yo-yo 0.1. You reach in and scope up a bundle of 5 toys. What are the chances of getting two cars, one top, and two yo-yos?

Solution. To find the probability, evaluate the following:

$$\frac{5!}{2!1!2!}(0.4)^2(0.5)^1(0.1)^2 = 0.024$$

---

[*] The abbreviation "iff" is shorthand for "if and only if" and was coined by the mathematician Paul Halmos.

Notice how the coefficient of factorials resembles the letter-arrangement problems (like MIS-SISSIPPI) we examined earlier.

Neither the binomial nor Poisson can model a situation in which trials need to be completed until the first success occurs.[*] Instead, we'll need to use the *geometric distribution*.

Definition 51.-3

Geometric distribution. A distribution that models probability experiments with the following properties: (1) There are two options, success or failure; (2) There is independence between the trials; (3) The probability of success stays the same trial to trial; and (4) The first success occurs on the kth trial.

The probability mass function of the geometric distribution relies on the general multiplication rule for independent events.

$$P(X = k) = (1 - p)^{k-1} p$$

Example 51.-9

Consider an unfair (or weighted) coin with a probability of 70% of landing on heads.

(a) What are the chances of the first head coming up on the second flip?

$$P(2) = (1 - 0.7)^1 \cdot 0.7 = 0.21$$

(b) What are the chances of the first head coming up on the fifth flip?

$$P(5) = (1 - 0.7)^4 \cdot 0.7 = 0.006$$

(c) What are the chances of the first head coming up on the fourth or fifth flip?

$$P(4) + P(5) = (1 - 0.7)^3 \cdot 0.7 + (1 - 0.7)^4 \cdot 0.7 = 0.019 + 0.006 = 0.025$$

---

[*] Real-world example: Repeatedly shooting a basketball until you make your first free throw. Another example: Calling someone over and over again until he or she finally agrees to go out with you—although, come to think of it, each call wouldn't be independent, which—as we're about to see—is a necessary but insufficient condition of being modeled geometrically. Unless it's a situation out of the movie *50 First Dates*.

(d) What are the chances of the first head coming up on the kth flip?

$$P(k) = (1 - 0.7)^{k-1} \cdot 0.7$$

While we can make use of the usual formulas for the mean and standard deviation of a geometric random variable, there is a simpler set of formulas we can use.

Let k represent the number of trials until the first success. Since we don't know when that first success will come—we might have to wait a lifetime for it—we'll use infinity as an upper limit of summation for the mean.

$$\mu = \sum_{k=1}^{\infty} k(1-p)^{k-1} p$$

Mathematically, since p is a constant, the formula can be rewritten as

$$\mu = p \sum_{k=1}^{\infty} k(1-p)^{k-1}$$

which, when expanded, becomes

$$\mu = p\left(1 + 2(1-p) + 3(1-p)^2 + \cdots\right)$$

Next, multiply both sides of the equation by $1 - p$:

$$(1-p)\mu = p\left((1-p) + 2(1-p)^2 + 3(1-p)^3 + \cdots\right)$$

Now, let's subtract the previous two equations, which gives us

$$\mu(1 - (1-p)) = p\left(1 + (1-p) + (1-p)^2 + (1-p)^3 + \cdots\right)$$

Next, you have to recognize that $1 + (1-p) + (1-p)^2 + (1-p)^3 + \cdots$ represents an infinite geometric series, with $r = 1 - p$—that is, the multiplicative rate of growth equal to $1 - p$.

The sum of any infinite geometric series is given by

$$\frac{1}{1-r}, \; |r| < 1$$

So, we see that

$$\mu(p) = p\left(\frac{1}{1-(1-p)}\right)$$

Dividing out p on both sides, we're left with

$$\mu = \frac{1}{p}$$

And the standard deviation of a geometric random variable is

$$\sigma = \sqrt{\frac{1-p}{p^2}}$$

Example 51.-10

Consider an unfair coin with a probability of 70% of landing on heads. Find the mean number of coin flips needed before landing the first head; also calculate the standard deviation.

Solution. The mean is $\mu = \dfrac{1}{0.7} = 1.429$, and the standard deviation is $\sigma = \sqrt{\dfrac{1-0.7}{0.7^2}} = 0.782$.

The geometric distribution is actually a special case of the *negative binomial distribution,* which can address more general questions such as: What is the probability that the fifth head of the unfair (70% heads) coin will occur on the twelfth toss? So instead of being restricted to the first success, you can calculate the probability of obtaining the rth success.

The probability mass function for the negative binomial distribution, where the rth success occurs on the xth trial, is

$$P(X = x) = \binom{x-1}{r-1} p^r (1-p)^{x-r}$$

Look closely at the formula: it is, effectively, the binomial distribution. And we've thus come full circle.[*]

---

[*] Just like the geometric distribution (looking for the "first success") is a special case of the negative binomial distribution (looking for the "rth success"), the *exponential distribution* (looking for the "first success") is a special case of the Poisson distribution (looking for the "rth success").

Example 5l.-11

Consider an unfair coin with a probability of 70% of landing on heads. What is the probability that the third head comes up on the tenth flip?

Solution. Using the formula, $P(X=10) = \binom{10-1}{3-1}0.7^3(1-0.7)^{10-3} = \binom{9}{2}0.7^3 \cdot 0.3^7 = 0.003$.

§5m. *Covariance*

Recall that the correlation coefficient quantifies the strength and direction of the linear relationship between two quantitative variables. Generalizing that definition slightly, instead of "two quantitative variables," let us instead focus on "two random variables." We will now introduce the *covariance* as a mediating tool to help us understand the generalization.

Definition 5m.-1

Covariance. A measurement of the dependence between two random variables; the correlation coefficient is, in fact, a special case of the covariance.

The formula for the covariance between two random variables X and Y is

$$\text{cov}(X,Y) = \sigma_{XY} = E[(X - E(X))(Y - E(Y))]$$

Example 5m.-1

Here is a joint probability distribution of discrete random variables X and Y.

		X	
		1	2
	3	0.20	0.60
Y	4	0.10	0.10

The $E(X) = 1.7$ and $E(Y) = 3.2$. Find the covariance of X and Y.

Solution. Working through every possible set of probabilities using the covariance formula, we see that

$$(1-1.7)(3-3.2)(0.20)+(1-1.7)(4-3.2)(0.10)+(2-1.7)(3-3.2)(0.60)+(2-1.7)(4-3.2)(0.10)=-0.04$$

So $\text{cov}(X,Y)=\sigma_{XY}=-0.04$.

Unsurprisingly, there is a "shortcut" formula we can use for the covariance. Start with $\text{cov}(X,Y)=\sigma_{XY}=E[(X-E(X))(Y-E(Y))]$. Then, expand out what's inside the brackets to obtain $E[XY-XE(Y)-YE(X)+E(X)E(Y)]$. Next, distribute the expected value to see that $E(XY)-E(X)E(Y)-E(Y)E(X)+E(X)E(Y)$. Thus, the "shortcut" formula for covariance is

$$\text{cov}(X,Y)=\sigma_{XY}=E(XY)-E(X)E(Y)$$

Example 5m.-2

Use the covariance "shortcut" formula to calculate the covariance of the joint probability distribution shown in the previous example.

Solution. Working through the "shortcut" formula, we see that

$$\text{cov}(X,Y)=(1)(3)(0.20)+(1)(4)(0.10)+(2)(3)(0.60)+(2)(4)(0.10)-(1.7)(3.2)=-0.04$$

When investing in the stock market, a maxim you'll repeatedly hear is "diversify, diversify, diversify." Covariance helps to explain why. Suppose you own two companies in the same business sector. Economic events affecting one company are likely to affect the other in kind; here, your portfolio will have a large positive covariance, since both companies' fortunes rise and fall together. But if, on the other hand, you pick two companies for your portfolio that have a large negative covariance, then when one company brings in great returns, the other does not, and vice versa. Harry Markowitz built his modern portfolio theory (MPT) around this basic idea, earning a Nobel as a result.

The correlation coefficient—of which the population symbol isn't r (recall that r denotes the sample correlation coefficient), but ρ (the Greek letter pronounced "rho")—can be expressed in terms of the covariance and the standard deviation of two random variables X and Y:

$$\rho_{XY}=\frac{\text{cov}(X,Y)}{\sigma_X\sigma_Y}$$

Like the sample correlation coefficient r, the population correlation coefficient ρ can be no larger than 1 and no less than -1, with the strength of the linear correlation between the two random variables set by how close $|\rho|$ is to 1.

Example 5m.-3

Reconsider the joint probability distribution used in the two previous examples. The standard deviation of X is 0.458, while the standard deviation of Y is 0.4. Find the correlation coefficient ρ between the two random variables X and Y, and also interpret your result.

Solution.

$$\rho_{XY} = \frac{\text{cov}(X,Y)}{\sigma_X \sigma_Y} = \frac{-0.04}{0.458 \cdot 0.4} = -0.218$$

A correlation coefficient of –0.218 indicates a weak, negative linear relationship between the random variables X and Y.

Observe that if the covariance between two random variables is equal to zero, then the correlation coefficient is equal to zero as well—indicating no linear relationship. If two random variables are independent, then the covariance (and correlation coefficient) is equal to zero, although the converse is not always true. But if the correlation between two variables is *not* equal to zero, then the random variables cannot be independent of each other.

And, finally, just like there is a connection between the correlation and the covariance, the slope of the regression line can be found utilizing covariance as well: $b = \dfrac{\text{cov}(X,Y)}{\sigma_X^2}$.

§5n. *A Momentary Look at Moments*

Calculation of the mathematical expectation of a series of integer powers for base x result in what are called *moments*. That is, $E(X)$, which is the mean of the random variable X, is the first moment; $E(X^2)$ is the second moment; $E(X^3)$ is the third moment; and so on.

The mean of a random variable, given by $E(X)$, is a function of a single moment (the first); the variance of a random variable, given by $E(X^2) - \mu^2$ (i.e., the "shortcut" formula), is a function of both the first and second moments.

If the random variable is continuous, and the probability density function is provided, the first moment, $E(X)$ (which is the mean), can be obtained by integrating the following:

$$\mu = E(X) = \int_{-\infty}^{\infty} x \cdot f(x)dx$$

The second moment can be found with this integration:

$$E(X^2) = \int_{-\infty}^{\infty} x^2 \cdot f(x)dx$$

Meaning that the variance, using the "shortcut" formula $E(X^2) - \mu^2$, can be calculated.

Example 5n.-1

Suppose that X is a continuous random variable with a probability density function of

$$f(x) = \frac{x^2}{3}$$

over the interval $0 < x < \sqrt[3]{9}$. Find the mean and variance of X.

Solution. The mean can be found using the first moment:

$$\mu_X = E(X) = \int_0^{\sqrt[3]{9}} x\left(\frac{x^2}{3}\right)dx = \frac{1}{3}\left[\frac{x^4}{4}\right]_{x=0}^{x=\sqrt[3]{9}} = \frac{1}{3}\left[\frac{x^4}{4}\right]_{x=0}^{x=9^{\frac{1}{3}}}$$

$$= \frac{1}{3}\left[\frac{9^{\frac{4}{3}}}{4} - \frac{0^4}{4}\right] = \frac{1}{3}\left[\frac{9^{\frac{4}{3}}}{4}\right] = \frac{9^{\frac{4}{3}}}{12} = \frac{\sqrt[3]{9^4}}{12} = \frac{\sqrt[3]{6561}}{12}$$

Thus, the mean is $\mu_X = E(X) = \dfrac{\sqrt[3]{6561}}{12}$.

The variance can be obtained by finding the second moment, and then utilizing the "shortcut" formula:

$$E(X^2) = \int_0^{\sqrt[3]{9}} x^2\left(\frac{x^2}{3}\right)dx = \frac{1}{3}\left[\frac{x^5}{5}\right]_{x=0}^{x=\sqrt[3]{9}} = \frac{1}{3}\left[\frac{x^5}{5}\right]_{x=0}^{x=9^{\frac{1}{3}}}$$

$$= \frac{1}{3}\left[\frac{9^{\frac{5}{3}}}{5} - \frac{0^5}{5}\right] = \frac{1}{3}\left[\frac{9^{\frac{5}{3}}}{5}\right] = \frac{9^{\frac{5}{3}}}{15} = \frac{\sqrt[3]{9^5}}{15} = \frac{\sqrt[3]{59049}}{15}$$

Plugging in the first and second moments into the "shortcut" formula, we see that the variance is

$$\sigma_X^2 = E(X^2) - \mu^2 = \frac{\sqrt[3]{59049}}{15} - \left(\frac{\sqrt[3]{6561}}{12}\right)^2$$

When we looked at uniform distributions, a continuous type of distribution, earlier, the mean and variance (and, by extension, standard deviation) of uniformly distributed random variables were reported without proof. It's time to remedy that.

Example 5n.-2

Find the mean and variance of a continuous uniformly distributed random variable with the limits a to b.

Solution. The probability density function for a continuous uniformly distributed random variable is $f(x) = \dfrac{1}{b-a}$ since, recall, the base of the rectangular uniform distribution has a length of $b-a$ and the area underneath any probability function must be equal to one, so since the area of a rectangle is base x height, we confirm that

$$(b-a) \cdot \left(\frac{1}{b-a}\right) = 1$$

Which leaves us, first, to find the mean (the first moment):

$$\mu_X = E(X) = \int_a^b x\left(\frac{1}{b-a}\right)dx = \frac{1}{b-a}\left[\frac{x^2}{2}\right]_{x=a}^{x=b}$$

$$= \frac{1}{b-a}\left[\frac{b^2}{2} - \frac{a^2}{2}\right] = \frac{1}{b-a}\left[\frac{b^2-a^2}{2}\right] = \frac{1}{b-a}\left[\frac{(b-a)(b+a)}{2}\right] = \frac{a+b}{2}$$

And, second, to find the second moment:

$$E(X^2) = \int_a^b x^2\left(\frac{1}{b-a}\right)dx = \frac{1}{b-a}\left[\frac{x^3}{3}\right]_{x=a}^{x=b}$$

$$= \frac{1}{b-a}\left[\frac{b^3}{3} - \frac{a^3}{3}\right] = \frac{1}{b-a}\left[\frac{b^3-a^3}{3}\right] = \frac{1}{b-a}\left[\frac{(b-a)(b^2+ab+a^2)}{3}\right] = \frac{b^2+ab+a^2}{3}$$

Finally, using the "shortcut" formula will get us the variance:

$$\sigma^2 = E(X^2) - \mu^2 = \frac{b^2 + ab + a^2}{3} - \left(\frac{a+b}{2}\right)^2 = \frac{b^2 + ab + a^2}{3} - \frac{a^2 + 2ab + b^2}{4}$$

$$= \frac{b^2 - 2ab + a^2}{12} = \frac{(b-a)^2}{12}$$

In general, we can use *moment-generating functions* (mgf's), to find the mean and variance of random variables. If X is a discrete random variable with a probability mass function $f(x)$ then $M(t) = E\left(e^{tx}\right) = \sum e^{tx} \cdot f(x)$ is the moment generating function of X. Evaluating the first derivative of the moment generating function at $t = 0$ will give us the mean of random variable X; finding the first and second derivatives of the mgf at $t = 0$ will result in the variance of random variable X.[*] On the other hand, if X is a continuous random variable, the moment generating function (if it exists) is given by $M(t) = \int_{-\infty}^{\infty} e^{tx} \cdot f(x)dx$, thus requiring a bit of integral calculus to evaluate.

In fact, there is a key connection between the mgf and moments: taking the nth derivative of the mgf at $t = 0$ results in $E(X^n)$—the nth moment "about the origin" (so called because zero is plugged into the mgf). This connection holds no matter if the random variable is discrete or continuous.

Example 5n.-3

Recall that, some pages back, when we introduced the normal distribution, we also relayed the distribution's probability density function:

$$f(x) = \frac{1}{\sigma\sqrt{2\pi}} e^{\frac{-1}{2}\left(\frac{x-\mu}{\sigma}\right)^2}$$

Let's rewrite this pdf with *exp* notation; the "exp," signifying e (Euler's number), is a convenient shorthand when the exponents of Euler's number get out of hand.

---

[*] For instance, instead of working through the derivations of the mean and variance formulas of binomial and geometric random variables that we completed earlier, using a bit of calculus to find the derivatives of the moment generating functions for binomial and geometric random variables would have also produced the mean and variance formulas. Those details, however, are well beyond the scope of this primer.

$$f(x) = \frac{1}{\sigma\sqrt{2\pi}} \exp\left\{-\frac{1}{2}\left(\frac{x-\mu}{\sigma}\right)^2\right\}$$

It can be shown, with *much* work (which we don't want to do here), that the moment generating function of a normal random variable X is given by $M(t) = \exp\left\{\mu \cdot t + \frac{\sigma^2 t^2}{2}\right\}$, meaning that

$$M(t) = \int_{-\infty}^{\infty} e^{tx} \cdot \left(\frac{1}{\sigma\sqrt{2\pi}} \exp\left\{-\frac{1}{2}\left(\frac{x-\mu}{\sigma}\right)^2\right\}\right) dx = \exp\left\{\mu \cdot t + \frac{\sigma^2 t^2}{2}\right\}$$

Use the mgf to find the mean and variance of a normal random variable X.

Solution. To find the mean, we need to calculate the first derivative of the mgf (with respect to t) and then set $t = 0$.

$$M'(x) = \exp\left\{\mu \cdot t + \frac{\sigma^2 t^2}{2}\right\} \cdot \left(\mu + \frac{2t\sigma^2}{2}\right)$$

$$M'(0) = \exp\left\{\mu \cdot 0 + \frac{\sigma^2 \cdot 0^2}{2}\right\} \cdot \left(\mu + \frac{2 \cdot 0 \cdot \sigma^2}{2}\right) = 1 \cdot (\mu + 0) = \mu$$

Unsurprisingly, we have shown that the mean of a normal random variable X is equal to μ.

To find the variance, we need to first calculate the second derivative of the mgf.[*]

$$M''(x) = \exp\left\{\mu \cdot t + \frac{\sigma^2 t^2}{2}\right\} \cdot \sigma^2 + \exp\left\{\mu \cdot t + \frac{\sigma^2 t^2}{2}\right\} \cdot (\mu + \sigma^2 t) \cdot (\mu + \sigma^2 t)$$

$$M''(0) = 1 \cdot \sigma^2 + 1 \cdot \mu \cdot \mu = \sigma^2 + \mu^2$$

Using the familiar "shortcut" formula will get us the variance of a normal random variable X.

$$\sigma^2 = E(X^2) - \mu^2 = \sigma^2 + \mu^2 - (\mu)^2 = \sigma^2$$

---

[*] Needed here is the *product rule of differentiation*: $(f \cdot g)' = f'g + g'f$ or, as my calculus teacher used to tell us to remember it as, "Hippity-hop plus hoppity-hip."

Part 2

Φ

The Axis of Collection

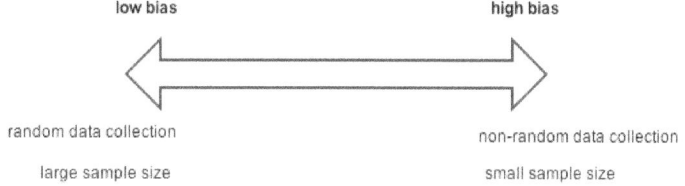

low bias   high bias

random data collection   non-random data collection

large sample size   small sample size

§6. Sampling, Studies, &c.

ɸ

Introduction to Sampling. Survey Design. Studies. The Trouble with Experiments.

§6a. *Introduction to Sampling*

Since it is oftentimes impractical to conduct a census, a sample of the population is obtained instead to learn as much as possible about the population. The sample that is collected must be representative of the population.

Up until now, we have summarized and examined data without questioning its validity. But if the sampling method used to collect a set of data has *bias*, this means that a portion of the population has been systematically excluded. We need to become adept at recognizing different kinds of biases.

Definition 6a.-1

Convenience sample. A sample obtained by a researcher who gathers information from those individuals most readily assessable.

Definition 6a.-2

Voluntary response sample (or **self-selected sample**). Respondents decide for themselves whether to participate in the study.

Definition 6a.-3

Nonresponse bias. Individuals selected for the study either refuse to participate or are not accessible.

Definition 6a.-4

Response bias. An umbrella term of cognitive biases, which include the wording of survey questions, presentation/appearance of the interviewer, and interviewees' reluctance to divulge truthful survey answers to questions of a sensitive nature.

Definition 6a.-5

Selection bias. When the sampling of a population is not sufficiently random, and the sample thereby excludes some portion of the population.

Definition 6a.-6

Sampling frame. A list of individuals from which the sample is taken.

Example 6a.-1

A sample of 178 high school students is asked by an interviewer how many times per day they brush their teeth.

(a) What is the population of this study?

The population is who (or what) we're trying to learn about. Here, we are attempting to learn about all high school students.

(b) What is the sample of this study?

The sample is individuals from whom we've managed to collect data. Here, we have collected information about 178 high school students.

(c) Suppose the mean of the data collected is 2.7 (in other words, $\bar{x} = 2.7$). Explain why you have reason to suspect that this number is not representative of the population of all high school students. Also address the following: What is this kind of bias called?

You might be tempted to say the error lies with the number itself: You can't brush your teeth 2.7 times per day. But that number's an *average*, and it's well within the bounds of reasonability.

Instead, the issue lies with the magnitude of the number. People are likely to inflate the number of times per day that they report brushing their teeth, perhaps to avoid embarrassment because they're self-conscious. This issue falls under the umbrella of response bias (people presenting the best version of themselves is called *social desirability bias*).

In addition, we don't know how the sample of 178 students was selected, or if every student queried actually answered the question.

(d) Suppose it is revealed that the interviewer who asked the questions was a gray-haired bearded tall man dressed in a white lab coat and formal wear. Explain why this could this have impacted the students' answers. Also, what is this kind of bias called?

An "authority figure"—the classically dressed scientist—may have intimidated the respondents to answer in a certain way. Again, this is classified as response bias.[*]

(e) Assume that the brushing-your-teeth question was asked in the following way: "The American Dental Association claims that brushing your teeth three times per day is the key to good dental hygiene. Given that recommendation, how many times per day do you brush your teeth?" What kind of bias is present in the question? Reword the question to eliminate as much of this bias as possible.

The bias is in the question wording, and this is yet again termed response bias. The question's wording, as it stands now, leads the respondents to answer a certain way. Rewriting the question thusly would eliminate that prompting: "How many times per day do you brush your teeth?"

(f) Let's say that some students refused to answer the brushing-your-teeth question in person. Each of these students was later sent a card via mail containing the question, but none of them filled the card out or sent the card back to the researchers. (These students who didn't respond to the question are *not* part of the sample of $n = 178$.) What kind of bias is this? Why might this bias impact the results of the study?

This is an example of nonresponse bias; those people who refused to answer the question may have had something in common—perhaps bad hygiene?—but they have been excluded from the sample, thus throwing into doubt the sample's representativeness of the population of all high school students. Nonresponse bias is a type of *nonsampling error*, or an error that has nothing to do with how the sample was chosen.

(g) Suppose it is revealed that the sampling frame was students who have recently taken the AP American History examination. Explain the bias with this sampling frame. Also state the direction of the bias (i.e., will the mean from this sample be an overestimate or an underestimate of the true population percentage)?

This is an example of selection bias, since the AP students may have qualities which skew the sample as compared to the host population. The direction of the bias will likely be higher than the truth—the mean will probably be an overestimate of the true number

---

[*] This also ties into an "expectancy effect" known as the *Pygmalion effect*, named after the Greek myth. See *Forty Studies that Changed Psychology* for more details.

of times per day all high school students brush their teeth.[*]

(h) The book *House Advantage* by Jeffrey Ma, one of the original members of MIT's famed blackjack team, contains an excellent example of selection bias worth quoting here:

> Another similar and dangerous bias is that of selection. The classic example of selection bias comes to us from World War II. American military personnel, when evaluating returning planes damaged during warfare, noticed that "some parts of planes were hit by enemy fire more often than other parts." Analyzing the pattern of bullet holes in the returning planes, they decided to have these areas reinforced to withstand enemy fire better. Seems logical enough, but there is a clear problem with their analysis.[†]

What was wrong with the American military personnel's analysis of the damaged planes?

They only examined the planes that survived, when they should have examined the planes that didn't.

(i) Instead of the sampling frame being those who took the AP American History exam, assume that a flyer was posted in several high schools across the country asking for participants to complete a survey about dental hygiene. Specifically state what kind of sampling bias is present in this method of data collection.

Voluntary response bias, since people have chosen, on their own volition, to be in the sample.

(j) Now assume that the interviewer collected all of the survey responses by simply driving to the high school closest to his house and questioning students there. Specifically state what kind of sampling bias is present in this method of data collection.

Convenience sampling.

Example 6a.-2

An online MSNBC poll during Christmastime asked users the following question: "How many online purchases have you made so far this holiday season?" The poll, which 19,231 people re-

---

[*] Although, I suppose you could make a counterargument claiming that since these AP students are so focused on their studies, they tend to neglect their hygiene.

[†] Do historians, when gathering facts about a time period, engage in a form of selection bias as well? In his classic text *What is History?*, Edward Carr argues that they do: "The historian is necessarily selective. The belief in a hard core of historical facts existing objectively and independently of the interpretation of the historian is a preposterous fallacy, but one which is very hard to eradicate."

sponded to, found that 63% of people surveyed clicked "3 or more."

(a) What's the population of this survey?

Holiday shoppers.

(b) What's the sample of this survey?

The 19,231 people who responded to the survey.

(c) A disclaimer on the bottom of the poll states that the results of the survey are "non-scientific"—in other words, they are not representative of the population. Explain why, mentioning different types of relevant sampling biases.

Most salient with online polls is the problem of voluntary response—people have chosen to click and take the survey. Not everyone has internet access, although this is much less of a concern than in the past.

The average response collected from the survey is likely to be an overestimate of the true population figure, since those people interested enough to answer a survey about online shopping are probably avid online shoppers.

Let's pause to note the distinction between voluntary response, which this online sampling method has in spades, and nonresponse. Voluntary response and nonresponse tend to be conflated for a logical reason: in both cases, people are making some sort of choice: in the former to participate on their own accord, in the latter to not participate. There's a subtle distinction between them, however. Although both biases harm the integrity of the data collected, with nonresponse, individuals have been *chosen in advance*—presumably randomly—to participate in the sample; they nonetheless decline (or are unable) to participate. With voluntary response, though, the individuals who participate *make up the sample to start with*, leaving us, ipso facto, with what are in effect worthless collected data. (No matter how sophisticated your statistical methods applied to worthless data, the inferences drawn are certain to be worthless, too.)

(d) Suppose another online MSNBC poll is conducted in the same way the next year, and 243,051 people answer it. Would you feel more comfortable with these results because the sample size is so much larger? Briefly explain.

No. If the sampling method is biased to start with, increasing the sample size does nothing to eliminate the bias.* If the sampling method is random in nature, though, bigger samples most often result in more representative data sets.

---

* A caveat: if somehow *all* holiday shoppers decided to answer the online poll, then we'd have conducted a census—and the entire population would be represented. Clearly, something like this occurring is extremely unlikely.

Not only is it true that the bigger the sample the better—as long as the data are collected randomly—it is also the case that, as long as the sample size is small relative to the size of the population (no greater than 10% of the population size), the number of individuals in the population makes no difference when measuring the accuracy (or "precision") of the sample. All that matters is the sample size itself.

Only with a variant of *random sampling* can we, probabilistically, set ourselves up for the best possible chance of capturing a sample that is representative of its host population. The most basic type of random sample is called the *simple random sample*, abbreviated *SRS*.

Definition 6a.-7

Simple random sample (SRS). A sampling method that permits every possible sample of size n from a population to have an *equal* chance of being selected.

An SRS can be obtained in a number of ways: mixing and selecting slips of paper in a hat, rolling dice, using the random number table (located at the back of this primer), or using a computer or calculator.

Example 6a.-3

Consider a population of 20 college students. We want to know the amount of money each student spent on textbooks for the semester. Because this population has only 20 individuals, we can easily find the mean; it turns out to be $\mu = \$260.25$. The standard deviation is $\sigma = \$50.83$.

What follows is the complete population data set of each student's expenses on textbooks for the semester; each student has been given an ID number.

Student ID#	01	02	03	04	05	06	07	08	09	10
Amount ($)	267	258	342	261	275	295	222	270	278	168

Student ID#	11	12	13	14	15	16	17	18	19	20
Amount ($)	319	263	265	262	333	184	231	159	230	323

(a) Use the table of random numbers to select an SRS of four students. Each student can be chosen for the sample at most once. Find the sample mean of their textbook expenses.

Choose several cells randomly from the table. Suppose these cells were selected:

11061	83674	34594	28056	85196

We will select two digits at a time, travelling left to right; if a set of digits is between 01 and 20, and doesn't repeat a previously chosen two-digit number, we will consider that student's ID# (and consequently that particular student) a part of the random sample. When four valid two-digit numbers have been chosen, our sample is complete.

11 = student; 06 = student; 18 = student; 36 = no student; 74 = no student; 34 = no student; 59 = no student; 42 = no student; 80 = no student; 56 = no student; 85 = no student; 19 = student.

Thus, our sample is of students 11, 06, 18, and 19. The sample mean of these four students' textbook expenses is $250.75.

(b) Use the graphing calculator to select an SRS of four students. Each student can be chosen for the sample at most once.

Use the randInt feature (found by pressing the $\boxed{\text{MATH}}$ key, then scrolling to the PRB menu). Querying for ten two-digit numbers from 1 to 20 should be enough to ensure no repeats: randInt(1,20,10)

Understanding the distinctions between populations and samples goes beyond simply identifying them, or finding out how they were collected; it also extends to assembling numerical information obtained from the population—called *parameters*—and the sample—called *statistics*. Let's formally define these terms.

Definition 6a.-8

Parameter. A number that originates from a population, excluding the population size N.

Definition 6a.-9

Statistic. A number that originates from a sample, excluding the sample size n.

Of course, noting that population and parameter start with the same letter (as do sample and statistic) helps us to remember which term goes with which.

Example 6a.-4

Identify each as a parameter or a statistic.

(a) In a study of all Harvard students, the mean height of all students at Harvard.

Parameter, since the number—the mean—comes from a study of *all* Harvard students.

(b) In a study of American college students, the mean height of all students at Harvard.

Statistic, because the population is now all college students, not just students at Harvard.

(c) In a study of American high school students, the proportion of high school students who have a driver's license.

Parameter.

(d) In a study of SAT scores, the median SAT score of students at Madison High School.

Statistic.

(e) The 67% of American college students who plan to vote in the next election.

Parameter, since the 67% is a number coming from the entire population of voters.

(f) The 2.4 children that the "average" U.S. family has.

Parameter.[*]

(g) The proportion of Native American students in Montgomery County, in a study about ethnic and racial composition in Maryland.

Statistic.

Example 6a.-5

A sample of 178 high school students is asked by an interviewer how many times per day they brush their teeth. Identify the parameter and the statistic in the study.

The statistic is the 2.7. The parameter is "how many times per day *all* high school students brush their teeth." Of course, we don't have the actual value of the parameter.

It's important to note that we rarely know a parameter's value—the best information we have usually lies with the statistic. The statistic estimates the parameter's value; as long as this *point estimate*, or this single sample value given by the statistic, was calculated off of a random sample, we can have confidence that the statistic targets the true population parameter value well.

---

[*] Replacement fertility rate, or the number of births per woman needed to maintain a population, is 2.1 for industrialized countries, and higher for developing countries. For instance, the U.S. is slightly below replacement, hovering at around 1.9. Japan's fertility rate, though, is significantly below 2.1; to understand how that's affecting their demographics, consider this: more adult diapers than children's diapers are sold there. For an intriguing thought experiment of what would happen to the world's industrialized societies if the fertility rate plummeted to zero, see the haunting film *Children of Men*, based on the book of the same name by P.D. James.

ȸ

Sometimes real-world sampling runs into problems: it is not always practical or feasible to obtain a "straightforward" SRS. Perhaps the population is too large, and a sampling frame cannot be readily constructed. Or maybe it is simply too expensive to randomly sample the entire population. There are alternatives that still preserve randomness but scrap the characteristic unique to a simple random sample: that every size sample n must have an equal chance of being selected from the host population.

The Nielsen ratings, which measure the popularity of television shows, have had to work around the SRS issue for many years. Begun by Arthur Nielsen, a market analyst, the Nielsens branched out from advertising and radio market analysis in the 1930s to television in the 1950s. Shows live and die based on their ratings, so the sampling methods utilized have had to be as representative as possible. A census of all television watching families is, of course, infeasible. So how does Nielsen obtain their data?

In two ways: viewer diaries and set meters. Viewer diaries are just that—diaries that television watchers keep over a set time period (usually a week or two). These diarists have been targeted randomly by Nielsen based on a host of demographic criteria.[*] According to their website, "Nielsen's TV families represent a cross-section of representative homes throughout the U.S."

Set meters are electronic monitors affixed to televisions in the homes of randomly selected families. Set meters can record viewing habits in real-time, as well as account for recorded shows (classified as "time-shifted viewing") and some online television-watching activity.

There are problems, though. First, not everyone selected for a Nielsen sample agrees to participate (nonresponse bias).

Second, people's viewing habits may change because they know that they are being observed; this is called the *Hawthorne effect*, named after the Hawthorne Works factory in which the phenomenon was first observed: in an experiment centering on productivity due to workspace light intensity, workers who were shown more attention (and light) performed better than those who weren't.

Third, bars, airports, and other public places aren't included in the sample, despite the many people watching community televisions at these sites.

Fourth, very few Americans, as a percentage of the total number of television viewers, participate in the surveys.

So, as the statisticians at the Nielsen Company certainly realize, randomly sampling a population is tricky business. Here is a listing of the most common types of random sampling methods, besides the SRS.

Definition 6a.-10

Sampling with replacement. After an observation is selected, it is placed back into the population so it can be selected again; people are not often surveyed by sampling with replacement.

---

[*] I myself was chosen to keep a Nielsen diary, back in 2002, right around the time the show *Felicity* was ending. Every couple of pages I pleaded for the show not to be canceled, to no avail.

Definition 6a.-11

Sampling without replacement. After an observation is selected, it is not placed back into the population; people are most often surveyed by sampling without replacement.

Definition 6a.-12

Stratified sampling. When observations in a population are grouped based on a common characteristic—these groupings are called *strata*—and an SRS is taken from each stratum; observations within strata should be as homogenous as possible.

Stratified sampling often arises when a researcher wishes to match the relative proportions of homogenous groups from a population in her sample. For instance, if population X contains 5% of stratum y, then, by using a stratified method, only 5% of the randomly collected data can come from stratum y—allowing for a proportionally representative sample that an equivalently large SRS likely wouldn't produce (e.g., an "unlucky," but appropriately conducted, SRS might contain no individuals from stratum y).

Definition 6a.-13

Systematic sampling. When observations in a population are arranged in some kind of order, beginning at some randomly selected starting point, every kth observation is selected to be part of the sample.

Definition 6a.-14

Cluster sampling. A population is divided into *clusters* that are similar in makeup to the population, only writ small; all observations from randomly selected clusters are selected to be part of the sample.

Definition 6a.-15

Multistage sampling. A random sampling method that proceeds in stages, combining other sampling methods together along the way.

Example 6a.-6

Determine what type of sampling method was used in each of the studies below. Some have more than one answer, so be sure to include all sampling methods that are relevant.

(a) People have lined up to buy tickets to a rock concert. A researcher decides to ask questions regarding such topics as age, smoking habits and income level to every fifth person in line. The starting person will be selected at random from the first five people in line.

This is systematic sampling (along with sampling without replacement).

(b) In the population of all undergraduate college students, a sociologist groups different graduation classes together—freshman, sophomores, juniors, and seniors—and conduct a survey on their drinking habits.

Stratified sampling.

(c) A researcher collects information from a large high school. She randomly selects five homerooms, and gathers data from every student in each of these homerooms.

Cluster sampling.

(d) Snapple decides to test every 20th bottle rolling off the factory's line. After a bottle is tested, it is disposed of.

Systematic sampling (and sampling without replacement).

Back in 1954, when Darrell Huff wrote his best-selling *How to Lie with Statistics*, the definition of a simple random sample wasn't quite what it is today. Here's how he put it: "The test of the [simple] random sample is this: Does every name [or individual] in the whole group have an equal chance to be in the sample?"

Since then, things have changed. It is no longer the case that an SRS is defined this way—as every individual in the population having an equal chance to be in the sample—since this sampling method isn't exclusive to an SRS.

Here's why. Consider a high school with 1,000 students: 500 males and 500 females. Let's take a stratified-by-gender sample from this school, of ten males and ten females, both randomly selected, for a combined sample of twenty students.

The chance of any particular male being in this sample is 1/50, and the chance of any particular female being in this sample is also 1/50. So *every single student in the population has an equal chance to be part of the sample*, yet this sample is *not* an SRS—it's a stratified sample.

The modern, revised definition of an SRS—that any particular subset of n students has an equal chance to be part of the sample—doesn't conform to this male-female example. To wit: the sample of twenty students, constructed according to the stratified sampling criteria detailed above, cannot have eleven males, or fifteen females, or no females: the sample must instead have precisely ten males and ten females. So not every subset of twenty students out of the 1,000 students, of which there are $\binom{1000}{20}$, can be selected, thereby disqualifying the sample from being considered an SRS.

Example 6a.-7

Suppose we wish to obtain a random sample of 70 Madison High School students. There are (let's assume) 70 classes being held during second period, so we have each individual class use a table of random numbers to randomly select one student. In total, 70 students—one randomly selected from each of the 70 classes—constitute the random sample. (Assume that every Madison High School student has a class during second period.)

(a) Is this random sample an SRS? Explain.

No, since in order to be classified as an SRS, every sample of size 70 has to have an equal chance of being selected. No class during second period can contribute more than one student to the sample.

(b) Does every single Madison student have an *equal* chance of being selected? Justify your answer.

No. For example, suppose Sam is in a class of ten students during second period. He has a greater chance of being selected for the sample than Zoe, who is in a different class of thirty students that class period.

Example 6a.-8

The Southfork Ranch, which appears as the backdrop for the long-running television drama series *Dallas*, encompassed over 200 acres and is abutted by a body of water on one side and a superhighway on the other. Suppose we can only afford to sample the soil of five randomly chosen acres. Would a simple random sample work best here? Or could we choose the acres in a more efficient way?

Solution. The problem with an SRS here is the possibility of getting an "unlucky" sample—a sample that although randomly selected, ends up clustering all five chosen acres far away from both the highway (whose proximity to land might affect the soil) and the water (likewise), for instance.

Therefore, some sort of stratification method that divides the ranch's property up into homogenous plots of land of, say, five strata of twenty contiguous acres each—a stratum of "adjacent to the highway," another stratum of "adjacent to the water," a third of "far away from both land and water," and two more somewhere physically in between—and then randomly sampling one acre from the twenty in each of the strata would result in a much more effective, and representative, sampling plan.

The Vietnam War—which consumed much of America's resources and many of the best of its citizens during the mid-1960s to the mid-'70s—was predicated on the idea of containing communist expansion in Southeast Asia. Because of a large demand, coupled with a limited supply, of troops needed to fight the war, the Selective Service System of the United States instituted a draft. The draft order was determined by a lottery.

Each of the days of the year, numbered from 1 to 365 (plus 366 for leap years), was written on a piece of paper and placed inside an opaque capsule. These capsules were then dumped into a large bowl, mixed up, and drawn successively. Suppose the capsule containing 2 was drawn first. Thus, all men of eligible draft age with a birthday of January 2 were called to serve first. The order of the capsules drawn determined the draft order by birthday.* And the lottery draft was shown on live television—it was the ultimate in reality TV before the decades-later official birth of the genre.

But in 1970, something went wrong. Look at a scatterplot of the draft order/rank (lower numbers = called up earlier) vs. birthday of draftees.

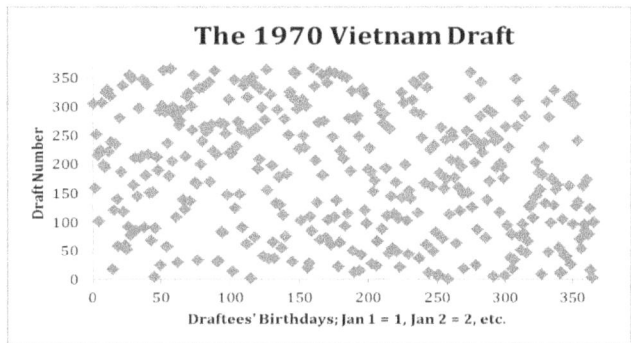

There seems to be two triangular clusters of data: one at the top left, the other at the bottom right, with a diagonal strip light on observations between them. The clusters reveal that if your birthday was early in the year—say, in January or February—you tended to have a high draft number, meaning you were less likely to actually be shipped out to Vietnam. But if your birthday was much later in the year, such as in November or December, you were likely to have a low draft number—and those with low draft numbers were called up to be part of those first (unlucky) waves of soldiers. As Norton Star, author of the article "Nonrandom Risk: The 1970 Draft Lottery" (in the *Journal of Statistics Education*), explains it, "[W]ith hindsight one can see how the attempt at randomization broke down. The capsules were put in a box month by month, January through December, and subsequent mixing efforts were insufficient to overcome this sequencing." When reaching into the transparent bowl, the later-in-the-year birthdays were up top. Statisticians almost immediately noticed, and efforts were made to correct the problem for the next year.

For comparison's sake, take a look at a scatterplot of the next year's war draft (shown at the top of the next page). The dots are much more evenly distributed than the previous year's because the capsules were thoroughly mixed.

---

* Unlike a professional sports draft, where being selected early on usually means millions of dollars for the lucky (and athletically talented) draftees, the eligible men in the war draft wanted nothing less than to be drafted early.

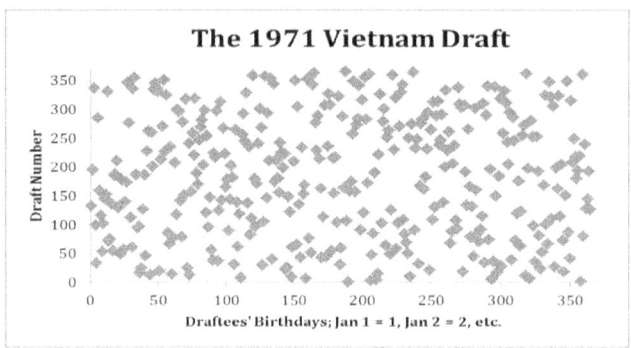

The 1971 Vietnam Draft

§6b. *Survey Design*

Besides the thorny issues of response bias, which accounts for respondents' hesitancy to truthfully answer questions of a sensitive nature (such as those centered on drugs, alcohol, hygiene, sexual practices, theft, and the like), the *wording of questions* is of vital importance—and can, in some cases, dramatically affect the responses.

Sometimes it's *wording ambiguity* that's the culprit. In 1993, the year that the Academy Award-winning film *Schindler's List* was released, there was renewed interest in learning about the Holocaust. Infamously, at around this time, the Roper Organization created a survey for the American Jewish Committee (AJC), of which one of its questions was "Do it seem possible or does it seem impossible to you that the Nazi extermination of the Jews never happened?" The question is confusing, replete with double negatives. Sixty-five percent responded that it (the Holocaust) is "impossible it never happened," while 22% responded that it is "possible it never happened." Was nearly a fourth of the American population Holocaust deniers?

In an effort to find out whether the responses were legitimate or muddled because of the question's wording, the Gallup Organization asked a similar question, but with a preface: "The term Holocaust usually refers to the killing of millions of Jews in Nazi death camps during World War II." Then, word for word, the same question as Roper's poll was asked. Respondents answered similarly, despite the additional information: 65% "impossible…never" and 33% "possible…never." Simultaneously, Gallup asked a different randomly chosen group of people a clearly worded question about the Holocaust, with the same preface: "The term Holocaust usually refers to the killing of millions of Jews in Nazi death camps during World War II. Do you doubt the Holocaust actually happened, or not?" This time, more than nine in ten expressed no doubt. (Roper would make amends, asking their own clearly worded version of the question several years later; again, around nine in ten were "certain" that the Holocaust occurred.)

The behavioral economists Daniel Kahneman and Amos Tversky conducted a number of interesting psychological experiments in the 1970s and '80s, some of which involved survey question wording. Try answering these questions:[*]

---

[*] The following questions were culled from Kahneman's book *Thinking, Fast and Slow*.

Which program would you prefer to combat a flu epidemic?
a. Program A would save two hundred people.
b. Program B would have a one-third chance of saving six hundred and a two-thirds chance of saving nobody.

***

You are shopping for a new camera. The ABC store and the XYZ store are one mile apart. You discover that the ABC store sells the same model as the XYZ store, but charges $200, which is $10 cheaper than XYZ. Would you drive the extra mile for the discount?

Now, suppose you are shopping for a new car. The ABC dealership and the XYZ dealership are one mile apart. You discover that the ABC dealership sells the same car as the XYZ dealership, but charges $30,800, which is $10 more than the XYZ dealership. Would you drive the extra mile for the discount?

Despite the arithmetic equivalence of answers (a) and (b) in the first question above, very few people tend to select choice (b).[*] As Todd G. Buchholz, in *New Ideas from Dead Economists*, explains,

> People like certainty. Behavioral economists have discovered that people hate to lose and sometimes feel paralyzed into standing in place. Stock market investors hate to sell their shares at a small loss, even when they are warned that big losses could come. They may become emotionally attached to their stocks, their homes, and their jobs.

As for the other two questions (about the camera and the car), notice that $10 is $10 in either case. But most people would drive the extra mile for the camera, but not the car. These questions threw into doubt the traditional "rational expectations" model of economics, helping to lay the foundation for the behavioral economists' take on the human condition.[†]

Even the opt-in, opt-out model of survey wording was studied. When obtaining your driver's license in the United States, you have to opt-in to become an organ donor. But wording makes all

---

[*] Kahneman and Tversky also asked people to interpret probabilities, most famously with the "Linda Problem": "Linda is 31 years old, single, outspoken, and very bright. She majored in philosophy. As a student, she was deeply concerned with issues of discrimination and social justice, and also participated in anti-nuclear demonstrations." So is Linda more likely to be (a) a bank teller, or (b) a bank teller active in the feminist movement? Although most respondents answered (b), (a) is more likely because of the *conjunction rule*: the probability of multiple events conjoined must be less than (or equal to) the probability of any single one of those conjoined events. An ability to deal properly with *base rates* (the populations of various groups) is critical to statistical numeracy.

[†] The behavioral economists, led by Kahneman and (the late) Tversky, through many creative experiments with sometimes unwitting subjects, arrived at a great number of so-called cognitive biases, such as the *sunk-cost fallacy*—when you believe you've invested too much in something to simply cut your losses—and the *availability heuristic*—when you evaluate situations based upon examples that come most readily to mind. Besides *Thinking, Fast and Slow*, the most thorough primer on these ideas is called *Predictably Irrational*, by Dan Ariely.

the difference. Two adjacent countries that have similar populations, Germany and Austria, also query people about becoming donors when distributing driver's licenses. Germany uses an opt-in system, while Austria uses opt-out. Only 12% agree to become organ donors in Germany, while a staggering 99% agree in Austria.

Another issue to consider when constructing (or examining) surveys is the idea that human memory is fallible, especially when it comes to time and recall; consider the *Rashomon effect*, named after the famous Japanese film *Rashomon*, where witnesses to the same crime or accident all relay different stories describing the event.

If a survey question reads, "How many times did you go to your doctor in the last ten years?" the respondent may be hard-pressed to recall. Human memory doesn't do particularly well with timeframes. There's even a name for this cognitive bias: the *telescoping effect*. People tend to think that some events, further in the past, happened more recently, and vice versa.

Even *question order* can affect responses. The Pew Research Center website notes that

> [w]hen determining the order of questions within the questionnaire, surveyors must be attentive to how questions early in a questionnaire may have unintended effects on how respondents answer subsequent questions. Researchers have demonstrated that the order in which questions are asked can influence how people respond; earlier questions—in particular those directly preceding other questions—can provide context for the questions that follow (these effects are called "order effects").[*]

Several Pew Research polls found these *order effects*. One, conducted in 2003, queried Americans about gay marriage versus civil unions; the results are summed up with the heading, "More People Favor Civil Unions When Asked After Gay Marriage." Another poll, conducted in 2008, asked two key questions back-to-back in different orders: "All in all, are you satisfied or dissatisfied with the way things are going in this country today?" and "Do you approve or disapprove of the way George W. Bush is handling his job as president?" When the Bush approval question was asked first, the percentage who expressed disapproval about the "way things are going" was much higher than when the question order was reversed, due to *verbal priming*.

Surveys can even be used to mislead. In his book *The Man Who Couldn't Stop: OCD and the True Story of a Life Lost in Thought*, David Adam describes a series of surveys that were distributed to Catholic friars and nuns by psychologists in 2002. Although the scientists wanted to investigate religious belief and obsessive symptoms, they were coy—"[t]hey said only that they were interested in how people think." Adam continues:

> Dozens of the nuns and friars came forward to help and the scientists sent them questionnaires to assess their personalities and to judge how obsessive-compulsive they were. They repeated the exercise with two other groups: citizens actively involved in church activities, and university students who said they had no interest in religion. The psychologists found that the friars and nuns, together with the regular churchgoers, were more likely to report thoughts and behaviours consistent with OCD.

---

[*] Found at http://www.people-press.org/methodology/questionnaire-design/question-order/

According to Sarah E. Igo's *The Averaged American: Surveys, Citizens, and the Making of a Mass Public*, survey research began in the U.S. between the world wars and was led by Elmo Roper and George Gallup, all in an effort to construct "rational citizens" in opposition to the totalitarianisms brewing across the Atlantic: "pollsters' democratic rhetoric relied on a notion of individuals able to speak and know their own minds." But the project very quickly turned into a great game of averages: the "scientific sampling [was employed] to better hear 'the man on the street' but instead created an averaged-out and abstracted public opinion that severed attitudes from their source."[*]

This concept of *l'homme moyen*, the "average man," though, has roots much earlier. Roughly one hundred years prior to Roper's and Gallup's first samples, mathematician Adolphe Quetelet compiled "vast numbers of statistics covering not only physical characteristics such as height and weight, but also 'moral characteristics' such as the propensities for individuals to commit crimes or to become drunk, [proposing] to be able to develop the idea of the representative individual in a given society at a given time… [to reveal] the regular laws of society, a 'social physics,'" as explained by Victor Katz.

Clearly we've only scratched the surface of survey history and survey design. There is much else—such as designing Likert items and scales, accounting for the *halo effect*, and writing unbiased questions—that is beyond the scope of this primer.

§6c. *Studies*

Surveys can be considered a type of *observational study*. Observational studies, while sometimes our only possible source of information, are not as useful as *experiments*. (Recall that John Stuart Mill realized that while observations are useful in establishing connections between variables, it is only with experiments that causation can be deduced.) But nothing's worse than an *anecdotal study*. Let's front-load as many definitions as possible.

Definition 6c.-1

Observational study. A method of the collection and examination of data obtained from individuals in which the imposition of treatments and/or controls is outside the purview of the researcher(s).

Definition 6c.-2

Experimental study. A method of the collection and examination of data obtained from individuals in which the imposition of treatments and/or controls is imposed by the researcher(s).

While it would be nice to always be able to construct experiments to test variables, in practice

---

[*] The preceding quotes from Igo's text were taken from Crawford's *The World Beyond Your Head*.

some experiments are either impractical or impossible to conduct. Determining safe levels of exposure to radioactive fallout after a nuclear explosion, the impact of foreign language study on SAT verbal scores among American high school students, and the dangers of pregnant women taking antidepressants to their future offspring are examples.

Definition 6c.-3

Anecdotal study. A method of the collection and examination of data that relies on anecdotes and is thus usually not rigorous enough to draw valid conclusions from.

Just because your friend tells you that a certain diet pill worked wonders for her waist doesn't necessarily mean it will work well for you—or others. Such anecdotal evidence is usually worthless.

But, to be fair, anecdotal evidence can sometimes give direction for future research; but, an anecdotal study in and of itself cannot withstand the scrutiny of the scientific method. As much as possible, we should avoid generalizing from a single individual or observation (except in the case of necessity, as in an "*N* of 1" design). Recall English economist Charles Davenant's admonition that conclusions "must not [be] argue[d] from single instances, but from a thorough view of many particulars." As the saying goes, the plural of anecdote is data.

Example 6c.-1

Suppose we wish to learn if a new diet pill is as robust as the drug maker claims. How could we go about testing the pill?

Solution. Perhaps we could randomly select people who have already (or are currently) taking the diet pill, and interview them, along with checking their weights periodically; we might also want to obtain a randomly selected sample of people who haven't (and aren't) taking the diet pill.

To obtain more control over the situation, though, we might instead construct an experiment. The brief outlines: have some people take the diet pill, others not; observe results over a set period of time.

Of course, we have to be much more rigorous and precise in the design of such an experiment. For example, we can't have just anyone participating—it makes most sense to select individuals who are *in need* of a diet pill. (Perhaps we could utilize BMI to screen our subjects.) Also, we probably won't have the luxury of random selection for the study; randomly selecting people from a population at large isn't only unethical, it's illegal. Instead, we'll have to rely on volunteers. [*] And, finally, we'll need strict ways to conduct, control, and measure the results of the ex-

---

[*] A persistent problem with using volunteers, though, is that it is difficult to draw conclusions that generalize to the host population. Also, college students frequently fill the ranks of volunteers in experimental studies, largely because of convenience, since researchers generally perform these studies in institutions adjacent to or part of college

periment—including making use of *placebos* and *double blindness*.

Definition 6c.-4

Comparison group. A group of individuals as similar as possible to those in the experimental group(s), except they do not receive the treatment(s); often called a *control group.*

Perhaps Francis Bacon was the first to make use of a control group; he conducted an experiment testing the efficacy of the germination of wheat seeds in various fluids as compared to no treatment at all.

Definition 6c.-5

Experimental units. The individuals in which treatments (or not) are applied in the study; if the experimental units are people, they are oftentimes instead referred to as *participants* or *subjects* (although animals in experiments are usually referred to as subjects as well; note that referring to human beings as subjects has generally fallen out of favor).

Definition 6c.-6

Placebo. An "empty drug" or sugar pill that looks, smells, tastes, feels, seems identical to the real treatment but contains no active therapeutic effects; note that a placebo does not have to take the form of a pill.

Although the most common form a placebo takes is a "fake" pill, placebos treatments come in many varieties. For instance, participants could be given false information in an experiment; or, when testing the efficacy of acupuncture, needles can be randomly placed into participants' skin rather than at traditional acupuncture points (acupoints). Placebos can even take the form of buttons—which are called, appropriately enough, placebo buttons—such as the "Close" button on most elevators, which usually only works when a key is inserted, or the no-longer-functional crosswalk signals at pedestrian crossings in major cities.

Not all studies are amenable to placebos, however. Take a study about the effectiveness of exercise: Can "placebo exercise" be constructed? Or an SAT test prep class: Can a "placebo test prep class" be made? Or even a study investigating the link between cigarette smoking and lung cancer: Could "placebo cigarettes" *really* fool anyone—even nonsmokers?[*]

---

campuses; in addition, such volunteers are usually paid for their efforts, incentivizing participation. But college students can't always be taken to represent the population at large (for instance, only a certain kind of college student would choose to participate in a study).

[*] Much more on the smoking-cancer link later on.

Definition 6c.-7

Placebo effect. The expectation of a beneficent effect, despite there being no active therapeutic treatment, results in a positive response felt by the participant.

Have you ever taken an over-the-counter pain reliever (such as ibuprofen) and, almost immediately, experienced pain relief?[*] Such medicine can't travel through the bloodstream that quickly; the lessened pain is due to the placebo effect. Although it's not clear how the placebo effect works—and it's difficult to construct experiments simply testing the effect itself—a complex mix of expectations and, perhaps, the body's release of endorphins drives the beneficent effect. But since only human beings have been shown to experience the placebo effect, placebos are not necessary with experiments involving animals or material objects.

Definition 6c.-8

Nocebo effect. The expectation of harmful effect, despite there being no active therapeutic treatment, results in a negative response felt by the participant.

Imagine taking a pill labeled "poison," despite the pill containing no active ingredient. Because of negative expectations, you might experience the nocebo effect through the psychosomatic expression of symptoms. (And you don't have to go to such melodramatic lengths as labeling a pill "poison"—people experience many symptoms from medications that have no basis in biochemistry.)

Definition 6c.-9

Blindness. When either the subjects or the researcher is not privy to which subjects receive which treatments; usually, though, blindness is taken to mean that the subjects are unaware.

Definition 6c.-10

Double blindness. When both the subjects and the researcher are not privy to which subjects receive which treatments.

If both the researcher and the participants aren't sure who's assigned to which experimental group, then how can an experiment be conducted? Perhaps a third researcher, one who will not participate in the implementation of the experiment or monitoring of the participants, has that information; alternatively, the assignments might be coded into a computer for later analysis.

---

[*] A recent study about pain relievers like ibuprofen showed that such anti-inflammatory drugs may help to relieve depression.

Though blindness or double blindness is not always possible or feasible—reconsider the experiments testing exercise's efficacy, the SAT test prep course, and smoking to see why—whenever possible, to reduce bias and strengthen the conclusions drawn from the study, blindness of some sort should be effected.

The American doctor Henry K. Beecher, in a 1955 paper titled "The Powerful Placebo," introduced placebos and double-blind study methods to a wide audience.

Definition 6c.-11

Explanatory variable(s). The independent variable(s) that may have an effect on the response variable(s); in an experiment, this is the variable(s) we wish to control for, and usually comes early in time.

Definition 6c.-12

Response variable(s). The dependent variable(s) that may be affected by the explanatory variable(s); in an experiment, this is the variable(s) we measure, and usually comes later in time.

A well-designed experiment will permit us to conclude a cause-and-effect relationship between the explanatory and response variables. Although many important conclusions can be drawn from well-designed observational studies, causation is not one of them.

Definition 6c.-13

Treatment(s). The different conditions of the explanatory variable(s).

Definition 6c.-14

Completely randomized design. An experiment in which participants are randomly assigned to the treatments; the randomization can be conducted with a physical object, such as a coin or die, or with a random number table, or via software on a computer or calculator.[*]

Definition 6c.-15

Lurking variable. An unmeasured or unmonitored variable in the "background" of a study.

---

[*] Statistician Ronald Fisher arrived at the idea of randomized experiments; epidemiologist Austin Bradford Hill came up with randomized clinical trials.

Definition 6c.-16

Confounding variable. A variable "tangled" up with the explanatory variable(s), with its effects on the response variable unable to be measured independently.

Although there are subtle distinctions between lurking and confounding variables, in practice we can consider them synonymous. Comparison groups, randomization, and blindness all conspire to reduce the impact of lurking and confounding variables as much as possible.

Example 6c.-2

Again, suppose we wish to learn if a new diet pill is robust. Design a completely randomized experiment that makes use of comparison and blindness.

Solution. Call for one hundred volunteers suffering from obesity. Using a coin, assign them randomly to one of two experimental groups: the treatment group (given the diet pill), and the control group (given a placebo). (The explanatory variable is "Type of Pill," the treatment conditions are active pill and placebo, and the response variable is "Weight.") Make sure that the groups contain fifty participants each; keep all participants in the dark as to which treatment condition they are receiving.

Over the course of the next six months, have all one hundred participants report for a weekly weighing, conducted by a researcher who has no knowledge whether the participants are taking the active pill or the placebo. Record participants' weights (and perhaps some other physiological factors) and analyze the results.

An *experimental design graphic*, as shown below, can map out this process.

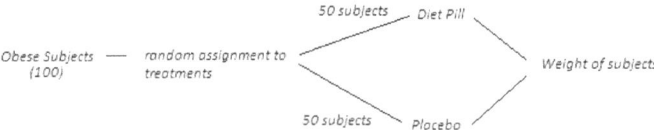

The graphic plots the experiment through time, left to right.

Example 6c.-3

At a conference for Advanced Placement courses at Madison High School, a presenter stated the following: "A recent study found that high school students who take at least one AP course are 35% more likely to have a higher GPA in college than those who don't. That's why getting our students to take AP courses is so important for their future success."

(a) The study the presenter referred to is either an observational study or a controlled experiment. Which is the study much more likely to be? Briefly explain.

Observational study. We're not "forcing" students to take AP courses.

(b) Who are the participants in this study?

High school students.

(c) What is the explanatory variable?

"Taken at least one AP course."

(d) What is the response variable?

"College GPA."

(e) Can you suggest an alternative explanation for why those who have taken at least one AP course are 35% more likely to have a higher GPA in college than those who don't—besides just the fact that they've taken at least one AP course?

Perhaps these students are smarter or more motivated to begin with.

(f) Look again at what the presenter said: "…That's why getting our students to take AP courses is so important for their future success." Explain what's wrong with the presenter's statement.

The statement implies cause and effect. If this really were the case, simply affix the initials "AP" in front of *every* course that students take in high school (AP Gym, AP Lunch, AP Study Hall)—which would consequently result in the GPA bump in college.

(g) Do you suppose that there's a cause-and-effect relationship between taking at least one AP course and doing well in college?

Perhaps. But causation can't be shown by a single observational study alone.

Example 6c.-4

A study is conducted to test a pill of 500 mg of amoxicillin orally twice daily to treat infections. In this study, one hundred people who have infections are randomly selected to participate. They are divided into two groups—those who take amoxicillin and those who take a placebo (sugar pill)—and, after three weeks of taking the medication daily, doctors examine if the people's infections have improved.

(a) Explain why is this an experiment and not an observational study.

Because researchers are imposing conditions on the participants.

(b) Who are the participants in this study?

People with infections.

(c) What are the explanatory and response variables?

"Pill type"; "Status of infection."

(d) What are the treatments?

Amoxicillin and placebo.

(e) Why should those in the control group be given a placebo instead of no pill at all?

To account for the placebo effect, so that improvement in people's infections can be safely attributed to the medicine rather than to an unintended psychological response.

(f) Why is double-blindness preferable here?

So the researcher won't bias the results of the study.

(g) Can a cause-and-effect relationship be shown in this study?

Yes, since the experiment is well-design, completely randomized, and has double blindness.

Example 6c.-5

Suppose we want to conduct an experiment to test whether a pill helps to lower high blood pressure, both with and without an exercise regime. You recruit eighty adult volunteers.

(a) Who are the participants in this experiment?

Adult volunteers with high blood pressure.

(b) What are the explanatory variables?

We are testing both the impact of "Pill Type" and "Exercise Type."

(c) Draw a graphic illustrating the experimental design.

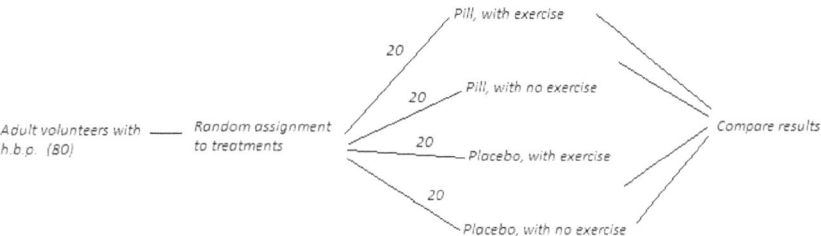

Notice that the four treatment groups are stocked with equal numbers of participants, all assigned randomly.

Perhaps researchers suspect that adults with high blood pressure respond differently to the pill because of their gender. In that case, a *randomized block design* would be more appropriate than a completely randomized design.

Definition 6c.-17

Randomized block design. An experiment in which experimental units are divided into homogenous groups called *blocks* prior to the random assignment of treatments; this design is analogous to stratified sampling, in that variability is reduced by the grouping of individuals: the maxim "similar within, but different between" applies.

Example 6c.-6

Again, suppose we want to conduct an experiment to test whether a pill helps to lower high blood pressure. You recruit eighty adult volunteers, half of which are male. Draw a graphic of the experimental design.

Solution. The graphic looks like this:

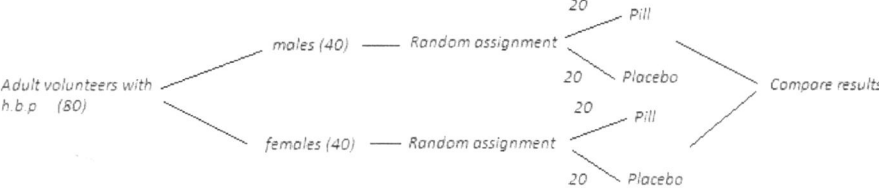

Of course, the random assignment comes *after* the blocking assignment, otherwise we'd be randomly assigning the volunteers a gender—which would be an entirely different sort of experiment.

Reducing the variability in an experiment helps to take chance variation as much out of the results as possible. The ultimate in reducing variation, though, is called a *matched pairs design.*

Definition 6c.-18

Matched pairs design. An experiment in which experimental units receive both treatments, usually in a randomized order (if possible; that way, *order effects* don't confound the results); each experimental unit, then, serves as his/her/its own control.

Example 6c.-7

Suppose we wish to test the claim that Coke tastes better than Pepsi. Describe how we could conduct a matched pairs experiment to test this claim; also draw a suitable graphic.

Solution. Let's gather sixty volunteers for this experiment. It doesn't make any sense to have thirty of them try Coke, the other thirty try Pepsi, and then "compare"—taste is very subjective.

Having each subject try *both* treatments gives everyone a comparison group: him- or herself. The Pepsi and Coke will be distributed to subjects in identical opaque paper cups. In addition, to ensure that the order of the trials isn't a lurking variable that may confound the results, the order will be randomized (with a coin, random number table, or computer/calculator).

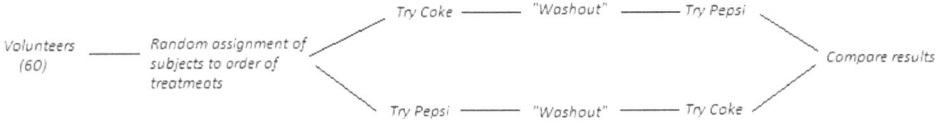

Notice that there is a "washout" period between the trials, meaning a slight pause and a glass of water to mix in one's mouth and spit out.

This experiment has blindness—the participants don't know which of the two treatments they're trying first—and could potentially have double blindness as well. The lurking variables of subjective taste and prior health and genetics and family background and…well, they're all safely eliminated due to the matched pairs design.

Matched pairs experiments can also be performed with pairs of identical twins. Such study designs are especially efficacious when examining nature-versus-nurture topics. For example, suppose one identical twin was raised from near-birth in a household that promoted reading and the classics, while the other was brought up in an environment that didn't place such a primacy on the love of literature. Since the twins are genetically identical, the confounding variable of heredity is eliminated; nurture, and family environment, then, is all that's left to explain their differences.

In a TED Talk from September of 2008, author and cognitive scientist Steven Pinker relays an especially interesting result of the Minnesota Twin Family Study,[*] in which many pairs of twins separated at birth were later found and reunited as adults, that casts serious doubt on a tabula rasa, all-nurture theory of parenting:

---

[*] Minnesota = Twin Cities, twin studies.

My favorite example is a pair of twins, one of whom was brought up as a Catholic in a Nazi family in Germany, the other brought up in a Jewish family in Trinidad. When they walked into the lab in Minnesota, they were wearing identical navy blue shirts with epaulettes; both of them liked to dip buttered toast in coffee, both of them kept rubber bands around their wrists, both of them flushed the toilet before using it as well as after, and both of them liked to surprise people by sneezing in crowded elevators to watch them jump. Now—the story might seem too good to be true, but when you administer batteries of psychological tests, you get the same results—namely, identical twins separated at birth show quite astonishing similarities.

Such identical twin studies can also get to the root of health variation between twins, since environmental factors, rather than genetic differences, seem the most likely explanation of the development of illnesses.[*]

With all experiments (and observational studies), however, especially those that involve human beings, we have to be especially careful. As a text like *Research Design* (by John W. Creswell) makes clear, there are many ethical issues related to data collection. Participants to research should never be put at risk; to that end, an Institutional Review Board (IRB) needs to review and protect against any violations of the rights of participants in studies, only approving those studies that conform rigorously to safety guidelines. In addition, an *informed consent form*, detailing the procedures of the experiment, the right to withdraw, and the confidentiality of individual results, must be obtained by all researchers before the collection of data. Although shaky guidelines for conducting studies in the past allowed researchers to both go too far and occasionally make tremendous breakthroughs in science,[†] careful attention to these details isn't optional.

For instance, if we wish to investigate the effects on fetuses of pregnant mothers taking antidepressants, we can't simply randomly divide pregnant mothers into two groups, giving one group antidepressants and the other group placebos. An IRB, designated with the mission of protecting against human rights violations, would never approve such a study.

On the other hand, it is important for health practitioners, as well as pregnant women, to know if mothers are putting their fetuses at risk by taking antidepressants during gestation. A *retrospective study* provides us with a way to an answer.

Definition 6c.-19

Retrospective study design. A *post hoc* study, in which the causes of events that have already occurred are deduced.

So the only viable approach to the antidepressant-pregnancy study is for researchers to examine what happened to babies *already* born of mothers on antidepressants. Unsurprisingly, many kinds of medical research have to be conducted retrospectively.

---

[*] Many more such studies of human beings—and the creative designs to test them—are detailed in *Forty Studies that Changed Psychology: Explorations into the History of Psychological Research* by Roger R. Hock.

[†] Especially in the history medicine; read the book *Open Heart: The Radical Surgeons who Revolutionized Medicine* by David Cooper for a taste of just how far those limits were pushed.

William Halsted, a surgeon who practiced in the late nineteenth century, is best known for the radical mastectomy—radical, in this case, not meaning novel but "root," as in removing not only the breasts but also everything around them, including chest muscle and lymph nodes and other parts of the neck. Halsted had concluded that in the advent of breast cancer diagnoses, simple mastectomies weren't sufficient to prevent cancer from creeping back at the "margins."

But retrospective studies, five to ten years out (if those patients even survived), revealed something unexpected, though Halsted never really accepted the studies' clear-cut conclusions. As Siddhartha Mukherjee details in *The Emperor of All Maladies: A Biography of Cancer*,

> In the summer of 1907, Halsted presented more data to the American Surgical Association in Washington, D.C. He divided his patients into three groups based on whether the cancer had spread before surgery to lymph nodes in the axilla or the neck. When he put up his survival tables, a pattern became apparent. Of the sixty patients with no cancer-afflicted nodes in the axilla or the neck, the substantial number of forty-five had been cured of breast cancer at five years. Of the forty patients with such nodes, only three had survived.
>
> The ultimate survival from breast cancer, in short, had little to do with how extensively a surgeon operated on the breast; it depended on how extensively the cancer had spread before surgery.

The retrospective study relies on what's already occurred; the *prospective study* takes the opposite approach.

Definition 6c.-20

Prospective study design. A "beforehand" study, in which set groups of individuals are observed and followed through time.

Definition 6c.-21

Longitudinal study. In effect, a type of prospective study, with an additional requirement that the same group of individuals (called a "cohort") is measured repeatedly and followed through time (perhaps for decades).

Example 6c.-8

An unethical researcher wishes to conduct an experiment to determine if smoking two packs of cigarettes a day causes lung cancer.

(a) If the researcher forces everyone in the treatment group to smoke two packs per day, what would the control group have to do to preserve blindness?
Smoke "fake" or "placebo" cigarettes.

(b) List some problems with conducting this experiment.

Besides the fact that it would never be approved by an IRB because it is potentially very harmful to its participants: (1) a placebo cigarette seems impossible to make; (2) subjects would have to be tracked for many decades; and (3) the incredible number of hereditary and environmental factors would pile up over the years, leaving any conclusions drawn at the end of such a long and problematic study suspect at best anyway.

It is with smoking that we press up against the absolute limits of experimental design and research. If correlational studies are all that's available, but only with well-designed completely randomized experiments can we demonstrate cause-and-effect relationships, how can smoking be shown to *cause* cancer?

"The same correlation could be drawn to the intake of milk." So said U.S. surgeon general Leonard Scheele in 1948, when dismissing a potential causative link between smoking and lung cancer. By then, smoking had become so ubiquitous, cigarettes so commonplace, that a many-fold increase in the number of lung cancer cases caused few to bat an eyelash toward America's (and the world's) all-consuming nicotine habit.

The epidemiologist Austin Bradford Hill and medical researcher Richard Doll, as well as the medical student Ernst Wynder and pulmonary surgeon Evarts Graham, were the first to perform serious, controlled studies examining a possible smoking-cancer connection. Wydner and Graham started by simply asking patients with lung cancer and those without their smoking history, looking to see if a higher proportion of people in the cancer group were smokers (this is called a *case-control study*). Hill and Doll did the same, gathering a multitude of interview data. The scientific community was mostly skeptical, believing that cause and effect could only be shown with infectious diseases like malaria. But the results were consistent: cigarette smoking was highly correlated with lung cancer.

The studies themselves, though, did not prove causation, since the results were examined *post hoc*—after the events in question had played out. Furthermore, the interviewees might not remember things correctly, or the interview questions themselves might be biased.

Here is where Hill, who invented the concept of the randomized experiment, had an idea: What if a group of people who don't already smoke (and don't already have cancer) were divided into two groups: those who will smoke (for a period of many years), and those who won't. Then researchers measure the frequency of lung cancer in each group—and can thus make an assessment about causation. Besides the fact that a placebo cannot be made—fake cigarettes to smoke, anyone?—this experimental design violates the "First, do no harm" tenet of the Hippocratic Oath, and is simply not feasible in practice regardless.

Instead, a prospective study, where changes through time could be observed, was needed. Such a longitudinal study would have to follow volunteers for many decades; eventually, members of the cohort—which had both smokers and non-smokers—would either get lung cancer or not. From that, a relative risk could be found that might answer this question: If you've smoked, how many times more likely are you to develop lung cancer than if you didn't smoke?

Doll and Hill mailed out questionnaires to thousands of doctors, who made up the cohort. After less than half a decade, the results were overwhelming: of those doctors who died of lung cancer, all were smokers. In addition, at around the same time, Wydner and Graham were able to subject mice to experiments with the ingredients in cigarettes, finding adverse effects on the animals.

Hill then said that for causation to be shown epidemiologically, at least these criteria needed to be met (now called the Bradford Hill criteria for causation): The association needed to be (1) strong; (2) consistent through time; (3) specific; (4) realistic; (5) responded to similarly in similar situations; and (6) backed up by experiments. With any particular association, the more criteria that are met, the more evidence there is in favor of causation, even sans direct controlled experiment.

Ronald Fisher, the statistician who arrived at the fundamentals of experimental design, believed that smoking and lung cancer were merely associated (he had a selfish reason: he loved smoking his pipe). He believe that there was a *publication bias*—that studies that demonstrated some sort of effect from smoking tended to be published, while those that didn't likely were rejected. Fisher also thought that a genetic predisposition might explain the link between smoking and cancer: those who were more likely to smoke were also more likely to get lung cancer.[*] Essentially, no matter what evidence was thrown at his brilliant mind, he stood firm: without experimental evidence, establishing cause and effect was impossible.

Jerry Cornfield, a contemporary of Fisher's and also a statistician, disagreed. He examined many—a hundred or more—retrospective observational studies of smokers. (Studies of studies, by the way, are termed *meta-studies*.) The evidence that lung cancer caused smoking was not only clear-cut and overwhelming, but also fit the Bradford Hill criteria for causation to a T.

Fisher died of cancer in 1962. Not too long after, the U.S. Surgeon General agreed that smoking causes cancer, and labeled cigarette packages accordingly.

§6d. *The Trouble with Experiments*

Newspaper articles that begin sentences with a sober "Studies show…" offer up a gloss of scientific certitude in a multifactorial world. Not all such "studies," though, are created equal. Even experimental studies, the most preferred kind by scientists, have their problems. Experiments, quite simply, are not always easy to conduct—or construct. Take food and dieting, for instance. In *The Gluten Lie: And Other Myths about What You Eat*, author Alan Levinovitz observes that

> [h]igh-quality studies of dietary practices are incredibly hard to design. How do you make a placebo piece of steak for your control group? Studies on the effect of diet and lifestyle in large populations are no less difficult. They depend on recollection and self-reporting, notoriously unreliable data. And even if the data were accurate—well, just tweak an equation, exclude a set of data points [this is called *cherry-picking*], isolate a different factor, and suddenly vegetarianism goes from increasing longevity to decreasing bone density.

Even prayer has been the subject of experimental scrutiny, with controversial results. Early

---

[*] This is a very strange argument to wrap your mind around. Fisher is rather counterintuitively claiming that there is some third variable, some lurking factor, that's causing *both* the inclination to smoke *and* the lung cancer. His mind spun fantastic tales to justify his bad habits.

last decade, Herbert Benson, cardiologist and director of the Mind/Body Medical Institute in Boston, performed a large blinded study on intercessory prayer by tracking a group of patients undergoing heart surgery and having separate groups of people pray for them to heal well after the procedures. The results? Those who were prayed for had approximately the same number of complications after surgery as those who were not prayed for. But, as Richard Sloan, a professor of medicine at Columbia, explains, "The problem with studying religion scientifically is that you do violence to the phenomenon by reducing it to basic elements that can be quantified, and that makes for bad science and bad religion."

An article in the *Atlantic* titled "Lies Damned Lies and Medical Science" by David H. Freedman profiles the meta-researcher John Ioannidis, who has "become one of the world's foremost experts on the credibility of medical research." According to Ioannidis, much medical research is "misleading exaggerated, and often flat-out wrong."

The low-hanging fruit, such as Ignaz Semmelweis's discovery of washing hands with chlorinated lime solutions in the mid-1800s or Florence Nightingale's assemblage of mortality data during the Crimean war, is long gone. Instead, many factors need to be considered when conducting clinical trials and other medical experiments. The reversals of conclusions from continuing research, though, are staggering in number. Freedman summarizes:

> "Randomized controlled trials," which compare how one group responds to a treatment against how an identical group fares without the treatment, had long been considered nearly unshakable evidence, but they, too, ended up being wrong some of the time. "I realized even our gold-standard research had a lot of problems," he says. Baffled, he started looking for the specific ways in which studies were going wrong. And before long he discovered that the range of errors being committed was astonishing: from what questions researchers posed, to how they set up the studies, to which patients they recruited for the studies, to which measurements they took, to how they analyzed the data, to how they presented their results, to how particular studies came to be published in medical journals.

Many health effects these studies show might even be "flukes," mere random noise amongst a blizzard of variables. But wrong ideas spread like an "epidemic," and, despite the best of intentions, hapless consumers and patients are left none the wiser.

§7. *From One, Many*

◁▷

Introduction to Sampling Distributions. Sampling Distribution of the Sample Proportion. Sampling Distribution of the Sample Mean.

§7a. *Introduction to Sampling Distributions*

We have summarized data sets using a variety of graphical displays: boxplots, histograms, stemplots, and the like. We've discussed the basics of probability (which underpin so much of statistics) including discrete probability distributions and continuous probability distributions. In addition, we have also worked with the most important continuous probability distribution: the normal distribution.

In this § we will consider *sampling distributions*, which serve as a segue for the two most important ideas of the primer: *confidence intervals* and *hypothesis testing*. Before defining sampling distributions, though, let's review a list of symbols necessary in order to work with them (the first four of which should already be familiar).

- The sample statistic for the mean is denoted by $\bar{x}$
- The population parameter for the mean is denoted by μ
- The sample statistic for the standard deviation is denoted by s or s_x
- The population parameter for the standard deviation is denoted by σ
- The sample statistic for the proportion is denoted by p
- The population parameter for the proportion is denoted by $\hat{p}$ (pronounced *p*-hat)

Note that it is *almost* always the case—with the exception of the population proportion—that Greek letters denote parameters. But there is variation in the symbol used for the population proportion. Although we will only use p in this primer, other textbooks and journal articles may use θ or π, keeping the pattern of Greek letters for population characteristics fully intact.

Example 7a.-1

Identify each of the following as a parameter or statistic, determining the symbol used to represent its value as well.

(a) The proportion of people in Madison High School with blue eyes.

Parameter: p

(b) The proportion of people in a single class in Madison High School with blue eyes.

Statistic: $\hat{p}$

(c) The average lifespan of U.S. adults.

Parameter: μ

(d) The mean age of students in a single classroom.

Statistic: $\bar{x}$

(e) The standard deviation of the number of teeth Pennsylvania residents have had filled.

Parameter: σ

(f) The standard deviation of the number of pencils in a small supply store at the local mall.

Statistic: s

Suppose that, instead of taking one random sample of size n from the population—as we examined last §, with our discussions of multifarious sampling techniques—we take *many* random samples of size n from the population. Specifically, suppose we take *every possible unique sample of size n* from the population of interest. Then we have arrived at a sampling distribution:

Definition 7a.-1

Sampling distribution. The distribution of the values when every possible unique sample of size n is gathered from a host population.

Technically, we haven't obtained a sampling distribution if we simply have "many" samples of the same size from the same population—we need to have *all* possible unique samples.

That presents a logistical problem, however, when the number of possible unique samples is very large. Consider the following examples.

Example 7a.-2

Suppose there are four people in our population—A, B, C, and D—and we wish to obtain a sampling distribution of size two. Calculate the number of unique samples, and list the samples out.

Solution. To find the number of samples, we need to evaluate $\binom{4}{2}$, which is equal to 6. The possible unique samples from this population of four people are AB, AC, AD, BC, BD, and CD.

Example 7a.-3

Suppose that there are ten people in our population—A, B, C, and so forth—and we wish to obtain a sampling distribution of size five. Calculate the number of unique samples in the distribution.

Solution. We need to find $\binom{10}{5}$, which is equal to 252 possible unique samples.

Typically, if there are an excessive number of unique samples from a host population, we will simply take "many" samples to get a rough idea of the shape, center, and spread of the sampling distribution.

Also, instead of simply listing out the unique samples in a sampling distribution, the mean or proportion (or, sometimes, variance) of *each* sample is found, and then aggregated.

Definition 7a.-2

Sampling distribution of the sample mean. The distribution of the sample means from each collected sample, when every possible unique sample of size n is gathered from a host population.

Definition 7a.-3

Sampling distribution of the sample proportion. The distribution of the sample proportions from each collected sample, when every possible unique sample of size n is gathered from a host population.

Definition 7a.-4

Sampling distribution of the sample variance. The distribution of the sample variances from each collected sample, when every possible unique sample of size n is gathered from a host population.

Example 7a.-4

Again, suppose there are four people in our population—A, B, C, and D—and we wish to obtain a sampling distribution of size two for the sample mean of these people's heights. Their heights are as follows: A is 70'', B is 65'', C is 72'', and D is 69''. Calculate all of the sample means, and then compare the mean of these sample means to the population mean.

Solution. The mean of sample AB is 67.5''; the mean of sample AC is 71''; the mean of sample AD is 69.5''; the mean of sample BC is 68.5''; the mean of sample BD is 67''; and the mean of sample CD is 70.5''.

The mean of this exhaustive set of sample means is 69''; this is the mean of the sampling distribution of the sample mean. The population mean—found by taking the arithmetic average of all four people's heights—is also 69''.

The mean of the means of all unique samples from a population will equal the mean of all the individuals in the population. Let's see if this same property holds with sample proportions.

Example 7a.-5

Again, suppose there are four people in our population—A, B, C, and D—and we wish to obtain a sampling distribution of size two for the sample proportion of males. A, B, and C are male, while D is female. Calculate all of the sample proportions, and then compare the mean of these sample proportions to the population proportion.

Solution. The proportion of males in sample AB is 1; the proportion of males in sample AC is 1; the proportion of males in sample AD is 0.5; the proportion of males in sample BC is 1; the proportion of males in sample BD is 0.5; and the proportion of males in sample CD is 0.5.

The mean of this exhaustive set of sample proportions of males is 0.75; this is the mean of the sampling distribution of the sample proportion. The population proportion of males is also 3/4 = 0.75.

Thus, the same property holds with sample proportions: the mean of all of the sample proportions—in other words, the mean of the sampling distribution of the sample proportion—is equivalent to the population proportion. (The property holds for variances as well.)

When the sample statistics *target* the population parameters, as they do here for the mean, proportion, and variance, we term them *unbiased estimators*.

Definition 7a.-5

Unbiased estimator. A statistic with the mean of its sampling distribution equal to the associated parameter value.

Not all sample statistics are unbiased estimators, however. Consider the range.

Example 7a.-6

Again, suppose there are four people in our population—A, B, C, and D—and we wish to obtain a sampling distribution of size two for the sample range of heights. Recall their heights: A is 70'', B is 65'', C is 72'', and D is 69''. Calculate all of the sample ranges, and then compare the mean of these sample ranges to the population range.

Solution. The range of sample AB is 5''; the range of sample AC is 2''; the range of sample AD is 1''; the range of sample BC is 7''; the range of sample BD is 4''; and the range of sample CD is 3''.

The mean of this exhaustive set of sample ranges is 3.667''; this is the mean of the sampling distribution of the sample range. But the population range is equal to 7''.

The large difference between the mean of the sampling distribution of the range and the population range demonstrates that the range is a *biased estimator*, since it does not target the associated parameter closely.

Definition 7a.-6

Biased estimator. A statistic with the mean of its sampling distribution not equal to the associated parameter value.

As you may suspect, range is not the only statistic that doesn't target the associated parameter well. Other biased estimators include the minimum and the maximum (which shouldn't be surprising, since range is calculated off of the minimum and maximum), the median, and the standard deviation.

Now that we have worked through examples with small populations, let's deal with a much larger population in which "many"[*] samples, rather than every possible sample, will be the rule of the day. Examining visual representations of these many samples, using the graphing calculator, will be useful.

Example 7a.-7

In order to illustrate different kinds of sampling distributions, first consider the discrete probabil-

---

[*] I'll stop with the scare quotes around the word *many* as long as you continue to keep this in mind: the set of samples we'll be collecting from large populations won't be exhaustive, so, technically, even thought we might call the distributions "sampling distributions," they really won't be so.

ity distribution of the random variable X of rolling a fair die.

(a) Fill in the probability distribution table below listing the events for random variable X, and the probability of each of these events for $P(X)$.

x	1	2	3	4	5	6
$P(x)$	1/6	1/6	1/6	1/6	1/6	1/6

(b) Using the discrete probability distribution, find the mean (or expected value), variance, and standard deviation of the random variable.

$$\mu_X = E(X) = \frac{1}{6}(1+2+3+4+5+6) = 3.5$$

$$\sigma_X^2 = \frac{1}{6}(1^2+2^2+3^2+4^2+5^2+6^2)-3.5^2 = 2.917$$

$$\sigma_X = \sqrt{2.917} = 1.708$$

Next, we will use the graphing calculator to conduct a simulation of rolling a fair die 5 times. Go to your home screen and $\boxed{\text{CLEAR}}$ it. Then press the $\boxed{\text{MATH}}$ key, scroll to the PRB menu, and select the RandInt(function. Your expression should be: RandInt(1,6,5). This is asking the calculator to select a random number from 1 to 6 five times—the equivalent of rolling a six-sided fair die five times. After pressing $\boxed{\text{ENTER}}$, five random rolls should appear on your screen. (Note: to ensure numbers closer to true randomness, you may wish to first pick your own initial random number and store that random number into the variable rand, located in the PRB menu—this is called a random number seed.)

But this is *not* a sampling distribution. A sampling distribution of the mean in this experiment, for instance, would consist of the distribution of the mean number of rolls of all unique samples of sample size 5 of rolling a die.

(c) How many different possible samples of rolling a die 5 times are there?

$$6^5 = 7,776$$

That's a lot of samples! The calculator cannot be expected to take *all* such samples and display them; but we can take *enough* samples—that is, many samples—to accurately simulate a sampling distribution using the calculator.

Here's how to do so. First, open up the seq(function, which found by typing $\boxed{\text{2}^{\text{nd}}}$, $\boxed{\text{LIST}}$, and then scrolling to the OPS menu. Next, find the mean(function, which is found by pressing $\boxed{\text{2}^{\text{nd}}}$,

LIST, and then scrolling to the MATH menu. Next, type RandInt(1,6,5), as you did before. After closing the parentheses, press the X key, comma, 1, comma, 200. Finally, store this expression into a new list called ROLL. Look at the leftmost screenshot below to see what should appear on your calculator screen. When you are satisfied it does, press the ENTER key and wait about two minutes for a list of numbers to appear underneath.*

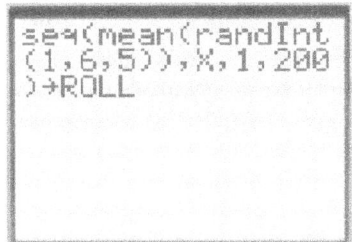 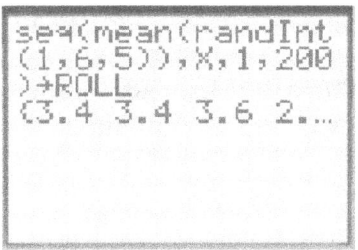 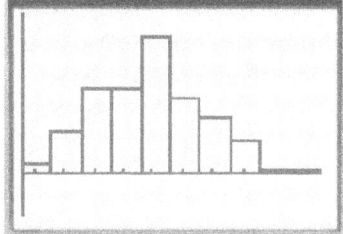

The top center screenshot shows the first several numbers in the list created when running the simulation. Of course, *your* list will probably contain different numbers, since each calculator is "rolling" a die randomly (or at least pseudorandomly).

(d) What do each of the numbers in your ROLL list represent? Be very specific.

The mean of the values of the 5 randomly generated die rolls.

(e) Using the 1-Var-Stats function, find the sample mean of your ROLL list. Compare this sample mean with the mean of the random variable X that you calculated in part (b) above. What can you say about how well the sampling distribution of means approximates the theoretical (population) mean of all dice?

Although your answer may vary slightly since *all* possible unique samples weren't generated by the calculator, the value should be around $\bar{x} = 3.48$, which is very close to the population mean, $\mu_X = 3.5$.

(f) Would you say that the sample means target the value of the population mean?

Yes. Although $\bar{x}$ likely will not exactly equal μ_X—since the calculator didn't generate *all possible samples* of rolls from the population—the values should be very close.

(g) View a histogram of the ROLL list. (It should closely resemble the histogram in the rightmost screenshot above.) How would you classify the general shape of this histo-

---

* Note: you may have an seq menu that opens up instead. To get the seq(function to appear on your home screen from the seq menu, type the number 1 in for every option on the menu, then highlight and press ENTER overtop of the Paste command. From there, replace all of the 1's with the correct commands, as in the leftmost screenshot on this page shown above.

gram?

The histogram looks relatively symmetrical and mound-shaped.

We will find that the sampling distribution of the sample means tends to be a normal distribution, becoming closer and closer to normal as the sample size increases.

Now, let's run through the same process again, only this time with a sampling distribution of the variances of the simulated die. Go back to your home screen and type this in: seq(variance(randInt(1,6,5)),X,1,200)→ROLL. The variance(function is located right below where the mean(function is. After waiting several minutes for the new ROLL list to appear, address the following questions:

(h) What does each of the numbers in your ROLL list represent? Again, be very specific.

The variance of the values of the 5 randomly generated die rolls.

(i) Using the 1-Var-Stats function, find the sample mean of your ROLL list. Compare this sample mean with the variance of the random variable X that you calculated in part (b) above. What can you say about how well the sampling distribution of variances approximates the theoretical (population) variance of all dice?

Although your answer may vary slightly since *all* possible unique samples weren't generated by the calculator, the value should be close to about $\bar{x} = 2.843$, which is very close to the population mean, $\sigma_X^2 = 2.917$.

(j) Would you say that the sample variances target the value of the population variance?

Yes. Although $\bar{x}$ likely will not exactly equal σ_X^2—since the calculator didn't generate *all possible samples* of rolls from the population—the values should be very close.

(k) View a histogram of the ROLL list. How would you classify the general shape of this histogram?

Skewed to the right.

As for sampling distributions of proportions, we could view a histogram of a sampling procedure in which we rolled a die five times and found the proportion of odd numbers (and completed this procedure 200 times). We would discover that the histogram would resemble a normal distribution, centered on the value of the population proportion (which is 0.5 in this case).

(l) Would you say that the sample mean, the sample variance, and the sample proportion are unbiased estimators—that is, that they target the value of the population parameter well?

Yes, since the expected value of their respective sampling distributions (theoretically) equals the mean of the associated parameters. ▮

Given a choice of a variety of sample statistics with which to estimate a population parameter, try to choose ones that have the least bias—i.e., that target the value of the parameter as closely as possible—as well as low variability—i.e., that are as minimally spread out as possible.

Because (1) the sample mean and the sample proportion are unbiased estimators of the population mean and proportion, and (2) sampling distributions of means and proportions tend to resemble normal distributions (and we have some experience working with normal distributions in prior §'s), going forward we will focus exclusively on two sampling distributions: that of proportions and of means.

§7b. *Sampling Distribution of the Sample Proportion*

The *Central Limit Theorem*, or CLT, will allow us to synthesize our knowledge of probability to help predict the center, spread, and shape of sampling distributions of all stripes. First, though, we'll need to connect the sampling distribution of the proportion to the binomial distribution.

Recall that the mean and standard deviation of a binomial random variable X were given by

$$\mu_X = np \text{ and } \sigma_X = \sqrt{np(1-p)}$$

Now, the value of $\hat{p} = \dfrac{x}{n}$, since $\hat{p}$, the sample proportion, represents some ratio of successes to the total number of observations in a sample. That ratio can be rewritten equivalently as $\hat{p} = \left(\dfrac{1}{n}\right)x$, so we are really multiplying the random variable X by a constant—$1/n$—to get a new random variable of $\hat{p}$. Thus, the new mean for the random variable of $\hat{p}$ is

$$\mu_{\hat{p}} = \frac{1}{n}np = p$$

The standard deviation derivation is slightly more complicated to obtain. See below:

$$\sigma_{\hat{p}} = \frac{1}{n}\sqrt{np(1-p)} = \sqrt{\frac{np(1-p)}{n^2}} = \sqrt{\frac{p(1-p)}{n}}$$

So we have the mean and standard deviation of the binomial random variable of $\hat{p}$. The normal distribution therefore can be used to approximate this binomial distribution if the following technical conditions (also called assumptions or requirements) hold:

$$np \geq 10 \text{ and } n(1-p) \geq 10, \text{ provided the "10% Condition" is met}$$

Recall that the 10% Condition states that, when sampling without replacement, the probability formulas assuming independence can be utilized so long as the percentage of population data sampled doesn't exceed 10%.

Laplace was the first to prove the Central Limit Theorem. Many years later, Galton expressed the results of the CLT colorfully when he said, "Whenever a large sample of chaotic elements are taken in hand and marshalled in the order of their magnitude, an unsuspected and most beautiful form of regularity proves to have been latent all along...." If the Greeks had known of the CLT, Galton continued, they would have "personified" and "deified" it.

Example 7b.-1

A bag of 8 oz. frozen vegetables has 100 vegetables. Suppose the factory claims that 25% of all vegetables packaged into the bags are frozen peas.

(a) What symbol would be used to represent the true population proportion of frozen peas?

p

(b) Suppose we bought a bag of frozen vegetables, and found that 28 were frozen peas.

a. In this sample of 100 vegetables, what proportion are frozen peas?

$28/100 = 0.28$

b. Is this proportion a parameter or a statistic?

A statistic, since it comes from a sample.

c. What symbol would be used to represent this proportion?

$\hat{p} = 28/100 = 0.28$

d. Suppose another bag of frozen vegetables is purchased. What's the best guess we could make for the proportion of frozen peas contained inside this new bag?

0.25, since that's the true population proportion of frozen peas.

(c) What does the symbol $\mu_{\hat{p}}$ refer to in context?

The mean of the sampling distribution of the proportion of the frozen peas from the population of all frozen vegetable bags.

(d) What does the symbol $\sigma_{\hat{p}}$ refer to in context?

The standard deviation of the sampling distribution of the proportion of the frozen peas from the population of all frozen vegetable bags.

(e) Is the 10% condition justified in this problem situation? Explain.

Yes. Each bag contains 100 vegetables—so 100 is our sample size. And 100 is surely much, much less than 10% of the population of *all* frozen vegetables produced at the factory.

Suppose we buy 700 bags' worth of frozen vegetables. Notice that this is the same as taking 700 simple random samples of 100 frozen vegetables each.

(f) Will all 700 bags (or samples) have the same proportion of frozen peas in them? Why or why not?

No. In this random process, there is sampling variability bag to bag.

(g) Find the mean of the sampling distribution of the proportion.

Imagine that we rip open each of the 700 bags, and calculate the proportion of frozen peas in each bag (proportions varying due to sampling variability):

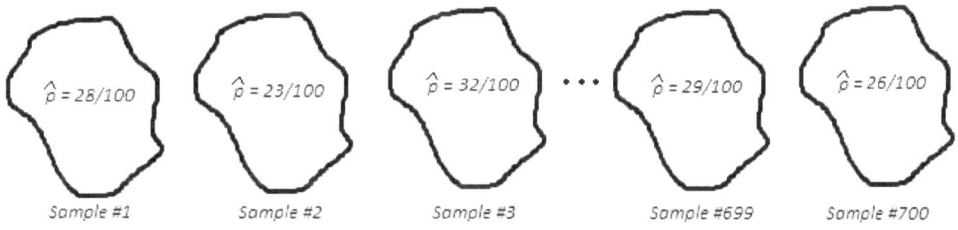

| Sample #1 | Sample #2 | Sample #3 | Sample #699 | Sample #700 |

Then, we take the sample proportion $\hat{p}$ of all 700 bags, add them together and divide by 700, finding the arithmetic mean. This mean should be very close to the theoretical mean—which is 0.25. Expressed properly using symbols, $\mu_{\hat{p}} = 0.25$.

(h) Find the standard deviation of the sampling distribution of the proportion.

Likewise, if we found the standard deviation of all 700 of the sample proportions, the result should be very close to $\sigma_{\hat{p}} = \sqrt{\dfrac{0.25(1-0.25)}{100}} = 0.043$, the theoretical standard deviation of this sampling distribution.

(i) Does the sampling distribution resemble a normal curve? Check, using the technical conditions.

$100 \cdot 0.25 \geq 10$ and $100 \cdot (1-0.25) \geq 10$. Since both are satisfied, the normal distribution can safely be used here to approximate the binomial distribution.

(j) What if we had had a sample size of 20 instead of 100 (in other words, what if each bag contained 20 vegetables instead of 100). Would the sampling distribution resemble a normal curve?

Since $20 \cdot 0.25 = 5 < 10$, we do not know if the sampling distribution is normal in shape.

(k) What is the absolute smallest sample size of frozen vegetables for which we could safely assume that the sampling distribution resembles a normal distribution?

Solve for n: $n \cdot 0.25 \geq 10$, $n \geq \dfrac{10}{0.25} = 40$. Note that if p had been greater than 0.5, we would have solved for n with $n(1-p) \geq 10$ instead, since *both* technical conditions need to be satisfied to assume the sampling distribution is normal in shape.

(l) Back to our bags of 100 vegetables each. Recall the percentages of the Empirical Rule: 68, 95, and 99.7. How many frozen vegetable bags (or samples) would you expect to be one, two, and three standard deviations away from the mean?

Because the technical conditions have been satisfied (thus indicating normality), we can use the Empirical Rule to estimate the number of bags (or samples) one, two, and three standard deviations from the mean.

- One standard deviation from the mean: $700 \cdot 0.68 = 476$

- Two standard deviations from the mean: $700 \cdot 0.95 = 665$

- Three standard deviations from the mean: $700 \cdot 0.997 = 698$, or nearly all samples

(m) Draw a sketch of the sampling distribution, labeling values one, two, and three standard deviations from the mean.

Using the mean and standard deviation calculated in parts (g) and (h), respectively, we construct the following sketch:

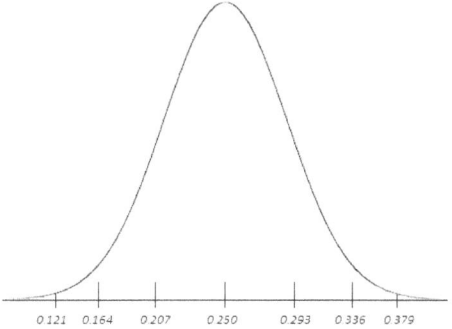

0.121   0.164   0.207   0.250   0.293   0.336   0.379

We are justified in using the normal distribution because the conditions for normality have been met.

(n) Which sample proportion is more likely to be obtained from the frozen vegetables' population: a $\hat{p} = 0.20$ from an SRS of $n = 20$, or a $\hat{p} = 0.20$ from an SRS of $n = 100$?

The bigger the sample size, the less the sample summary measures will deviate from the population summary measures—as long as the sample data were collected randomly. Thus, since $p = 0.25 \neq 0.20$, the smaller sample size of $n = 20$ is more likely to have the sample proportion $\hat{p} = 0.20$.

(o) Suppose people are dissatisfied with the proportion of peas in each frozen vegetables bag. A new packing procedure at the factory is claimed to increase the proportion of peas to 0.30. You buy a single bag of 100 vegetables and find that the proportion of frozen peas is higher, at $\hat{p} = 0.29$. Does this higher sample proportion lead you to believe that the factory has really increased the proportion of peas in each bag, or is the increased sample proportion in the bag you've just bought more likely the result of mere sampling variability?

Let's work through this question assuming that the factory did not, in fact, change their procedures. We need to see how likely or unlikely a value of the sample proportion $\hat{p} = 0.29$ is based on mere chance alone. We'll make use of a z-score to do so.

$$z = \frac{0.29 - 0.25}{0.043} = 0.930$$

Generally we consider any value more than two standard deviations away from the mean an "unusual value." Since $\hat{p} = 0.29$ is not even one standard deviation from the mean, we do not have evidence that the factory's process has changed; the deviated sample value is more likely the result of sampling variability in the packing of the vegetable bags.

Example 7b.-2

Write "True" or "False" next to each statement, and justify your answer.

(a) If $np \geq 10$ and $n(1-p) \geq 10$, then a sampling distribution of $\hat{p}$ is approximately <u>normal</u> with $\mu_{\hat{p}} = p$ and $\sigma_{\hat{p}} = \sqrt{\dfrac{p(1-p)}{n}}$.

This is true. For sampling distributions of the proportion, this is the central crux of the CLT.

(b) It is unrealistic to assume that we can obtain p.

Also a true statement. In the previous example, with the frozen vegetables, p was given as 0.25—but that was an artificial situation. Normally, we won't have the value of p with which to complete calculations, except in special situations (since, after all, to obtain a parameter usually requires a census of the population).

(c) Even if $np < 10$ and/or $n(1-p) < 10$, it is *still* indeed the case that both $\mu_{\hat{p}} = p$ and $\sigma_{\hat{p}} = \sqrt{\dfrac{p(1-p)}{n}}$, but we cannot assume that the sampling distribution of $\hat{p}$ is approximately normal.

This is true. With the binomial distribution formulas, we can always find the mean and standard deviation of the random variable of $\hat{p}$, but there are additional requirements with respect to sample size for the shape to be considered approximately normal.

(d) If we take many SRSs (simple random samples) of size 1 from the same population, the sampling distribution will be *effectively identical* in shape to the population distribution.

This is true as well. Think of it this way: many random samples of size 1 from the same population is effectively equivalent to taking one large random sample. And that large random sample will be very visually similar to the host population.

(e) A sampling distribution of $\hat{p}$ with SRSs of $n = 30$ has <u>more</u> variability than a sampling distribution of $\hat{p}$ with SRSs of $n = 15$.

Finally, a false statement! The bigger the sample size, the *less* the variation (and variability) around the mean.

Example 7b.-3

What is the *smallest* sample size n for which we could assume that the sampling distribution of the proportion is approximately normal?

Solution. When $p = 0.5$, it doesn't matter which of the two inequalities we solve for n for—thus minimizing n:

$$n \cdot 0.5 \geq 10$$

$$n \geq \frac{10}{0.5} = 20$$

Another way to arrive at the answer is to use algebraic substitution (recall that with two equations and two unknowns, both unknowns can be found—if there are not an infinite number of solutions, or no solutions, for them):

Distribute the left side of $n(1 - p) \geq 10$: $n - np \geq 10$

Substitute in $np \geq 10$ into the result: $n - 10 \geq 10$

Solve for n: $n \geq 20$

The preceding example brings us to the notion of the *skew* of sampling distributions. The smaller the sample size, the closer the sampling distribution visually appears to resembling the host population distribution. If the true value of the population proportion p is 0.5, then the population distribution exhibits no skew whatsoever; thus, the *smallest possible* sample size—a sample size of 20, as we've just calculated—is needed to "push" the sampling distribution toward resembling a normal shape.

But as the population proportion p moves away from 0.5, the host population becomes skewed. The closer p is to 0, the more skewed right the population is—since the mass of the data lies on the left (closer to 0) side of a graphical display—and vice versa. And the more skewed the host population is, the bigger the sample size needed to "overcome" that skewness to achieve normality.

For instance, if $p = 0.1$, meaning the population is very right-skewed, the minimum sample size needed to assume the sampling distribution is normal in shape is

$$n \cdot 0.1 \geq 10$$

$$n \geq \frac{10}{0.1} = 100$$

And if $p = 0.01$, indicating an extremely skewed right population distribution, an even bigger minimum sample size is needed to ensure normality: 1,000.

We've already hinted at it, but it should seem obvious by now that we can calculate probabilities of obtaining certain SRSs off of sampling distributions that resemble normal distributions—by simply finding z-scores and then looking them up on the normal distribution table.

Example 7b.-4

It is known that 28% of U.S. 16-to-18 year olds watch *The Daily Show*.

(a) What symbol would represent the 28%?

$p = 0.28$

(b) Suppose a random sample of fifty 16-to-18 year olds is taken. Will the sample proportion $\hat{p}$ of those who watch the show be 0.28?

Not necessarily, since there is sampling variability.

Suppose many samples of size $n = 50$ are taken from the population.

(c) What's the mean of these samples?

$\mu_{\hat{p}} = 0.28$

(d) What's the standard deviation of these samples?

$$\sigma_{\hat{p}} = \sqrt{\frac{0.28(1-0.28)}{50}} = 0.063$$

(e) What's the chance than less than 27% of fifty 16-to-18 year olds watch the show? Find $P(\hat{p} < 0.27)$.

First, we need to see if a normal approximation to the sampling distribution is appropriate here:

Since $50 \cdot 0.28 \geq 10$ and $50 \cdot (1-0.28) \geq 10$, the sampling distribution resembles a normal distribution.

Next, we'll find a z-score and look the value up on the normal distribution table:

$$z = \frac{0.27 - 0.28}{0.063} = -0.16,\ 0.4364$$

(f) What's the probability that of fifty teens more than 32% watch the show? Find $P(\hat{p} > 0.32)$.

$$z = \frac{0.32 - 0.28}{0.063} = 0.63,\ 1 - 0.7357 = 0.0537$$

Example 7b.-5

What is the probability that a fair coin comes up tails more than 65 times in 100 flips?

Solution. Since the sample size is sufficiently large to assume a normal distribution shape to the sampling distribution (check the technical conditions to see), we can immediately work with a *z*-score:

$$z = \frac{0.65 - 0.5}{\sqrt{\dfrac{0.5 \cdot (1 - 0.5)}{100}}} = 3$$

Looking up the probability gives us:

$$1 - 0.9987 = 0.0013$$

So there is only a 0.13% chance that you'd see more than 65 tails in 100 flips of a fair coin.

§7c. *Sampling Distribution of the Sample Mean*

Instead of finding the sample proportion from each SRS of a sampling distribution, we'll now turn to the sample mean.

Example 7c.-1

Reconsider the population of 20 college students. We want to know the amount of money each student spent on textbooks for the semester. Because this population has only 20 individuals, we can easily find the mean; it turns out to be $\mu = \$260.25$. The standard deviation is $\sigma = \$50.83$.

What follows is the complete population data set of each student's expenses on textbooks for the semester; each student has been given an ID number.

Student ID#	01	02	03	04	05	06	07	08	09	10
Amount ($)	267	258	342	261	275	295	222	270	278	168

Student ID#	11	12	13	14	15	16	17	18	19	20
Amount ($)	319	263	265	262	333	184	231	159	230	323

(a) Create a dotplot of the population of textbook expenses; describe the shape of the distribution.

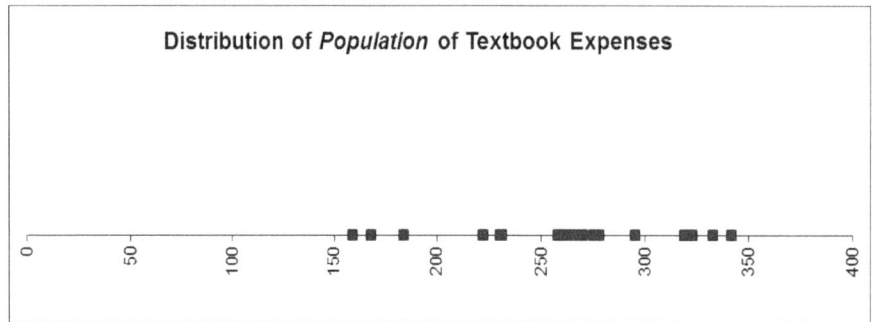

The shape of the distribution, obtained with Excel, is somewhat left-skewed.

(b) Pull random samples of size 11 (with replacement), finding the mean of each sample of students' textbook expenses. Do this process 300 times, for a total of 300 SRSs. Plot these sample means on a dotplot.

Using the RANDBETWEEN(1,20) function in Excel to select 11 random students (based on their ID numbers) in one cell apiece, and then the VLOOKUP function to pair the ID number with the associated textbook expense (as well as the ROUND function to eliminate any decimal dollar amounts), the following dotplot of 300 samples was obtained:

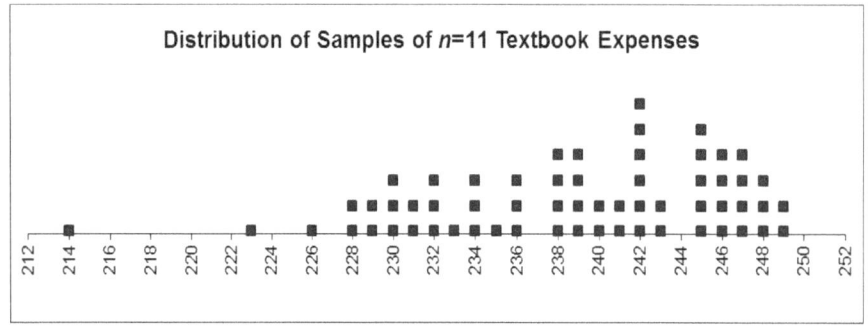

(c) Pull random samples of size 22 (with replacement), finding the mean of each sample of students' textbook expenses. Do this process 300 times, for a total of 300 SRSs. Plot these sample means on a dotplot.

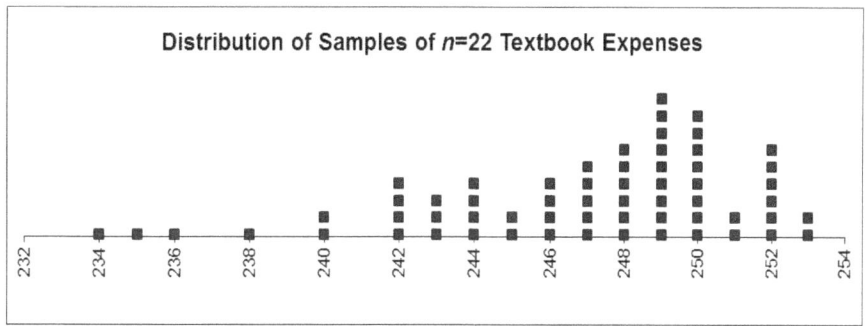

(d) Comment on your findings.

Generally, as the sample size of each SRS increases, the distribution becomes "smoother" and, theoretically, should begin resemble a normal distribution; in addition, and more noticeably, the distribution of ever-increasing sample sizes is less and less spread out, as compared to the population. While the population distribution must include *all* of the data points—even the extreme values—the sets of small samples need not have *all* of the extremes present in every sample. Hence, the distributions of samples, which are not as "stretched out" by the extremes, are less spread out than the population distribution.

With sampling distributions of sample proportions, the technical conditions for normality call for ten "successes" and ten "failures"; with sampling distributions of sample means, the condition is a lot simpler to check:

*Either the sample size is **at least 30**, or the host population is normal.*

But we still need to arrive at formulas for calculating the mean and standard deviation of the sampling distribution of the sample mean.

Let X_1, X_2, ... , X_n be n independently drawn observations from a distribution. This distribution has a mean of μ and a variance of σ^2.

Now, we want the *sample mean* of these independent observations:

$$\overline{X} = \frac{X_1 + X_2 + \cdots + X_n}{n}$$

That's the same as asking for

$$E(\overline{X}) = E\left(\frac{X_1 + X_2 + \cdots + X_n}{n}\right)$$

But we can remove the constant $1/n$ out of the parentheses:

$$E(\overline{X}) = \frac{1}{n} E(X_1 + X_2 + \cdots + X_n)$$

Which is the same as the sum of the expected value of each independently drawn observation:

$$E(\overline{X}) = \frac{1}{n} (E(X_1) + E(X_2) + \cdots + E(X_n))$$

But the expected value of each independently drawn observation is equal to μ.

$$E(\overline{X}) = \frac{1}{n} (\mu + \mu + \cdots \mu)$$

$$E(\overline{X}) = \frac{1}{n} (n\mu)$$

$$E(\overline{X}) = \mu$$

Thus, we can now claim that the mean of the sampling distribution of the sample mean is $\mu_{\bar{x}} = \mu$.

To find the *sample standard deviation* of these observations, consider that

$$\mathrm{var}(\overline{X}) = \mathrm{var}\left(\frac{X_1 + X_2 + \cdots + X_n}{n} \right)$$

We can remove the $1/n$ constant, but we have to square the constant when doing so.

$$\mathrm{var}(\overline{X}) = \frac{1}{n^2} \mathrm{var}(X_1 + X_2 + \cdots + X_n)$$

Since the terms are independent, we can say that

$$\mathrm{var}(\overline{X}) = \frac{1}{n^2} (\mathrm{var}(X_1) + \mathrm{var}(X_2) + \cdots \mathrm{var}(X_n))$$

But the variance of each independently drawn observation is equal to σ^2.

$$\mathrm{var}(\overline{X}) = \frac{1}{n^2} (\sigma^2 + \sigma^2 + \cdots + \sigma^2)$$

$$\text{var}(\overline{X}) = \frac{1}{n^2}(n\sigma^2)$$

$$\text{var}(\overline{X}) = \frac{\sigma^2}{n}$$

And because the standard deviation is the square root of the variance, we can claim that the standard deviation of the sampling distribution of the mean is $\sigma_{\bar{x}} = \dfrac{\sigma}{\sqrt{n}}$, as long as either the observations are independently drawn, or we're making the assumption that they're *effectively* independently drawn because the 10% Condition is satisfied (where we can assume independence even when there's sampling without replacement, since the sample size is relatively small compared to the host population size).

Example 7c.-2

A toy company makes lots of different types of toys. It randomly packages 30 of these toys in a box. One such box from the factory is selected. In it, the mean weight of its thirty toys is 5 pounds, with a standard deviation of 1 pound.

(a) Is the "5 pounds" in the problem setup a parameter or a statistic? Why?

A statistic, because it's the arithmetic average from a sample.

(b) Is it the case that any box selected from the factory has its mean toy weight equal to 5 pounds? Explain.

No—sampling variability.

(c) Define, in words, the parameter we wish to find in the problem.

The mean weight of *all* toys in the factory.

(d) What symbol would the parameter be represented by?

μ

(e) If $\mu = 4$ pounds, does the sample result above make sense?

Perhaps. The 4 pounds seems close to the sample mean weight of 5 pounds, although we're not given a population standard deviation so we don't know how close. (Although our best guess for the population standard deviation should be 1—the sample standard deviation.)

(f) If $\mu = 9$ pounds, does the sample result above make sense?

Probably not.

We take 600 SRSs (simple random samples) of boxes, which have 30 toys each in them. This is the same as taking 600 SRSs of size 30 each of toys. The parameters (the true population values) for all toys are $\mu = 4.5$ and $\sigma = 0.5$.

(g) What is the mean of the sampling distribution of the sample mean?

$\mu_{\bar{x}} = 4.5$

(h) What is the standard deviation of the sampling distribution of the sample mean?

$\sigma_{\bar{x}} = \dfrac{0.5}{\sqrt{30}} = 0.09$

(i) Are the technical conditions for sampling distributions of the sample mean satisfied?

The sample size is at least 30, and the sample size is also certainly less than 10% of the population size.

(j) How many of the 600 boxes are one, two, and three standard deviations away from the mean?

Because the technical conditions have been satisfied, we can use the Empirical Rule to estimate the number of boxes (or samples) one, two, and three standard deviations from the mean.

- One standard deviation from the mean: $600 \cdot 0.68 = 408$

- Two standard deviations from the mean: $600 \cdot 0.95 = 570$

- Three standard deviations from the mean: $600 \cdot 0.997 = 599$, or nearly all samples

(k) Draw the sampling distribution of $\bar{x}$, labeling the values of one, two, and three standard deviations from the mean.

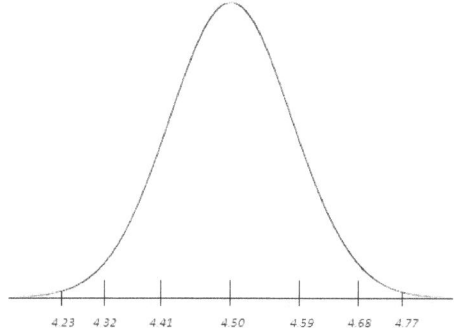

4.23 4.32 4.41   4.50   4.59   4.68 4.77

(l) If there were 40 toys per box instead of 30, what would happen to the sampling distribution?

The distribution would be taller and less spread out.

(m) What if there had been 20 toys per box instead of 30?

The distribution wouldn't necessarily resemble a normal distribution; it depends on what the population distribution looks like. Regardless, though, the distribution would be more spread out with a sample size of 20 because of the increased variability.

Like with sampling distributions of sample proportions, with sample means we can calculate probabilities of obtaining certain SRSs off of sampling distributions that resemble normal distributions—by simply finding z-scores and looking them up on the normal distribution table.

Example 7c.-3

The average height of American males is 70 inches, with a standard deviation of 2 inches. Assume that heights are normally distributed.

(a) Find the probability that a randomly selected male is greater than 71 inches tall. In other words, find $P(x > 71)$.

$1 - 0.6915 = 0.3085$

(b) Find $P(x > 73)$.

$1 - 0.9332 = 0.0668$

(c) Find $P(71 < x < 73)$.

$0.9332 - 0.6915 = 0.2417$

Now suppose that 30 males are selected at random.

    (d) Find the probability that the mean height of these 30 males exceeds 71 inches. In other words, find $P(\bar{x} > 71)$.

$$z = \frac{71-70}{2/\sqrt{30}} = 2.74,\ 1-0.9969 = 0.0031$$

    (e) Find $P(\bar{x} > 73)$.

$$z = \frac{73-70}{2/\sqrt{30}} = 8.23,\ 1-1 = 0^+,\ \text{or effectively zero}$$

    (f) Find $P(71 < \bar{x} < 73)$.

$$1-0.9969 = 0.0031$$

    (g) Explain why your answers to questions (a) and (d), (b) and (e), and (c) and (f) are so different.

    The average height of 30 random selected men has much less variability about the population mean than the height of a single randomly selected man.

    (h) Could you find the probabilities asked for in questions (a) to (c) above if n had been less than 30?

    Yes, since the population of heights was stated as normal, probabilities can be found no matter what n is used.

    All these ideas about sampling distributions approaching a normal distribution can be bundled up into one term: the Central Limit Theorem (CLT). The CLT tells us that for a population with any distribution, the distribution of the sample means (and proportions) approaches a normal distribution as the sample size increases. Recall the requirements for normality for a sampling distribution of the sample mean: the sample size must be at least 30 if the population from which the samples were taken was not normal, or the sample size can be anything if the population from which the samples were taken was normal. And the requirements for normality for a sampling distribution of sample proportions are $np \geq 10$ and $n(1-p) \geq 10$. Don't forget to always check the 10% condition to ensure the effective independence of selection.

    In general, we can state the following about any sampling distribution (of either the proportion or of the mean):

- If you have a sampling distribution of samples of size $n = 1$, the sampling distribution will closely resemble the population distribution.
- As your sample size n increases, the sampling distribution comes closer and closer to resembling a normal curve (this is the basic premise of the Central Limit Theorem).
- If your population is already normal, your sampling distribution will be normal no matter the n.

As long as we have the value of the population proportion p or the population mean μ with which to calculate the mean and standard deviation of the sampling distribution, probability calculations are a piece of cake.

But what happens if we don't have or can't obtain the population parameters p or μ? This will be addressed in the next §.

Part 3

⚬

The Axis of Interpretation

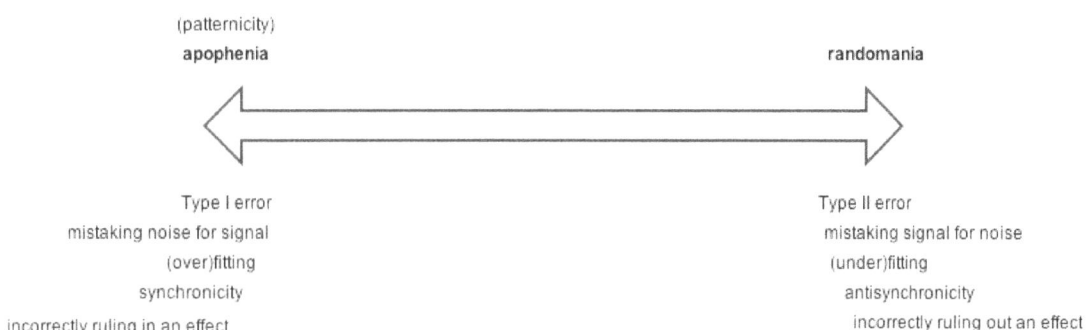

(patternicity)
apophenia   randomania

Type I error   Type II error
mistaking noise for signal   mistaking signal for noise
(over)fitting   (under)fitting
synchronicity   antisynchronicity
incorrectly ruling in an effect   incorrectly ruling out an effect

§8. *Not Lacking in Confidence*

∞

Introduction to Confidence Intervals. Confidence Intervals of Population Proportions. Confidence Intervals of Population Means.

§8a. *Introduction to Confidence Intervals*

In *Proofiness: How You're Being Fooled by the Numbers*, author Charles Seife relays a silly but telling anecdote about a natural history museum guide who, when asked by a teenager how old a particular Tyrannosaurus rex skeleton is, says, "Sixty-five million and thirty-eight years old."

Incredulous, the teenager asks, "How could you possibly know that?"

"Simple! On the very first day that I started working at the museum, I asked a scientist the very same question. He told me that the skeleton was sixty-five million years old. That was thirty-eight years ago."

When we trust a measurement beyond the point which it should be trusted, like the aging museum guide does,[*] we commit what Seife calls *disestimation*: "Disestimation is an act of taking a number too literally, understating or ignoring the uncertainties around it." Carl Wunderlich, the German physician who first measured body temperature by claiming to have collected a million data points for his calculation, although, as Siefe notes, he may have constructed his measurements out of whole cloth—and settled on the nicely rounded number of 37 degrees Celsius (98.6 degrees Fahrenheit)—or, at the very least, made body temperature readings from different points of the body, affecting his results greatly. Not only does temperature vary depending on where it's taken (mouth, arm pit, etc.), it also varies throughout the day as a result of circadian rhythm. Yet the seeming preciseness of the 98.6 degree figure has maintained a stranglehold on medicine for well over a century.

Who perpetuates disestimation most often today? The media, especially during election time. When reporting election polling—even from reputable polling firms, like Gallup or Zogby—the margin of error is frequently not mentioned, and the calculation of *confidence levels* are almost never discussed. To understand polling and other types of estimation, we need to unpack the

---

[*] And like the American public did during the McCarthy hearings, as discussed in the "Foundations" §.

mathematically multifarious *confidence intervals*.

If we want to estimate a parameter with a statistic—whether the statistic is an unbiased estimator (the ideal) or not—it seems we're in effect shooting in the dark.

Reconsider the set of textbooks expenses from the previous §. The true population mean—the parameter—of these book prices was $\mu = \$260.25$. Not knowing that, though, suppose we gather an SRS of five students and find the mean of their textbook expenses:

Student #5: $275; Student #9: $278; Student #12: $263; Student #18: $159; Student #20: $323

The sample mean of their expenses is $\bar{x} = \$259.60$

Despite the sample mean $\bar{x}$ being an unbiased estimator of the population mean μ, as we demonstrated in the previous §, there's no guarantee that $\bar{x}$ is sufficiently "close"—or the same—as μ. A single summary measure obtained from a population could be equal to any value due to variability.

Constructing sampling distributions, or at least many samples of some size n from the same population, was of the ways we found to counteract this variability; although a single random sample's summary measure might be close to or far from the parameter value, the aggregate of many SRSs likely cluster near to or exactly at the value of the parameter.[*] In the previous §, whenever we tested this idea—such as with the textbook expenses' example—we had the value of the targeted parameter at hand. But, when working with real-world data, we'll almost never have the parameter accessible: it might be infeasible or even impossible to obtain. Therefore, we'll have to have confidence that our methods of obtaining sampling distributions, with a few minor tweaks, will estimate the unknown parameters of interest well.

But, to cover any inherent error we'll encounter, let's move away from using a *point estimate* for a population parameter.

Definition 8a.-1

Point estimate. A statistic used to estimate a parameter.

A point estimate of the textbook expenses, for instance, is $\bar{x} = \$259.60$. But instead of relying on a single value to target some unknown true value (in the form of a parameter), we'll use an *interval estimate* to capture a range of values in which the parameter could conceivably reside.

---

[*] If an estimator is unbiased, and we construct a true sampling distribution—i.e., obtaining every possible sample of size n from the host population—then the aggregate of the samples' measures will target the parameter exactly.

Definition 8a.-2

Interval estimate. A range of values, built off of an estimate from a statistic and a margin of error, that potentially capture the parameter.

We can never be completely sure that the interval constructed off of sample data captures the associated parameter, unless a census of the population is conducted to find the parameter value outright.[*] There is, of course, uncertainty associated with the accuracy of an interval. We can, however, have a degree of *confidence* that the interval contains the parameter—how much confidence, however, depends on how, precisely, the interval estimate is calculated. This type of interval estimate is called, unsurprisingly, a *confidence interval*, named by its originator, mathematician Jerzy Neyman, back in 1934—although the first such interval estimates were made hundreds of years before by Nicolaus Bernoulli, nephew of Jacob, while adding to and completing his uncle's papers; later Abraham de Moivre refined the process when he fully described the normal distribution.

Definition 8a.-3

Confidence interval. An interval calculated from sample data; it takes the form *estimate* $\pm$ margin of error .

The degree of "confidence" used in an interval's calculation is called the *confidence level*.

Definition 8a.-4

Confidence level. A measure of the percentage of all possible samples that include the population parameter.

In order to fully unpack the ideas of confidence intervals, we'll need to try some examples. Although confidence intervals can be constructed off of any population parameter, the focus here will be on sample proportions first, and then on sample means.

§8b. *Confidence Intervals of Population Proportions*

When a researcher obtains data most easily accessible to him or her, recall that this is termed convenience sampling; the sample is not random, so we probably can't generalize these sample results to learn about the population from which the sample was

---

[*] There is actually one other way to be sure, though it's a bit of a cheat; we'll briefly highlight the method in subsequent pages.

taken.

Nonetheless, I recently gathered data from a college statistics class I teach; the lesson was on introducing confidence intervals, and my young charges were asked to respond to this question: *Are you an introvert or an extrovert?*[*] Eighteen out of thirty of them said that they were introverts. Despite this being a nonrandom convenience sample, let's work through some basic confidence interval of proportion ideas making use of this data.

Example 8b.-1

(a) How would you symbolically represent the proportion of introverts in the class?

$$\hat{p} = \frac{18}{30} = 0.6$$

(b) Is this sample proportion of introverts a point estimate?

Yes, since it is a value obtained from a sample that estimates a parameter. Since the value was obtained in a nonrandom manner, though, the point estimate likely does not target the true population value.

(c) Calculate the *standard error*.

First, notice that we cannot calculate a standard deviation here. Recall the formula for the standard deviation of a sampling distribution of the proportion: $\sqrt{\frac{p(1-p)}{n}}$. We don't have p, the true population proportion of introverts, since that's the value we're estimating using the sample proportion.

Since the sample proportion is a point estimate (albeit a poor one here because of the convenience sampling) of the population parameter, let's use that value in the calculation instead. Thus, our new formula is $\sqrt{\frac{\hat{p}(1-\hat{p})}{n}}$. What seems like a bit of hand waving allows us to press forward. But we cannot call $\sqrt{\frac{\hat{p}(1-\hat{p})}{n}}$ the "standard deviation"; because there's error inherent in using a statistic in place of a parameter, we'll instead call the formula the "standard error":

---

[*] These terms were defined by me, perhaps simplistically, as follows: an introvert gains energy when he is alone, and loses it when around other people; an extrovert gains energy when around others but loses it when alone.

$$SE = \sqrt{\frac{\hat{p}(1-\hat{p})}{n}}$$

$$SE = \sqrt{\frac{0.6(1-0.6)}{30}} = 0.089$$

(d) Construct a *95% confidence interval* for p, the true proportion of students who are introverts.

In order to construct a 95% confidence interval, we first need to find a z^* (pronounced "z star") critical value for 95% confidence. Visually, using the normal curve, we see the following:

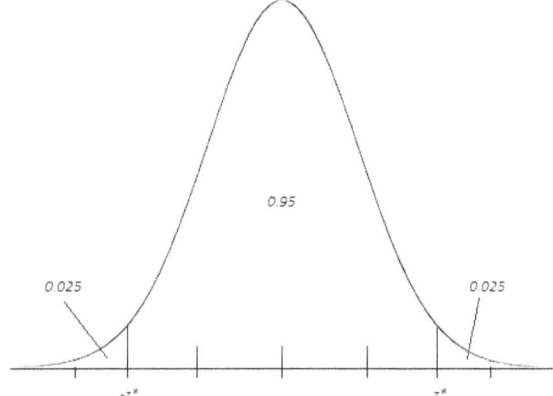

The middle 95% of the normal curve is, according to the Empirical rule, about two standard deviations away from the mean. However, if we look up 0.025 "backward" using the normal distribution table—meaning, we look up the *proportion/area* rather than the *z*-score—the *z*-score reveals itself to be −1.96, which is not quite two standard deviations below the mean.

Nonetheless, the z^* critical value for 95% confidence is 1.96. From this number, we will construct a confidence interval by using *estimate* ± margin of error :

$$\hat{p} \pm ME$$

$$\hat{p} \pm z^* \cdot SE$$

$$\hat{p} \pm z^* \cdot \sqrt{\frac{\hat{p}(1-\hat{p})}{n}}$$

$$0.6 \pm 1.96 \sqrt{\frac{0.6(1-0.6)}{30}}$$

$$0.6 \pm 1.96 \cdot 0.089$$

$$0.6 \pm 0.174$$

(*lower bound, upper bound*)

$$(0.426, 0.774)$$

(e) Do you feel sure that the interval $(0.426, 0.774)$ contains p, the true population proportion of introverts?

Even if our data had been collected randomly—which it wasn't—we can't be sure short of taking a census of the host population. Instead, we can be *confident*—specifically, 95% confident—that we have captured the true population proportion p.

(f) Find the width and half-width of the confidence interval.

The width is simply $0.774 - 0.426 = 0.348$. The half-width is half that: $0.348 / 2 = 0.174$. Another name for the half-width is the margin of error, or *ME*.

(g) Find the midpoint of the confidence interval.

You could do the following: $\dfrac{0.426 + 0.774}{2} = 0.6$. But that's unnecessary. Notice that the midpoint of the confidence interval is $\hat{p}$, since an equal amount (the *ME*) is being added and subtracted from $\hat{p}$ to construct the interval: $\hat{p} \pm ME$.

(h) Are the technical conditions (also called assumptions or requirements) met for this confidence interval procedure?

These conditions should have been checked *prior* to constructing the interval. After all, in order to obtain z^* we made the assumption of normality—an assumption that perhaps was unwarranted.

The conditions are $n\hat{p} \geq 10$, $n(1 - \hat{p}) \geq 10$, SRS, and the 10% Condition (if sampling without replacement). The 10% Condition is certainly met here; our sample is very small relative to the size of all possible students. And the successes and failures are both at least ten. (Notice that we use $\hat{p}$ in place of p.) But this was not a random sample, since the data were obtained with convenience.

(i) Interpret the 95% confidence interval in two valid ways.

The straightforward interpretation first: We are 95% confident that p, the true population proportion of introverts, is between 0.426 and 0.774.

A confidence interval doesn't estimate a sample statistic, like $\hat{p}$; it also doesn't estimate the number of sample (or population) observations in the interval. A confidence interval estimates the value of a single population parameter—in this case, p.

For the second interpretation, recall the definition of the confidence level: the percentage of all possible samples that include the population parameter. We'll pick up off of that to claim that if all possible samples of size 30 were obtained from the population, and 95% confidence intervals were constructed off of each one, 95% of those samples would contain the true population proportion p.[*]

So which 5% of the samples *don't* contain p? We don't know. That's why we're 95%, rather than 100%, confident.

In addition, it is *incorrect* to bring the notions of probability into the mix, no matter how tempting. For instance, consider this interpretation: There is a 95% probability (or chance) of p lying inside the confidence interval. This confidence interval interpretation is wrong on its face, since p is a fixed population number, not some value that floats hither and thither, always plotting its escape from the interval's proverbial net. Either the confidence interval captures p, or it doesn't—there is no probability involved.

Likewise with the terms "sure" and "certain." Don't interpret a confidence interval by claiming that you are 95% sure/certain that p lies inside the interval. Unlike the word "confidence," which expresses things according to some degree, "sure" and "certain" are absolute rather than relative (or gradient) quantities, similar to the words "pregnant" and "perfect." You can't be 95% sure any more than you can be 95% pregnant or 95% perfect, because you're either sure or you're not (or pregnant or not, or perfect or not).

Here's the upshot: use the term "confident" and you can't go wrong, except if the technical conditions aren't met—which they aren't in our introvert example. So, because the data weren't collected randomly, there's actually very little we can safely conclude.

(j) Construct a 90% confidence interval for the introverts' sample data.

Draw a graph of the normal curve, this time focusing in on the middle 90% of the data. Each tail houses 5% of the remaining data, so, using the normal distribution table, look up 0.05 "backward" to find the $z*$ critical value, which ends up being 1.65 (ignore the negative value, since the curve is perfectly symmetrical about the mean). Then calculate the 90% confidence interval:

$$0.6 \pm 1.65 \cdot 0.089$$

$$0.6 \pm 0.147$$

---

[*] Note that this interpretation is consistent with the frequentist approach to probability.

$$(0.453, 0.747)$$

(k) Construct a 99% confidence interval for the introverts' sample data.

Now the $z*$ critical value is found by looking up 0.005 "backward" on the normal distribution table, giving us 2.58. Therefore, the interval is

$$0.6 \pm 2.58 \cdot 0.089$$

$$0.6 \pm 0.230$$

$$(0.370, 0.830)$$

(l) What do you notice about the widths of your 90%, 95%, and 99% confidence intervals?

As the confidence level increased, the width increased. If you want more confidence, the interval will get larger.

Think of it this way: while Barry Bonds surely was confident he could hit a homerun in a ballpark (a wide possible area encompassing left field, center field, and right field), he was probably less confident to which *particular* field he would hit the homerun. As the size of the area decreased, Bonds' confidence decreased with it.[*]

(m) Suppose we asked more students whether they were introverted or extroverted. In general, if we increase n—holding all other things equal—what happens to the size of our confidence interval?

A bigger sample size reduces variability, so the interval shrinks. (Also note that n is located in the denominator of the standard error, so increasing the sample size increases the denominator thereby decreasing the quotient.)

(n) Calculate a 0% confidence interval for the introverts' sample data.

The middle 0% of the normal curve nets us a $z*$ critical value of 0. Thus, a 0% confidence interval is simply our point estimate: $\hat{p} = 0.6$.

(o) Calculate a 100% confidence interval for the introverts' sample data.

At 100%, $z*$ is very large—but not so large as to exceed the minimum and maximum possible proportions, namely, 0 and 1. Therefore, any 100% confidence interval is $(0,1)$,

---

[*] Perhaps Babe Ruth's "Called Shot at Wrigley Field," in which he supposedly pointed to center field with his bat before hitting a homerun there, is the ultimate expression of confidence in a baseball game.

which is mathematically trivial: of course the percentage of *any particular characteristic in a population* is between 0% and 100%!

The next example unpacks a political poll from a past election season.

Example 8b.-2

In the famously contentious Democratic primary season of 2008, two candidates—Barack Obama and Hillary Clinton—struggled mightily to obtain the presidential nomination of their party. In a Gallup daily election poll for March 1st to 3rd, 2008, based on national Democratic and Democratic-leaning voters, 45% pledged their support to Obama.

On their website, Gallup stated their public opinion polling methodology as such:

> Gallup is interviewing no fewer than 1,000 U.S. adults nationwide each day during 2008. The results reported here are based on combined data from March 1–3, 2008, including interviews with … 1,282 Democratic and Democratic-leaning voters. For results based on these samples, the maximum margin of sampling error is ±3 percentage points. In addition to sampling error, question wording and practical difficulties in conducting surveys can introduce error or bias into the findings of public opinion polls.

(a) What is the sample proportion of this study? Express it using the correct symbol.

$$\hat{p} = 0.45$$

(b) Calculate the standard error.

$$SE = \sqrt{\frac{0.45(1-0.45)}{1282}} = 0.0139$$

(c) Find a 95% confidence interval for p, the true proportion of Democratic and Democratic-leaning voters who supported Barack Obama on March 1st to 3rd, 2008.

$$0.45 \pm 1.96 \cdot 0.0139 = (0.423, 0.477)$$

(d) The half-width of a confidence interval is the survey's margin of error. Calculate this half-width. Was Gallup correct to claim that their "maximum margin of sampling error is ±3 percentage points" for this public opinion poll?

Yes, since $\dfrac{0.477 - 0.423}{2} = 0.0272$, or 2.72%, which is less than 3%.

(e) Why does Gallup claim that "[i]n addition to sampling error, question wording and practical difficulties in conducting surveys can introduce error or bias into the findings of public opinion polls"? Doesn't the calculation of the margin of error handle these issues? Explain.

The calculation of the *ME* does not account for how the survey was conducted, and cannot "correct" for nonsampling errors such as undercoverage (systematically excluding a portion of the population) and nonresponse bias.

In the 2012 presidential election, the Rasmussen polling firm vastly overestimated support for the Republican candidate, Mitt Romney, because only individuals with landlines were surveyed—but the young, who tend to vote Democratic, disproportionately have only cell phones.

Now let's turn our attention to more advanced confidence interval ideas.

Example 8b.-3

Consider the following plot of twenty 95% confidence intervals obtained from twenty equally-sized random samples, all from the same population, where the true population proportion $p = 0.6$.

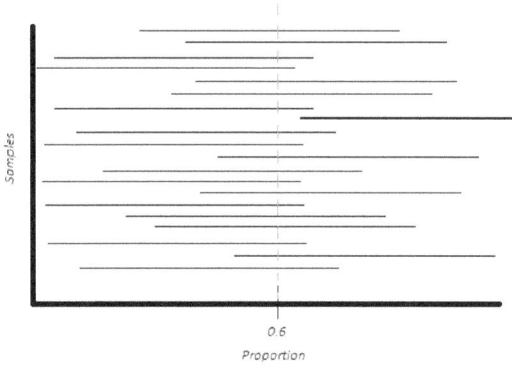

One of the twenty confidence intervals above does not capture the true population proportion (it is represented by the thicker black line). Explain, by correctly interpreting a 95% confidence interval, why this is logical.

Solution. Out of twenty samples, nineteen capture p—that's 95% of the samples. We would expect, over the long term, 95% of the intervals calculated off of samples of the same size from the same population to contain p.

Example 8b.-4

(a) If we know that $\hat{p} = 0.45$, explain why (0.352, 0.648) cannot be a correct confidence interval (no matter the confidence level).

Because $\hat{p} = 0.45$ is not the midpoint of the given interval.

(b) Suppose a confidence interval is (0.545, 0.755). Find $\hat{p}$. Also address this: what would we need to know in order to find the confidence level?

$\hat{p} = \dfrac{0.545 + 0.755}{2} = 0.65$. In order to find the confidence level, we would need to know the sample size n.

(c) Now consider the confidence interval of (0.179, 0.321). We know that $n = 100$. Find both $\hat{p}$ and the confidence level.

$\hat{p} = \dfrac{0.179 + 0.321}{2} = 0.25$. The half-width of this interval is $\dfrac{0.321 - 0.179}{2} = 0.142$, which is also the interval's margin of error. So we can solve for $z*$:

$$0.071 = z* \cdot \sqrt{\dfrac{0.25(1 - 0.25)}{100}}$$

$$z* = \dfrac{0.071}{\sqrt{\dfrac{0.25(1 - 0.25)}{100}}}$$

Thus $z*$ is equal to 1.64. Looking up 1.64 on the normal distribution table, we obtain 0.9495, or about 0.95. There is 95% of the area to the left of the $z*$ critical value and 5% of the area to the right, meaning there is 5% in *each* tail. With 5% in each tail, that leaves 90% in the middle of the curve. Hence, the confidence level is 90%.

(d) Suppose that two researchers, Carolyn and Jenna, decide to obtain two samples, one with a sample size of 250 and one with a sample size of 1000. Both decide to form 99% confidence intervals from their samples. Carolyn obtains the interval (0.320, 0.480), and Jenna gets (0.521, 0.601). Which sample size goes with which researcher? Explain.

The larger interval pairs with the smaller sample size, so since Carolyn's interval is bigger than Jenna's, she has the sample size of 250.

(e) Reconsider the previous problem. Which interval—Carolyn's or Jenna's—has a better shot at capturing the true population proportion p?

Assuming that the technical conditions were all met for both intervals, they have the same "chance" of capturing p since both intervals have the same confidence level, 99%.

Your graphing calculator has the built-in capability to compute confidence intervals of sample proportions automatically.

First, hit the $\boxed{\text{STAT}}$ button, then scroll over to the TESTS menu.

Now, the formal name for a confidence interval of sample proportions is a **One Sample z Confidence Interval for p**. We have been working with the z-table (the normal distribution table), so the option we select should mention Z in it. Also, we are finding confidence intervals of proportions, so Prop will figure into the option name. We want a confidence interval, and it appears that the calculator abbreviates intervals using Int. Finally, we are looking at one proportion at a time.

Therefore, select the option 1-PropZInt. Your calculator screen should look very similar to this one:

```
1-PropZInt
 x:0
 n:0
 C-Level:.95
 Calculate
```

Answer the following question using the 1-PropZInt calculator option.

Example 8b.-5

In a study published in *The Journal of Family Psychology* in 2008, researchers from the University of Colorado and Texas A&M asked 4,884 married women if they had been unfaithful to their husbands, of which 6% said yes. Find the margin of error for 95% confidence.

Solution. The calculator will not accept an input of 6% into "x:"; instead, take 6% of the 4,884 women to arrive at 293 women. Inputting the 293, the 4,884, and a C-Level of 0.95 gives us an interval of (0.053, 0.067).

However, we need to find the *ME*, which is the same as the half-width:

$$\frac{0.067 - 0.053}{2} = 0.007^{*}$$

We also need to know how to solve for the sample size *n* given a half-width, a $\hat{p}$, and a confidence level for a sample.

---

[*] Is it merely coincidence that the half-width of a survey about women cheating on their husbands is equal to 0.007, legendary philanderer James Bond's secret agent number? Yes.

Example 8b.-6

Suppose the confidence level is 95%, the margin of error (or half-width) is 0.03, and $\hat{p} = 0.6$. Find *n*, the sample size.

Solution. Start by plugging in everything we know:

$$0.03 = 1.96 \cdot \sqrt{\frac{0.6(1-0.6)}{n}}$$

$$\frac{0.03}{1.96} = \sqrt{\frac{0.6(1-0.6)}{n}}$$

$$\left(\frac{0.03}{1.96}\right)^2 = \frac{0.6(1-0.6)}{n}$$

$$\left(\frac{1.96}{0.03}\right)^2 = \frac{n}{0.6(1-0.6)}$$

$$n = 0.6(1-0.6) \cdot \left(\frac{1.96}{0.03}\right)^2 \approx 1025$$

Instead of repeating the algebra steps with each such problem, we can make a formula based off of the final step.[*] Thus, to directly solve for the sample size, simply use the following formula:

$$n = \hat{p}(1-\hat{p}) \cdot \left(\frac{z*}{ME}\right)^2$$

Before actually collecting any data at all, a researcher may wish to determine a sample size for a half-width of an interval. Since the researcher does not know ahead of time what *p* is, she will have to make a conservative-as-possible estimate of *p*. The most conservative estimate of *p*, and thus of $\hat{p}$, is 0.5. This will produce the largest possible *ME*.

---

[*] As an analogue, consider the quadratic formula: it is also a generalized version of the final step of solving a quadratic equation by completing the square.

Example 8b.-7

(a) We want to estimate some proportion to within ± 0.05 with 95% confidence. How big should n be?

$$n = 0.5(1-0.5) \cdot \left(\frac{1.96}{0.05}\right)^2 \approx 385$$

(b) Suppose that you want to estimate p, the true proportion of students at Madison High School who have a tattoo, with 90% confidence and a margin of error of no more than 0.10. Calculate how many students you'll have to survey.

$$n = 0.5(1-0.5) \cdot \left(\frac{1.65}{0.10}\right)^2 \approx 68$$

Because surveying individuals uses resources like time and money, it is important to do these sorts of sample size calculations ahead of time; that way, you'll know exactly how many people need to be surveyed—and not miscalculate by asking too few or waste time by asking too many.

§8c. *Confidence Intervals of Population Means*

Recall that the sample proportion $\hat{p}$ is the best point estimate for the population proportion p. We are able to construct confidence intervals for p by using $\hat{p}$.

Likewise, the sample mean $\bar{x}$ is the best point estimate of the population mean μ, and we want to construct confidence intervals for μ by using $\bar{x}$. Suppose we don't know what μ is (if we knew what this population mean equaled, why create a confidence interval for it?), but we *do know* what σ is equal to. This is an unrealistic situation; after all, how would we know the population standard deviation σ but not know the population mean μ, especially since σ requires μ in its calculation? Despite it being unlikely, let's consider this situation anyway.

If n is large ($n \geq 30$), and the sample mean $\bar{x}$ is from a random sample of the population, and we know the population standard deviation σ, then the **One-Sample z Interval for a Population Mean** is

$$\bar{x} \pm z^* \frac{\sigma}{\sqrt{n}}$$

As long as $n \geq 30$, the confidence interval we just created can be used since the Central Limit Theorem applies: as long as n is sufficiently large, the sampling distribution of $\bar{x}$ is approximately normal for any population—no matter the population distribution's shape.

But if $n < 30$, and it is reasonable to believe that the distribution of values in the population is normal, we can still construct a confidence interval for μ, as long as we know the value of σ.

Example 8c.-1

Suppose that the population standard deviation of the heights of American males is $\sigma = 2$. We measure the heights of a random sample of 35 men and find that their sample mean is $\bar{x} = 69$ inches. Construct and interpret a 95% confidence interval of the true mean of American males' heights.

Solution. We'll plug in everything we know into the formula:

$$69 \pm 1.96 \frac{2}{\sqrt{35}} = (68.337, 69.663)$$

We are 95% confident that μ, the true mean of American males' heights, is between 68.337 and 69.663 inches.

It is very unlikely, though, that we would know σ yet, somehow, not know μ. If, as is most common, both σ and μ are missing, we'll be forced to use two point estimates: one for σ and one for μ. As we've already seen, the best point estimate for μ is $\bar{x}$. Unsurprisingly, even though this particular point estimate is technically not unbiased, the best point estimate for σ is s.

But using two point estimates in the confidence interval instead of just one introduces a great deal of uncertainty—so much, in fact, that we'll have to discard the normal distribution in favor of a new distribution, called the *Student's t distribution* (or the *t distribution* for short).

The *t* distribution unfolds into many distributions, based on the sample size of the data set. At low sample sizes, the *t* distribution is stout, low, and flattish, with thick tails. But as the sample size increases, the *t* distribution gets taller, rounder at the top, and its tails thin out; in fact, the bigger the sample size, the more the *t* distribution starts resembling the normal distribution. No matter the sample size, though, all versions of the *t* distribution are perfectly symmetrical about the mean, median, and mode, which—like the normal distribution—are located at its center.

Instead of using sample size to distinguish between different variants of the *t* distribution, though, *degrees of freedom* is utilized. The degrees of freedom (abbreviated *df*) is always *one less than the sample size*. For instance, if the sample size is 10, the df is $10 - 1 = 9$.

A *t* distribution table can be found at the back of this primer. The leftmost column houses the degrees of freedom; the topmost row gives the *right* tail probabilities (recall that the normal distribution table only relays the probabilities below, or to the left, of associated *z*-scores). For instance, the *t** critical value for 95% confidence with 5 degrees of freedom is 2.571, since, with

95% confidence, each tail contains 2.5%, or 0.025, of the remaining area in the distribution.

If a particular df is not present on the table, always round down, except if the sample size is very large—in that case, use the infinity row. Say, for instance, that the sample size is 90: round down to use the 80 df row. But if the sample size is 2,000, default to the infinity row.

Example 8c.-2

(a) Find the t^* critical values and degrees of freedom of these confidence levels and sample sizes:

		Confidence Level				
n	df	80%	90%	95%	98%	99%
6	5	1.476	2.015	2.571	3.365	4.032
11	10	1.372	1.796	2.201	2.764	3.169
41	40	1.303	1.684	2.021	2.423	2.704
71	70 (row 60)	1.296	1.671	2.000	2.390	2.660
5000	4999 (row ∞)	1.282	1.645	1.960	2.326	2.576

(b) At a "large" df, the t distribution becomes what distribution?

The normal distribution. Looking on the infinity row, the t^* critical values are equal to the z^* critical values—but they have more precision. For instance, earlier we found that the z^* critical value for 90% confidence was 1.65, but the t distribution instead gives 1.645.

We will use the t distribution when dealing with confidence intervals of means and, later, with tests of significance of means. We will never use the t distribution when dealing with proportions—that's the exclusive province of the normal distribution—only with means.

Now that we've worked a bit with the t distribution, let's enrich our understanding by taking a quick detour exploring its history. The Guinness Brewing Company of Dublin, Ireland—yes, the same Guinness Brewing Company that originally published the annual Guinness Book of World Records—hired a recently minted degree holder in math and chemistry by the name of William Sealy Gosset in 1899 as part of an initiative to bring greater scientific rigor to the production of their beer.

Guinness ended up having many uses for such a well-rounded mathematical mind like Gosset's. He rose quickly through the company's ranks, solving all sorts of brewing quandaries using mathematical models. But there was a problem: Gosset wanted to publish his mathematical results for all the world to see, but it was against company policy to do so.

Nonetheless, Gosset secretly formed a friendship with Karl Pearson—the same Karl Pearson who derived the correlation coefficient—and began publishing under the pseudonym "Student."

Gosset would publish many papers in *Biometrika* as "Student"—including one entitled "The Probable Error of the Mean," in which he introduced the *t* distribution (or, as it is also still called, the Student's *t* distribution). As statistician David Salsburg explains in *The Lady Tasting Tea: How Statistics Revolutionized Science in the Twentieth Century*,

> All of Pearson's work assumed that the sample data was so large that the parameters could be determined without error. Gosset asked, What happens if the sample were small? How can we deal with the random error that is bound to find its way into our calculations? Gosset sat at his kitchen table at night, taking small sets of numbers, finding the average and the estimated standard deviation, dividing one by the other, and plotting the results on graph paper…. It did not matter where the data came from or what the true value of the standard deviation was…. As Frederick Mosteller and John Tukey [later] point[ed] out, without this discovery [of the *t* distribution], statistical analysis was doomed to use an infinite regression of procedures….

Later statisticians, such as Stanford's Bradley Efron, showed the strict requirements of population normality that Gosset assumed were necessary actually weren't—because of the robustness of the *t* distribution.

Just like we created confidence intervals for proportions, we can create confidence intervals for means. So we'll now be dealing exclusively with *quantitative* variables, rather than categorical ones.

First of all, let's recall how we obtained the standard deviation of a sampling distribution of sample means: $s = \dfrac{\sigma}{\sqrt{n}}$. There's a problem, though. Just like we didn't have p and had to use $\hat{p}$ in its place, we won't have the population mean μ. (We are trying to make a precise guess about what μ is—that's the whole point of a confidence interval: chipping away at the truth of a population that we only have a simple random sample from.)

But, recall, if we don't have the population mean μ, we probably don't have the population standard deviation σ, either! Let's use the sample standard deviation s in place of σ in our formula for standard deviation, which we will now call the standard error (*SE*) of the sample mean: $\dfrac{s}{\sqrt{n}}$. Replacing both the population mean and population standard deviation with point estimates forces our hand: because of all of the additional uncertainty, the *t* distribution, in place of the normal distribution, must now be used.

With all this in mind, the **One-Sample *t* Interval for a Population Mean** is equal to the following:

$$\bar{x} \pm \left(t^{*}_{\,n-1}\right)\dfrac{s}{\sqrt{n}}$$

Here are the technical conditions (or assumptions or requirements): the sample is an SRS, the sample size is at least thirty or the host population is normal, and the 10% Condition is satisfied (if sampling without replacement).

Just like with confidence intervals of sample proportions, with sample means there is always a *battle* between obtaining the highest confidence level possible with having the smallest, most precise confidence interval possible.

Example 8c.-3

In the book *The Assault on Reason* by Al Gore, Gore writes that "according to an authoritative global study, Americans now watch television an average of *four hours and thirty-five minutes every day*." Suppose that in this study, only 31 individuals were randomly surveyed and the sample mean (restated) is 275 minutes with a sample standard deviation of 16 minutes.

(a) What are the cases?

   Americans.

(b) Calculate the standard error of $\bar{x}$.

$$SE = \frac{16}{\sqrt{31}} = 2.87$$

(c) Find an appropriate t^* critical value for 95% confidence.

   The sample size is 31, so the degrees of freedom is 30; looking on the right tail probability of 0.025 column, $t^* = 2.042$.

(d) Is this critical value less than or greater than the corresponding critical value of z^*?

   t^* critical values are always greater than their associated z^* critical values, except when the sample size is infinity—in that case, they're equal.

(e) Check to make sure that the conditions for this confidence interval procedure are satisfied.

   The sample size is 31, which is at least 30, so it makes no difference whether the host population is normal; the sample was obtained randomly; and the sample size is much, much less than 10% of the population size (all Americans).

(f) Find a 95% confidence interval for the population mean of the average time that Americans spend watching television per day.

$$275 \pm 2.042 \cdot \frac{16}{\sqrt{31}} = (269.13, 280.87)$$

(g) Find the width, half-width (i.e., margin of error), and midpoint of the interval.

The width is $280.87 - 269.13 = 11.74$.

The half-width is 11.74/2 = 5.87.

The midpoint is $\bar{x} = 275$.

(h) Label each of the following "correct" or "incorrect" interpretations of the confidence interval. Briefly justify your answer.

    a. We can be certain that the true mean μ of average time Americans watch television per day lies inside the interval.

    Incorrect. We can be 95% confident, but not certain or sure.

    b. If random samples of size 31 were repeatedly taken from this population of Americans, and 95% confidence intervals were generated off of each sample, then approximately 95% of these intervals would contain μ.

    Correct. Notice the term "approximately"—the statement would have been incorrect had it read "exactly" instead. "Exactly" could only be used if *all possible samples* of size 31 were taken from the host population.

    c. The probability of μ lying inside the interval is 95%.

    Incorrect. Confidence intervals have nothing to do with probability. Either the parameter is contained within the interval, or it is not.

    d. 95% of Americans watch television between 269.13 and 280.87 minutes per day.

    Incorrect. The bounds of the confidence interval estimate the value of the mean of the population, not any particular data points from the population.

(i) If we wanted to obtain a 90% confidence interval for our sample of Americans, would the width of the interval be bigger or smaller than the 95% confidence interval already calculated above? Why?

It would be smaller. If you have less confidence, you don't need as big of a range of values.

(j) Calculate a 90% confidence interval. Was your prediction in part (i) confirmed?

$275 \pm 1.697 \cdot \dfrac{16}{\sqrt{31}} = (270.12, 279.88)$. Yes, the interval is smaller.

(k) Suppose we wish to decrease the width of the 90% confidence interval just calculated. Holding everything else constant, what would we need to do to the sample size?

Increase it.

(l) Again, suppose we want to decrease the width of the 90% confidence interval. Would decreasing or increasing the sample standard deviation decrease the width?

Decreasing the sample standard deviation. However, we can't just arbitrarily change the standard deviation of a set of sample data—s is a fixed number originating from a sample. But we can try collecting a new sample, and, while we're at it, increasing the sample size. An increased sample size will decrease the size of our confidence interval (and since the square root of the sample size is in the denominator of the standard error, the standard error will decrease too).

As is perhaps obvious by now, your graphing calculator has the built-in capability to compute confidence intervals of sample means automatically.

First, hit the STAT button, then scroll over to the TESTS menu.

Now, the formal name for a confidence interval of sample means is a **One Sample t Confidence Interval for** μ. We have been working with the t distribution, so the option we select should mention T in it. We want a confidence *interval*, and it appears that the calculator abbreviates intervals using Int.

Therefore, select the option TInterval. Your calculator screen should look very similar to one of these:

```
TInterval
 Inpt:DATA Stats
 List:L1
 Freq:1
 C-Level:.95
 Calculate
```

```
TInterval
 Inpt:Data STATS
 x:0
 Sx:0
 n:0
 C-Level:.95
 Calculate
```

Use the screen on the above left if you have the list of data already stored in the calculator; use the screen on the above right if you have the sample mean, the sample standard deviation, and the sample size of your data set.

Given a small sample of data (less than thirty observations), if the host population is not stated to be normal, the normality condition needs to be checked—either with a modified boxplot or a normal probability plot (NPP).

Example 8c.-4

The following is a data set of a random sample of healthy adult males' systolic blood pressure readings:

125, 136, 110, 131, 143, 140, 115, 112, 109, 142, 126, 134, 112, 140, 124

Construct an 87% confidence interval using the graphing calculator. Make sure that the assumptions for the procedure are met.

Solution. First, type the data into the calculator, calling the list BP. Then, examine a modified boxplot of data. You'll see no outliers—remember, the presence of any outliers would disqualify the host population of normality—and the boxes and the whiskers are roughly the same size.

Looking at an NPP, the minor bit of curvature—a bit of a departure from linearity—doesn't dissuade us from our conclusion: this host population is probably normal. So, along with the data being randomly obtained and the sample being markedly less in size than the size of the population, the conditions are all met.

What follows is a screenshot of the set up for constructing the confidence interval using the TInterval calculator function:

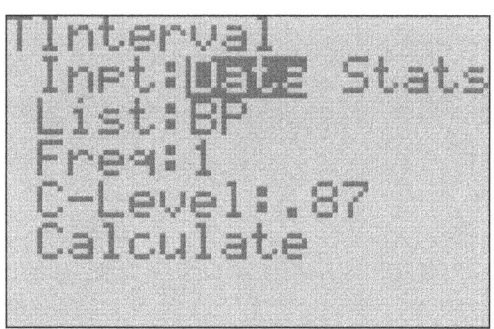

Select "Calculate" to obtain the interval: (121.41, 131.79).

If a population is not normal and the sample size is less than 30, we can't use any of the methods described above to obtain a valid confidence interval; instead, we have to either (a) use a nonparametric method (also called a distribution-free method, meaning we'll be making no assumptions about the shape of the population distribution) or (b) perform resampling (such as bootstrapping). We will examine these ideas at the end of the next §.

§9. *But is It Significant?*

⚭

Population Proportion Significance Testing. Population Mean Significance Testing. Committing Errors. Power. Comparing Independent Samples. Chi-Square. Correlation Coefficient Redux. Linear Regression Redux. Analysis of Variance. Nonparametric Methods. Logic vs. Probability.

§9a. *Population Proportion Significance Testing*

Confidence intervals are a form of *statistical inference* because they help us generalize and draw conclusions about a population based on a sample statistic. *Tests of significance*, also called *hypothesis tests*, also are a form of statistical inference: we will be testing the validity of claims about population parameters (the population proportion p and the population mean μ) using sample data culled from the populations of interest.

In order to do this correctly, we are going to have to combine our understanding of probability, the normal distribution, the Central Limit Theorem, and sampling distributions of the sample proportion and the sample mean. All of what we're going to accomplish here, however, is based on the rare event rule for inferential statistics. Recall the rule: An assumption probably isn't true if the likelihood of an observed event occurring based off of that assumption is improbable. Before continuing, let's define two terms:

Definition 9a.-1

Hypothesis. A claim about a population of interest.

Definition 9a.-2

Hypothesis testing. A statistical method for testing a claim.

Hypothesis testing was first used, albeit informally, by John Arbuthnot in his study of the distribution of boy-girl births in London in the eighteenth century. He noted that in each of eighty-two years of births in the registry, more boys than girls were born every year. Since the chances of that happening purely by chance were incredibly small, Arbuthnot rejected that there was an equal probability of boys and girls being born, instead leaning on a teleological argument to explain the distribution: divine providence, since the supply of males, who died at a faster clip, needed to be replenished.

Much later, significance testing was formalized mathematically by Fisher, who, along with Gosset, Neyman, and Pearson, put the procedure on firm probabilistic foundations. Since significance testing is such an important idea of this primer—perhaps *the* most important idea—we need to take some time to put the mathematical ideas into historical context. In *The Lady Tasting Tea: How Statistics Revolutionized Science in the Twentieth Century*, David Salsburg describes the impetus for the discovery of hypothesis testing, along with relaying the colorful[*] life of the father of hypothesis testing, Sir Ronald Fisher.

> It was a summer afternoon in Cambridge, England, in the late 1920s. A group of university dons, their wives, and some guests were sitting around an outdoor table for afternoon tea. One of the women was insisting that tea tasted different depending upon whether the tea was poured into the milk or the milk was poured into the tea. The scientific minds among the men scoffed at this as sheer nonsense. What could be the difference? They could not conceive of any difference in the chemistry of the mixtures that could exist. A thin, short man, with thick glasses and a Vandyke beard beginning to turn gray, pounced on the problem.

Putting the latent sexism of the situation aside—it seems that the "scientific minds" dismissed the lady's suggestion out of hand simply *because* she was a woman—Ronald Fisher, the bespectacled, bearded man, quickly spoke up.

"Let us test the proposition!" he exclaimed. Fisher, a British statistician, sketched out the design for an experiment in which the lady—whose name, by the way, was Muriel Bristol, Ph.D.—would be given sets of cups of tea: some with tea poured into milk, the rest with milk poured into tea. Fisher later published this experimental design in his most influential book, *The Design of Experiments* (1935).

Example 9a.-1

(a) What is the probability that the lady, just by guessing, will over time correctly identify which tea-milk infusion is which?

1/2 or 0.5.

---

[*] And controversial: Ronald Fisher, like Galton and Pearson before him, dabbled in eugenics. Gavin Kennedy argues that eugenics and the other social movements (e.g., Social Darwinism) springing from the Victorian age provided much of the impetus for the development of statistical techniques we still use today—sans references to eugenics and the like, of course. (In *The Mismeasure of Man*, Gould speaks of the "unholy alliance" between the theory of evolution and the tools of quantification that cast a "scientific" gloss on anthropological theories: the "allure of numbers, the faith that rigorous measurement could guarantee irrefutable precision, and might mark the transition between subjective speculation and a true science as worthy as Newtonian physics," drove an obsessive Victorian predilection to measure everything in sight.)

(b) Is this value a parameter or a statistic? Why? And what symbol would be used to represent this value?

The value is a parameter, because it's from a theoretically infinite population of tea-milk trials. The symbol used to represent the value, therefore, is p.

(c) What exactly does p represent?

p = the proportion of tea-milk mixtures correctly identified by the lady.

Even if we give the lady some cups to drink, how can we really tell if she knows the difference between them—or is merely guessing? Giving her one cup clearly isn't enough; two is too few; three isn't sufficient. Maybe ten? Or twenty? Or thirty?

At what point in such an experiment would it no longer be surprising that the lady was detecting a difference in the tea-milk infusion? At what point would we cease attributing her tasting expertise to mere chance alone? We must find a way to quantify this idea of what would be a surprising result in an experiment and what would be unsurprising—i.e., an experimental result likely due to chance.

This is where tests of significance, also called hypothesis testing, come into play. We need to set two hypotheses: the *null hypothesis* and the *alternative hypothesis*.

Definition 9a.-3

Null hypothesis. Represents our "best guess" of the population parameter; presupposes that there is "no effect" or "no difference" or maintains some sort of "status quo"; and is symbolically stated as H_0 (read as "H-naught").

Fisher himself said that the null hypothesis "is never proved or established, but is possibly disproved, in the course of experimentation. Every experiment may be said to exist only in order to give the facts a chance of disproving the null hypothesis."

Viewed another way, noted mathematics professor Ned Wolff has aptly called the null hypothesis the "boring hypothesis" because the null represents the assumption that things are just as they seem—boring!

Definition 9a.-4

Alternative hypothesis. Usually represents our "claim" about the population parameter; assumes that there is some sort of effect; and is symbolically stated as H_a or H_1.

The alternative hypothesis can have a greater-than (more than expected), a less-than (less than expected), or a not-equal-to (more or less) relationship with the conjectured value of interest. Sometimes the not-equal-to alternative is termed an "exploratory hypothesis."

An alternative hypothesis must be set so that if we end up rejecting the null hypothesis, we have a formal, alternate account of what is likely to be true (other than the divine providence explanation of Arbuthnot). Gossett said,

> It doesn't in itself necessarily prove that the sample is not drawn randomly from the population even if the chance is very small, say .00001; what is does is to show that if there is any alternative hypothesis which will explain the occurrence of the sample with a more reasonable probability, say .05 (such as that it belongs to a different population or that the sample wasn't random or whatever will do the trick) you will be very much more inclined to consider that the original hypothesis is not true.

Example 9a.-2

Reconsider the lady tasting tea.

   (a) State the null and the alternative hypothesis.

   Null: The lady correctly identifies 50% of the tea-milk mixtures.

   Alternative: The lady correctly identifies more than 50% of the tea-milk mixtures.

   (b) Restate the null and alternative hypotheses symbolically.

   $$H_0 : p = 0.5$$
   $$H_a : p > 0.5$$

In our lady tasting tea experiment detailed above, we ran a "greater-than" test for the alternative. This is called a one-sided or one-tailed test. Sometimes our alternative hypothesis will be of the not-equal-to variety. This is called a *two-tailed*, *two-sided*, *double-sided*, or *exploratory alternative hypothesis*, and we will see these types of tests later on.

When expressing the null and alternative hypotheses symbolically, make sure to *always* use the same numerical value for both, the same population parameter for both (in this case, p), an equals sign for the null hypothesis, and no equals sign for the alternative.

And when conducting any test of significance, we always assume that the null hypothesis is true and attempt to, by analyzing data, prove otherwise.

The question still stands, though: Can the lady tell the difference between tea in milk and milk in tea? Although we could argue one side or the other until we're blue in the face, collecting data is the only way to settle the question—at least with a high degree of probability. In the case of the lady tasting tea, an experiment would have to be conducted. Fisher did perform such an experiment, "mixing eight cups of tea, four in one way and four in the other, and presenting them to the subject for judgment in a random order," as he wrote in *The Design of Experiments*. We will modify the design a bit below in order to better illustrate the basics of hypothesis testing.

Example 9a.-3

Suppose that the lady sits for 24 trials of the tea experiment. Of these 24 trials, she correctly identifies the tea-milk mixture 18 times.

(a) Find the value of the sample statistic, and state what it represents.

$\hat{p} = 18/24 = 0.75$. The value represents the proportion of correctly identified tea-milk mixtures by the lady in the experiment.

(b) Is the sample statistic *consistent* with the claim (i.e., the alternative hypothesis)?

Yes, since $\hat{p} > 0.5$.

Analogous to sampling distributions of the proportion and confidence intervals of proportions, significance tests of proportions have technical conditions (or requirements, or assumptions for inference) that must be met for the procedure to be considered mathematically valid.

The technical conditions are $np_0 \geq 10$, $n(1 - p_0) \geq 10$, SRS, and the 10% Condition (if sampling without replacement); the p_0 refers to the *conjectured value of interest*.

(c) Are the conditions for inference met?

$24 \cdot 0.5 \geq 10$ is met, as is $24 \cdot (1 - 0.5) \geq 10$; the sample is, in effect, "random," since the observations are all being generated by the same individual (the lady).

(d) According to the Central Limit Theorem, the sampling distribution of the sample proportion will be a normal curve, centered about the conjectured value of interest (note that the null hypothesis is assumed to be true). Draw this sampling distribution, labeling the mean and one, two, and three standard deviations above and below the mean. Also find and label the sample proportion $\hat{p}$ on the sketch.

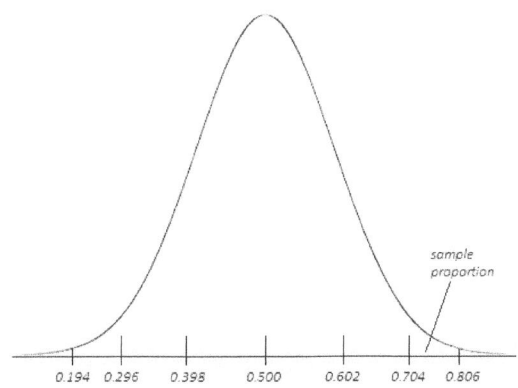

sample proportion

0.194   0.296   0.398   0.500   0.602   0.704   0.806

The mean and standard deviation are $\mu_{\hat{p}} = 0.5$ and $\sigma_{\hat{p}} = \sqrt{\dfrac{0.5(1-0.5)}{24}} = 0.102$, respectively.

In the context of hypothesis testing, a z-score is called a *test statistic*.

(e) Calculate the test statistic (z-score) for the lady's sample proportion.

$$z = \frac{0.75 - 0.5}{0.102} = 2.45$$

(f) Use the normal distribution table to find the probability that the lady obtains a sample proportion equal to or greater than 0.75 (assuming the null hypothesis is true). Mathematically stated, find $P(\hat{p} > 0.75)$.

Using the normal distribution table, look up the z-score of 2.45 and subtract that value from 1 (we subtract from 1 because the alternative hypothesis is of the "more-than" variety). So, $1 - 0.9929 = 0.0071$.

Just like we used a new term for the z-score, calling it a test statistic, we're going to label the probability under the normal curve we find based off of the test statistic a *p-value*.

Definition 9a.-5

p-value. The probability of obtaining a test statistic at least as extreme as the one actually observed, assuming that the null hypothesis is true.

Example 9a.-4

Interpret the *p*-value of 0.0071 in the lady tasting tea experiment.

Solution. If the null hypothesis were true, we would expect to see a sample result ($\hat{p} > 0.75$) this or more extreme 0.71% of the time.

Here's the interpretation of the *p*-value restated and expanded: Suppose for a moment that you believe the lady is a liar—she's simply a really good guesser, you think. How good of a guesser? Well, when she sat for 24 trials, she got 75% of them correct; the chance of that happening, being that she's just an amazing guesser, is 0.71%. According to the rare event rule, you should immediately challenge your she's-a-good-guesser assumption: the probability tied to that assumption—only 0.71%—is very, very low. Although it's *possible* that the lady guessed her way to such an improbable number of correct trials, in all likelihood she wasn't merely guessing and can in fact taste the difference (most of the time) between tea in milk and milk in tea.

How low would the *p*-value have to be for us to dismiss the she's-a-good-guesser argument outright? Fisher struggled with this question. In 1929, he wrote an article for *Proceedings of the Society for Psychical Research*, a publication examining extrasensory perception (ESP) and other qualities of clairvoyance. In it, Fisher set down an arbitrary rule of significance: "It is common practice to judge a result significant, if it is of such a magnitude that it would have been produced by chance not more frequently than once in twenty trials [i.e, 5% of the time]."

Here's what Fisher is saying: we all have to agree on a clear-cut way to identify and categorize what "unlikely" and "not so unlikely" means. First of all, understand that the smaller the *p*-value, the stronger the evidence against the null hypothesis being true; Fisher stated that a *p*-value of 0.05 gives us sufficient evidence—of the statistically significant sort—that the null hypothesis is not true.

But not all researchers use 5% as a *significance level*. We denote the significance level with the symbol α (read: alpha), which is treated as a cut-off value. The researcher, in advance of her experiment (and certainly before examining any data), has to designate an α level. Common values used for α are 0.10, 0.05, and 0.01. But sometimes, in a statistics problem or an actual research study, the α level isn't specified explicitly; in that case, assume Fisher's preferred significance level of $\alpha = 0.05$.

If the *p*-value ends up less than or equal to the significance level α, the *test decision* is to *reject* the null hypothesis. Otherwise, we *fail to reject* the null hypothesis. (This process of decision making is termed a *decision rule*.) Pay careful attention to the wording: we can never claim that the null hypothesis is true but can only say that based on the *p*-value the evidence isn't overwhelming enough to reject the null hypothesis.[*]

Example 9a.-5

Recall that the *p*-value of the lady tasting tea experiment is 0.0071.

(a) Based on the *p*-value, under which significance level(s) of α would the data be considered statistically significant?

Since the *p*-value is less than $\alpha = 0.10$, $\alpha = 0.05$, and $\alpha = 0.01$, the data would be considered statistically significant under the three most common significance levels. In fact, any significance level less than or equal to 0.0071 would constitute statistically significant data in the lady tasting tea experiment.

(b) At the 5% significance level, report the test decision.

We reject the null hypothesis at the 5% significance level.

(c) Explain what the test decision means in the context of the study.

---

[*] There are various mnemonic devices to help remember the test decision algorithm, like Triola's: "If the *p* is low the null must go…"

We have enough evidence to claim that the lady can correctly identify more than 50% of the tea-milk mixtures.

When reporting the test decision and its implications, use this format:

- First sentence: "We [reject/fail to reject] the null hypothesis at the [level of significance]% significance level."
- Second sentence: "We [have enough/do not have enough] evidence to claim [state the alternative hypothesis]."

If you reject the null, there is enough evidence to claim the alternative; if you fail to reject the null, there is not enough evidence to claim the alternative. Rejecting the null = statistically significance data = p-value $\leq \alpha$; failing to reject the null = not statistically significant data = p-value $> \alpha$.

Why do we "fail to reject" the null, rather than "accepting" the null? Fisher said, "For the logical fallacy of believing that a hypothesis has been proved to be true, merely because it is not contradicted by the available facts, has no more right to insinuate itself in statistical than in other kinds of scientific reasoning...." Recall that this stems back to what Karl Popper said about the scientific method: "A scientific idea can never be proven true, because no matter how many observations seem to agree with it, it may still be wrong." Failing to reject is a hedge against future observations, against the possibility of Hume's black swan lurking around the corner.[*]

It was the statisticians Jerzy Neyman and Egon Pearson (son of Karl Pearson) who worked together to systematize and codify hypothesis testing. For instance, it was Neyman and Pearson who coined the terms "null hypothesis" and "alternative hypothesis."

If not for Fisher's hypothesis testing (along with Gosset, Neyman, and Pearson's formalizations), which helped precipitate the flowering of evidence-based research, the discipline of psychiatry might still be waddling in the childhood-narrative eddies of Freudian psychodynamics.[†] As David Adam explains in *The Man Who Couldn't Stop*, psychiatry took a long time to catch up to empirical observations and the scientific method, and one need look no further than the categorizations of illnesses in the DSM-I (Diagnostic and Statistical Manual of Mental Disorders) and DSM-II to see.

By the publication of the DSM-III in 1980, the authors "looked not to Freud but to a different role model, one as far away from Freud as it is possible to imagine," named Emil Kraeplin, who took biology, not "childhood toilet habits," to be the cause of mental disorders. This is the "category approach" that, while still flawed, relies on observation and empirical evidence to sort pa-

---

[*] Consider how embarrassing it might be to live your life by hedging on everything. For example, at your wedding, imagine saying, "I don't not take you to be my wife/husband," instead of "I do."

[†] In the early 1900s, as Gould explains, "[P]sychology still wallowed in its reputation as a 'soft' science, if a science at all. Some colleges did not acknowledge its existence; others ranked it among the humanities and placed psychologists in departments of philosophy." This led psychologists, such as Robert Yerkes, to institute mass intelligence testing, finally giving the nascent science of psychology something to measure.

tients into groups. Psychiatric diagnosis finally caught up to the scientific method.

In June of 2005, Gallup conducted a poll about supernatural beliefs among adult Americans. One such question was "[Do you believe in] psychic or spiritual healing or the power of the human mind to heal the body?" Approximately 55% said yes; another similar question, centering on ESP, resulted in 41% who answered in the affirmative.

James Randi, former magician and famed debunker of self-stylized psychics such as Sylvia Browne and Uri Geller, has put up one million dollars to any individual "who can show, under proper observing conditions, evidence of any paranormal, supernatural, or occult power or event." Those "proper observing conditions"? A scientifically designed test of significance, co-designed by Randi and the applicant. (As of this writing, the money is still unclaimed.)

ESP tests customarily use Zener cards, developed by perceptual psychologist Karl Zener. Each Zener card has one of five symbols on one side—a circle, a cross, three wavy lines, a square, or a five-pointed star—and is blank on the other. There are twenty-five such cards in a standard deck. After shuffling the deck thoroughly, subjects are tested for clairvoyance by being presented with the cards (symbol not showing), one at a time, and asked to guess the symbols. We would expect a subject, by guessing alone, to get 20% (or 1/5) correct; the farther away from 20%, the more evidence of a supernatural power (or some sort of behind-the-scenes cheating).

You may already be familiar with Zener cards from one of the opening scenes of the classic comedy *Ghostbusters*, when Dr. Peter Venkman, played by Bill Murray, administers electric shocks to a male graduate student each time he makes a mistake guessing the Zener symbol, but fails to shock an attractive female graduate student, also participating in the experiment, despite her many faulty guesses.

Example 9a.-6

In an ESP test using a standard deck of twenty-five cards, how many Zener cards does a subject need to correctly identify to demonstrate statistically significant evidence of ESP at the 5% significance level?

Solution. The null and alternative hypotheses here are similar to the lady tasting tea's:

$$H_0 : p = 0.2$$
$$H_a : p > 0.2$$

We do not, however, have a value for $\hat{p}$. That will need to be calculated.

If data is significant at the 5% level, then its *p*-value will be 5% or less. However, in this case, the hypothesis test has a one-tailed greater-than alternative. Therefore, we need to look up the complement of 0.05—which is 0.95—on the normal distribution table "backward." This gives us a *z*-score of 1.65, which is also the value of the test statistic.

Let's solve for $\hat{p}$:

$$1.65 = \frac{\hat{p} - 0.2}{\sqrt{\dfrac{0.2(1-0.2)}{25}}}$$

$$\hat{p} = 0.2 + 1.65 \cdot \sqrt{\frac{0.2(1-0.2)}{25}} = 0.332$$

Thus, in order to offer statistically significant evidence of ESP at the 5% level, the subject would need to get at least $0.332 \cdot 25 = 8.3 \approx 9$ cards correct. (At even more stringent levels of significance, such as 1%, many more than nine cards would need to be identified correctly.)

Example 9a.-7

Suppose you are handed a gold coin. You suspect it's unfair (or weighted), so you flip the coin 100 times. It lands on heads 60 times. At the 1% significance level, do you have enough evidence to claim that the coin is unfair?

Solution. It makes the most sense to use a not-equal-to alternative hypothesis (recall, a not-equal-to alternative is also called a double-sided or two-sided or exploratory hypothesis) since, if a coin is unfair, it might be unfair in *either* direction—too many heads or too few heads.

Thus, the null and alternative hypotheses are

$H_0 : p = 0.5$ (the gold coin lands on heads 50% of the time)

$H_a : p \neq 0.5$ (the gold coin doesn't land on heads 50% of the time)[*]

There were 100 flips with 60 heads, so $\hat{p} = 0.6$. Checking the technical conditions, we find that $100 \cdot 0.6 \geq 10$, $100 \cdot (1 - 0.6) \geq 10$, and the sample is random (it is from a theoretically infinite population of coin flips from a single coin).

Next, calculate the test statistic:

$$z = \frac{0.6 - 0.5}{\sqrt{\dfrac{0.5(1-0.5)}{100}}} = 2$$

---

[*] Again, an exploratory (not-equal-to) alternative hypothesis makes the most sense here, simply because you don't know ahead of time how/if the coin will diverge from 50-50. When conducting studies, researchers use exploratory hypotheses most of the time since such methods allow for the possibility of being surprised by the data; in addition, by raising the threshold for statistical significance, a double-sided p-value is a more conservative approach. Computer statistical software packages, such as SPSS, only relay double-sided p-values to users.

And look up the p-value (off of the normal distribution): 0.9772. Now, with a two-sided test, the alternative hypothesis doesn't clue you in on whether or not to obtain a complement from the normal distribution probability (if the alternative is a greater-than, then you do; if it's a less-than, then you don't). So, here's the rule for two-sided tests: if the probability you snag from the normal distribution is greater than 0.5, subtract it from one; if not, leave it alone. Next, regardless of whether you subtracted from one, multiply the result by two—after all, this is called a "two-sided" test, which is like two tests for the price of one: testing both greater-than and less-than.

To find the p-value for this example, then, we do the following:

$$(1 - 0.9772) \cdot 2 = (0.0228) \cdot 2 = 0.0456$$

Even though this data would be significant at the 5% level, it is not significant at the 1% level. Therefore our test decision and conclusion read as follows: We fail to reject the null hypothesis at 1%. We do not have enough evidence to claim that the gold coin is unfair.

Your graphing calculator has the built-in capability to conduct tests of significance of a sample proportion—well, sort of: only the test statistic and the p-value will be reported.

First, hit the STAT button, then scroll over to the TESTS menu.

Now, another formal name for a hypothesis test of a sample proportion is a **Large Sample z Test for p**. We have been working with the z-table (the normal distribution table), so the option we select should mention **Z** in it. Also, we are testing *proportions*, so Prop will figure into the option name. We want to conduct a hypothesis *test*, so Test needs to be part of the option name. Finally, we are looking at one proportion at a time.

Therefore, select the option 1-PropZTest. Your calculator screen should look very similar to the screenshot shown below.

The p_0 above is the conjectured value of interest (for population proportion p).

Example 9a.-8

Reconsider the previous example with the gold coin. Redo the calculations for the test statistic and p-value, this time using the 1-PropZTest calculator option. Also interpret the p-value.

Solution. Enter $p_0 = 0.5$, $x = 60$, $n = 100$, and $prop \neq p_0$. The Calculate option will produce the summary measures; but the Draw option will do one better: in addition to giving the test statistic and p-value, the screen will also display the shaded p-value region on the sampling distribution. See below:

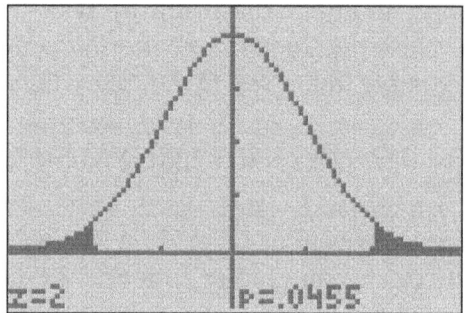

What's more, there's no need to do any conversion of the p-value to take into account the exploratory hypothesis: the calculator automatically accounts for the not-equal-to alternative.

The interpretation of the p-value is as follows: if the null hypothesis were true—and the gold coin was really a fair coin—then we would expect to see a sample result (which here is 60 out of 100 heads) this or more extreme 4.55% of the time.

Now that we've explored how to implement two methods of statistical inference—hypothesis testing and confidence intervals—we should link them together; they are, mathematically, two sides of the same coin (so to speak), both leaning on the frequentist approach to probability.

Example 9a.-9

Recall the test for ESP. Conduct a significance test at the 10% level for the following data: 9 out of 25 cards being correctly matched by a subject. In addition, find a 90% confidence interval and interpret the results.

Solution. The hypotheses for the ESP test are

$$H_0 : p = 0.2$$
$$H_a : p > 0.2$$

Assume that the data were gathered randomly; the remaining conditions are satisfied.

Using the graphing calculator, the test statistic is 2 and the p-value is 0.023, so we reject the null hypothesis and note that there is enough evidence to claim that the subject does better than merely guessing the Zener cards' symbols.

Again, using the calculator, we find the following 95% confidence interval: (0.20209, 0.51791). Thus, we are 95% confidence that p, the true proportion of Zener cards the subject correctly identifies, is between 0.20209 and 0.51791.

The confidence interval does not contain the conjectured value of p, which is 0.2; recall we rejected a null hypothesis that claimed the proportion was indeed 0.2. By virtue of 0.2 not being present in the interval, we know, automatically, that the null would be rejected (and vice versa). (If 0.2 had been present in the confidence interval, then there would not have been enough evidence to reject the null.)

This connection between confidence intervals and hypothesis testing is called *duality*, and usually works well if the following conditions are met:

- The significance level and the confidence level are complements (e.g., 10% and 90%, or 5% and 95%); and
- The hypothesis test is two-tailed.

In the example above only the first condition was met, yet there was still a correspondence between the confidence interval and the hypothesis test (if we had lowered the significance level, the confidence interval would have expanded, thus including 0.2 and causing a disagreement in the conclusions). But even if both conditions are met, duality isn't exact. Here's why: the standard error is used for confidence interval calculations— $\hat{p} \pm z * \sqrt{\dfrac{\hat{p}(1-\hat{p})}{n}}$ —while the conjectured value is used for the test statistic— $z = \dfrac{\hat{p} - p_0}{\sqrt{\dfrac{p_0(1-p_0)}{n}}}$. Nonetheless, working through both the hypothesis test and the associated confidence interval results in a more thorough analysis.

§9b. *Population Mean Significance Testing*

W e will now perform hypothesis testing using the population mean μ . There are many similarities to tests of significance with population proportions, of course, but there are also several important differences. For instance, we won't be using the normal distribution. The t distribution will be reintroduced, since it is extremely unlikely we would ever have the value of the population standard deviation σ .

With all that in mind, let's complete a **One-Sample t Test** problem.

Example 9b.-1

The following is a data set of a random sample of healthy adult males' systolic blood pressure readings:

125, 136, 110, 131, 143, 140, 115, 112, 109, 142, 126, 134, 112, 140, 124

A researcher wants to test the claim that men's normal resting systolic blood pressure is 120. Conduct a complete hypothesis test.

Solution. It makes the most sense to use a not-equal-to alternative hypothesis, since we don't know whether the collected data will fall greater or less than 120 (recall, a not-equal-to alternative is also called a double-sided or two-sided or exploratory hypothesis). Thus, the null and alternative hypotheses are

$$H_0 : \mu = 120 \text{ (the mean of men's resting systolic blood pressure is 120)}$$
$$H_a : \mu \neq 120 \text{ (the mean of men's resting systolic blood pressure is not 120)}$$

The technical conditions for this inference procedure are (1) either the sample size is at least 30, or the population is normal, (2) the sample data was obtained randomly, and (3) the 10% Condition (if sampling without replacement).

Although the sample size is small, recall that, at the end of the last §, we used a modified box-plot and an NPP to check the normality condition, which was satisfied. In addition, the sample is random, and the sample size is certainly much less than 10% of the population of all healthy adult males.

Thus, we can proceed with finding a test statistic for a sample mean, whose formula is analogous to the test statistic for a sample proportion:

$$t = \frac{\bar{x} - \mu_{\bar{x}}}{\frac{s}{\sqrt{n}}}$$

where the degrees of freedom is one less than the sample size.

The test statistic is equal to

$$t = \frac{126.6 - 120}{\frac{12.489}{\sqrt{15}}} = 2.047$$

From there, travel to the 14th row of the *t* distribution and try to find, as closely as you can, 2.047 (it is between two *t* scores). See the illustration on the top of the next page.

So the *p*-value seems like it is between 0.025 and 0.05—except, remember, with a two-sided test, the *p*-value needs to be multiplied by two. Therefore, the *p*-value is between 0.05 and 0.10. At a 5% level of significance—which, when none else is stated, we default to—we fail to reject the null and do not have enough evidence to claim that the mean of men's resting systolic blood pressure is not 120.

df	0.25	0.2	0.15	0.1	0.05	0.025	0.0
14	0.692	0.868	1.076	1.345	1.761	2.145	2.26
15	0.691	0.866	1.074	1.341	1.753	2.131	2.24

The 95% confidence interval, which is (119.68, 133.52), contains the conjectured value of interest (120), confirming the fail-to-reject test decision.

The test statistic in the example above is between two *t* scores on the *t* distribution table. But if the test statistic is, quite literally, off the chart, there are two possibilities, as illustrated by the diagram below.

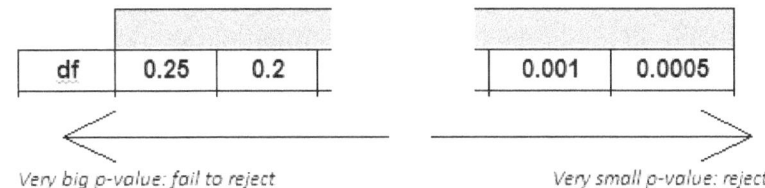

df	0.25	0.2		0.001	0.0005

Very big p-value: fail to reject   *Very small p-value: reject*

Rather unsurprisingly, the graphing calculator option T-Test can help us avoid using the *t* distribution chart altogether. Entering the list into the calculator and selecting T-Test (with $\mu_0 = 120$), the test statistic is given as 2.047 with a *p*-value of 0.059.

When a matched pairs experiment is conducted, two sets of data emerge from the study: two data points for each individual. When such data points are matched up (such as from before-and-after studies, for example), the data are considered *dependent* data. These *paired-difference problems* can be analyzed by using the *t* distribution, as long as we examine only the *differences* between the data pairs.

Example 9b.-2

A small random sample of adult males with elevated blood pressure are recruited for a study of a new blood pressure medication. Their resting systolic blood pressures are recorded before taking the medicine as well as after one month of daily doses. The data, with the differences calculated ("After" subtracted from "Before"), are shown at the top of the next page.

At the 1% level of significance, can you claim that the medicine had a significant effect on reducing blood pressure?

Subject	1	2	3	4	5	6	7	8
Before	135	150	164	140	149	138	182	176
After	121	134	132	115	135	140	152	135
Difference	14	16	32	25	14	-2	30	41

Solution. We will only be examining the differences' data set (since the data pairs are dependent). If a difference is positive, the medicine had a beneficial effect; if negative, the medicine had a harmful effect; if zero, the medicine had no effect.

The hypotheses, using the symbol μ_d to represent the mean of the *population* of differences of men's resting systolic blood pressure readings, are

$$H_0 : \mu_d = 0 \text{ (the medicine has no effect on men's resting systolic blood pressure)}$$
$$H_a : \mu_d > 0 \text{ (the medicine lowers men's resting systolic blood pressure)}$$

The technical conditions need to be checked. First, let's examine an NPP of the differences:

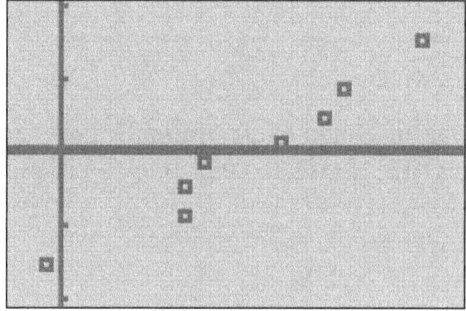

The plot appears relatively straight and linear, with some exceptions perhaps on the left side. Nonetheless, pressing on, we also note that the sample is random, and the sample size is very small compared to the population of all men with elevated blood pressure.

Using the T-Test calculator function, the test statistic is 4.463, with a *p*-value of 0.001. We would reject the null hypothesis. The associated 99% confidence interval of the differences confirms the test decision, since zero is not present in the interval: (4.588, 37.912). We have enough evidence to claim that the medicine lowers resting systolic blood pressure among men with elevated blood pressure.

§9c. *Committing Errors*

The test decision, whether rejecting or failing to reject the null hypothesis, is based on results from a sample. Because of sampling variability, and the possibility of obtaining an "unlucky" (non-representative) sample, there's no way to be absolutely certain that you've made the correct decision, unless you can somehow manage to collect data from the entire population—at which point a hypothesis test would be unnecessary because you'd already have the values of all parameters of interest. Two possible errors could be made: a *type I error* or a *type II error*.

Definition 9c.-1

Type I error. When a true null hypothesis is rejected; in other words, a "meaningful" pattern was falsely perceived (i.e., apophenia). The probability of this error is represented by α.

Definition 9c.-2

Type II error. When a false null hypothesis is not rejected; in other words, a truly meaningful pattern was dismissed (i.e., randomania). The probability of this error is represented by β.

Either a type I error or a type II error might be committed, but they cannot be made simultaneously. These errors can be mapped visually using a "decision matrix":

Decision	Null is true	Null is false
Fail to reject	Correct	Type II error
Reject	Type I error	Correct

Pascal's Wager (also called Pascal's Gambit) is a famous and well-worn analogue to hypothesis testing (and, by extension, type I and type II errors) devised by Blaise Pascal, the philosopher-mathematician and ardent religionist. "God is, or he is not," Pascal said. "Which way should we incline? Reason cannot answer."

So Pascal created a "payoff matrix" to attempt to solve the problem. He argued that one's belief in a deity was a lifetime bet where one had little to lose (the type I error would be leading a "good" life according to scripture, yet there was no need to after all) and much to gain—i.e., eternal salvation—while, by not believing, one had much to lose: a miserable afterlife spent in eternal damnation (the type II error). So, if necessary, fake your belief until it becomes second nature: "You would like to attain faith, and do not know the way; you would like to cure yourself of unbelief, and ask the remedy for it... by acting as if [you] believed, taking the holy water, having masses said, etc. Even this will naturally make you believe, and deaden your acuteness."

Needless to say, there are numerous holes in Pascal's argument. For example, would any deity be naïve enough not to penalize someone for falsely believing?

What's especially interesting about the wager, though, is the systematic use of the notions of probability outside of games of chance (such as the Problem of Points [POP]). The wager, not the POP, is the "first significant statement of the philosophy of probability in history—the old era had ended," according to Gavin Kennedy. Pascal rejects Descartes' deductive approach to proving the existence of God—recall Descartes' axiomatic method: "I found that the existence of the [Perfect] Being was comprised in the idea in the same way that the equality of its three angles to two right angles is comprised in the idea of a triangle…."—instead opting for a demonstration rooted in expected values and payoffs. Pascal's Wager later cued seventeenth century mathematician Christian Huygens to explore the notion of expected value with respect to games.[*] "Although in a pure game of chance," he wrote, "the results are uncertain, the chance that one player has to win or to lose depends on a determined value."

Hypothesis testing can also be compared to the U.S. legal system. The null hypothesis is, of course, innocence; the alternative is guilty. If a jury finds a defendant "guilty," and he really was guilty, then that's a correct decision; likewise a jury finding an innocent man "not guilty."[†] But if a jury finds an innocent man "guilty," or a guilty man "not guilty," then a mistake has been made—and justice has not been served. The former is a type I error, the latter a type II error. Which error you consider worse depends upon the kind of society you wish to live in.

However, putting aside questions of which error is worse, observe that the probabilities of the two errors have an inverse relationship with each other: as one goes down, the other goes up. Imagine living in a totalitarian state like North Korea: the merest whisper of wrongdoing might land you in jail. There, the probability of a type I error is very high, since convictions occurs with the scantest of evidence, locking up (probably) many innocent people. Now imagine the reverse scenario: a state in which the burden of proof is very high on the prosecution. In this society, lock-ups occur infrequently, if at all, so the probability of a type II error—letting the guilty walk free—is high. The probability of type I and type II errors, in any society (real or imagined), can't *both* be high. As the chance of committing a type I error rises, the chance of making a type II error falls (and vice versa). Expressed symbolically: as α increases, β decreases, and as α decreases, β increases.

When the significance level α is set in a hypothesis test, the probability of making a type I error is set as well. Understanding the inverse relationship between α and β answers another question: Why not just make α incredibly low—say, 0.00000001—for every significance test, setting the burden of proof extremely high? Because then β, the probability of a type II error, explodes. You must therefore always be judicious when setting a significance level.

Example 9c.-1

A software company wants to produce new anti-spam software to help prevent users from getting unwanted e-mail messages. Assume that spam filtering is like the U.S. judicial system: the e-mail message is "innocent" (i.e., not spam) until proven otherwise. Draw a decision matrix detailing all four possible decisions the anti-spam software can make about each e-mail message; in addition, describe the consequences of each error.

---

[*] And also Daniel Bernoulli, from the famous Bernoulli family of learned scholars, to expand the idea of expected value to capture the idea of *utility*—the personal valuation of risk.

[†] Notice the wording here: "not guilty" is just as much of a hedge as "fail to reject." ("Beyond a shadow of a doubt" is the de facto significance level.) U.S. juries, or judges for that matter, never proclaim a defendant "innocent."

Solution. The decision matrix is as follows:

Decision	E-mail isn't Spam	E-mail is Spam
Not Spam	Correct	Type II error
Spam	Type I error	Correct

The consequences of a type I error: an e-mail that is valid is sent to a spam folder and perhaps never even seen by the user. The consequences of a type II error: a spam e-mail is thrown to a user's inbox and has to be manually deleted.

Example 9c.-2

Suppose your friend Ruth claims that a coin is weighted, but you don't think so. You decide to test it. Your $\hat{p}$ value ends up being 0.42, at $\alpha = 0.05$. Describe the type I and type II errors.

Solution. If you make a type I error, you reject that the coin is fair when, in reality, it is fair. If you make a type II error, you fail to reject that the coin is fair (in other words, you claim that the coin is fair) when it actually is weighted (unfair).

Example 9c.-3

A new medical treatment for chronic heart disease shows great promise and is about to undergo clinical trials. Describe the type I and type II errors, and the consequences of each.

Solution. The type I error: concluding that the new medicine is effective when it really is not; people will be given ineffective medicine. The type II error: concluding the new medicine is ineffective when it is effective; a medical advance, perhaps with great promise, will be inadvertently discarded.

Medical science, especially with screening and treatment of patients, is the most ripe area for committing type I and type II errors. Screening otherwise healthy (and especially younger) populations for a variety of diseases, such as breast cancer, can lead to all sorts of diagnostic errors—and much needless pain and suffering—of those screened. These mistaken treatments—this persistent apophenia and, to a lesser extent, randomania—serve as reminders of the continued fallibility of the medical profession, even in our advanced technological age.[*]

---

[*] See the work of Gerd Gigerenzer for more.

§9d. *Power*

L ord Acton famously noted that "Power tends to corrupt, and absolute power corrupts absolutely." Not so in statistics, however. The more *power*, the better.

Definition 9d.-1

Power. The probability of correctly rejecting a false null hypothesis.

Since power is the probability of one of the two correct decisions, it shouldn't be surprising that increasing the sample size n likewise increases the power.

In addition, it's easy to understand why if we could somehow decrease the population standard deviation σ, the power would increase—there would be less spread, less variability, and an increased probability of making a correct decision. (In practice, of course, we can't alter the value of any parameter, σ included.)

Harder to conceptualize, however, is why the power increases if the conjectured value of interest is moved further away from the true population value, so an example is in order.

Example 9d.-1

Two researchers, Carolyn and Jenna, decide to study American adult females' heights. Before collecting any data, Carolyn sets two hypotheses:

$$H_0 : \mu = 65 \text{ inches}$$
$$H_a : \mu \neq 65 \text{ inches}$$

Jenna, though, has other ideas. She sets her hypotheses as

$$H_0 : \mu = 25 \text{ inches}$$
$$H_a : \mu \neq 25 \text{ inches}$$

Both Carolyn and Jenna collect random samples of size thirty of adult women with which to test their hypotheses. Which researcher has more power?

Solution. Clearly, Jenna has a much better chance of rejecting her null hypothesis. Her conjectured value of interest, 25 inches, is very far from what the true average height of American females is likely to be. Because Jenna's probability of rejecting the null is much, much higher than Carolyn's, Jenna has more power in her test.

Let's look again at a decision matrix of the errors and correct decisions, this time with the symbols inserted:

Decision	Null is true	Null is false
Fail to reject	Correct	Type II error (β)
Reject	Type I error (α)	Correct (POWER)

The power is the probability of rejecting a false null hypothesis; that's the same as saying that the power is the probability of the complement of rejecting a false null; which is the same as saying that the power is the probability of the complement of making a type II error. Thus, each column in the decision matrix above sums to one.

With all that in mind, we have arrived at a formula for power:

$$\text{Power} = 1 - \beta$$

Also, we now see that power travels in the same direction as α, but in the inverse direction of β. So, if we increase α, we also increase power—simply because we're increasing the chances of rejecting the null and (perhaps) making a correct decision by doing so.

Although we won't calculate power here—in order to do so, a hypothesized "alternative value" of the parameter must be set, and the power found off of that—but statisticians generally agree that a hypothesis test with a power greater than 0.8 is considered sufficient.

§9e. *Comparing Independent Samples*

Tests of hypothesis aren't restricted to examining one sample of independent data at a time. Consider an experiment testing a weight loss pill against a placebo: how big does the average weight differential between the two groups have to be? What constitutes statistically significant data in this case? And what about confidence intervals—how do they factor in to the analysis?

The clinical trials of pharmaceutical corporations rely on such tests of significance of independent samples to measure the efficacy of their chemical compounds. Putting aside the ethical issues of side effects (and other Big Pharma-related notions), does the significance testing on these proto-drugs have a strong measure of scientific validity? According to Melody Petersen, author of the book *Our Daily Meds: How the Pharmaceutical Companies Transformed Themselves into Slick Marketing Machines and Hooked the Nation on Prescription Drugs*, the answer is, disturbingly, no: statistical results are oftentimes cherry-picked:

> In 2002, Dr. Arif Khan, a psychiatrist in Bellevue, Washington, reviewed the data from the dozens of clinical trials that companies had performed to prove that Zoloft, Prozac, Paxil, and six other antidepressants actually worked.

> These drugs are now some of the most prescribed medicines in America. Dr. Khan and his colleagues found 52 completed trials of these drugs, which involved more than 10,000 patients. In more than half of these studies, the sugar tablet relieved the patients' depression just as well as, or better than, the antidepressant.

And even when the data are not cherry-picked, the effectiveness of certain types of cancer drugs are minimal at best, as Steven Levitt explains in *SuperFreakonomics: Global Cooling, Patriotic Prostitutes, and Why Suicide Bombers Should Buy Life Insurance*:

> Chemo is effective on some cancers, including leukemia, lymphoma, Hodgkin's disease, and testicular cancers, especially if these cancers are detected early. But in most other cases, chemo is remarkably ineffective. An exhaustive analysis of cancer treatment in the U.S. and Australia showed that the five-year survival rate for all patients was about 63 percent but that chemo contributed barely 2 percent to this result.
>
> There is a long list of cancers for which chemo has zero discernible effect, including multiple myeloma, soft-tissue sarcoma, melanoma of the skin, and cancers of the pancreas (what Steve Jobs died of), uterus, prostate, bladder, and kidney. Even with lung cancer, chemo is not very effective at extending life.

All of these disheartening conclusions rely on significance testing of multiple independent data sets. So how, exactly, are significance tests of two independent data sets implemented? Such inference procedures, like their simpler counterparts, rely on sampling distributions—in this case, sampling distributions of the *differences* of population parameters. Let's begin by examining categorical data (proportions) and then moving on to quantitative data (means).

When comparing the proportions from two populations, we'll make use of the following null hypothesis:

$$H_0 : p_1 - p_2 = 0 \text{ or, rearranged, } H_0 : p_1 = p_2$$

The alternative hypothesis can take one of three forms:

$$H_a : p_1 - p_2 > 0 \text{ or, rearranged, } H_a : p_1 > p_2$$
$$H_a : p_1 - p_2 < 0 \text{ or, rearranged, } H_a : p_1 < p_2$$
$$H_a : p_1 - p_2 \neq 0 \text{ or, rearranged, } H_a : p_1 \neq p_2$$

The test statistic is $z = \dfrac{(\hat{p}_1 - \hat{p}_2) - (p_1 - p_2)}{\sqrt{\hat{p}_c(1-\hat{p}_c)\left(\dfrac{1}{n_1} + \dfrac{1}{n_2}\right)}}$, but notice that since we always assume that

the null hypothesis is true when determining the *p*-value, $p_1 - p_2 = 0$. Also notice the presence of a new proportion, $\hat{p}_c$, called the *combined proportion*, which is a weighted average of the two

sample proportions: $\hat{p}_c = \dfrac{x_1 + x_2}{n_1 + n_2}$.

The technical conditions are a little more extensive this time. First of all, there needs to be some sort of randomization, whether that be with the samples being compared having been collected randomly, or the sample data having been obtained through the use of a randomized experiment. Next, there needs to be at least ten successes and ten failures from each group, giving us four sets of inequalities to check. In addition, the samples need to have been collected independently from one another.[*] And, finally, the 10% Condition needs to be checked (if sampling without replacement).

A confidence interval of the difference of the two means can also be constructed:

$$(\hat{p}_1 - \hat{p}_2) \pm z^* \sqrt{\frac{\hat{p}_1(1-\hat{p}_1)}{n_1} + \frac{\hat{p}_2(1-\hat{p}_2)}{n_2}}$$

Let's work through an example.

Example 9e.-1

A new pill to decrease the incidence of strokes among high-risk populations is tested. Fifty people suffering from high cholesterol and at a high risk of stroke are randomly given the medicine; the other 45 individuals in the study are given a placebo. The study lasts for one year. The researchers find that of the 50 subjects who took the medicine, 5 had strokes, but of the 45 subjects who took the placebo, 12 suffered strokes. Is there sufficient evidence that the medicine is effective at reducing the incidence of strokes?

Solution. Our two hypotheses, replete with more descriptive subscripts than "1" and "2," are

$H_0 : p_m - p_p = 0$ (The proportion of those who suffer strokes while taking the medicine is the same as the proportion of those who suffer strokes while taking the placebo)

$H_0 : p_m - p_p < 0$ (The proportion of those who suffer strokes while taking the medicine is less than the proportion of those who suffer strokes while taking the placebo)

The two sample proportions are $p_m = 5/50 = 0.1$ and $p_p = 12/45 = 0.267$, meaning the com-

---

[*] As psychologist Richard Nisbett explains in his book *Mindware*, comparing two new teaching procedures in two different classrooms of more than two dozen students each—procedure A in classroom 1, and procedure B in classroom 2—will result in a sample of size of only two: "This is because N equals the number of cases only when there is *independence of observations*. But in the case of a classroom of students or any group of people who interact with one another during the period of treatment and measurement of its effects, the individual behaviors are not independent of one another."

bined proportion is $\hat{p}_c = \dfrac{5+12}{50+45} = 17/95 = 0.179.$[*] Before finding a test statistic, p-value, and confidence interval, we have to check the conditions for inference.

This study had randomization in the assignment of subjects to the treatment groups.[†] Although $0.9 \cdot 50 = 45$, $0.267 \cdot 45 = 12.015$, and $0.733 \cdot 45 = 32.985$ are all greater than 10, $0.1 \cdot 50 = 5$ is not—so the normal approximation for the binomial distribution of the differences of the proportions cannot safely be classified as normal in shape (although some statisticians are a bit more lenient on this condition, saying that five successes is sufficient, so we'll press on). And the samples are certainly independent of each other, since no subject received both the pill and the placebo.

Although it might be enjoyable by some to calculate the test statistic and p-value by hand—feel free to do so if that strikes your fancy—let us instead lean on technology and use the graphing calculator.[‡] The calculator function for the **two-sample z test for the difference between two proportions** is 2-PropZTest. A sample calculator screenshot is shown below.

Using 2-PropZTest, we find a test statistic of –2.17 and a p-value equal to 0.017, significant at the 5% level. A 95% confidence interval for the data can be found using the 2-PropZInt calculator function: (–0.3203, –0.0130). Since zero, the conjectured value, is not in the interval at the 95% level, we reject the null hypothesis at the 5% level. In addition, we are 95% confident that the difference in the population proportion of strokes between the two groups is between –0.3203 and –0.0130. Thus, there is enough evidence to claim that the proportion of those who suffer strokes while taking the medicine is less than the proportion of those who suffer strokes while taking the placebo.

---

[*] Realize that adding up the two proportions and then dividing by 2 will not give a *weighted* average—unless, of course, the two samples sizes are the same.

[†] Although the study most certainly didn't have random selection, since the subjects for experiments of this nature are usually volunteers.

[‡] Computer statistical software, such as SPSS, Fathom, Minitab, and Excel, can perform all of the inferential procedures presented in this §. A quick online search will give you any particular instructions you may require.

When comparing the means from two populations, we'll make use of the following null hypothesis:

$$H_0 : \mu_1 - \mu_2 = 0 \text{ or, rearranged, } H_0 : \mu_1 = \mu_2$$

The alternative hypothesis can take one of three forms:

$$H_a : \mu_1 - \mu_2 > 0 \text{ or, rearranged, } H_a : \mu_1 > \mu_2$$
$$H_a : \mu_1 - \mu_2 < 0 \text{ or, rearranged, } H_a : \mu_1 < \mu_2$$
$$H_a : \mu_1 - \mu_2 \neq 0 \text{ or, rearranged, } H_a : \mu_1 \neq \mu_2$$

The test statistic is $t = \dfrac{(\bar{x}_1 - \bar{x}_2) - (\mu_1 - \mu_2)}{\sqrt{\dfrac{s_1^2}{n_1} + \dfrac{s_2^2}{n_2}}}$, but notice that since we always assume that the

null hypothesis is true when determining the p-value, $\mu_1 - \mu_2 = 0$. There are two ways to obtain the degrees of freedom for the test statistic: either by taking one less than the smaller sample size (this is termed the conservative calculation) or by using technology (a complicated formula calculating the df awaits those who dare to rush in).

The technical conditions are as follows. First of all, there needs to be some sort of randomization, whether that be with the samples being compared having been collected randomly, or the sample data having been obtained through the use of a randomized experiment. Next, both sample sizes either need to be at least thirty or both samples need to have come from normally distributed populations. In addition, the samples need to have been collected independently from one another. And, finally, the 10% Condition needs to be checked (if sampling without replacement).

A confidence interval of the difference of the two means can also be constructed:

$$(\bar{x}_1 - \bar{x}_2) \pm t^* \sqrt{\dfrac{s_1^2}{n_1} + \dfrac{s_2^2}{n_2}}$$

Let's work through an example.

Example 9e.-2

A new weight loss pill is tested. One hundred adult volunteers suffering from obesity are randomly given the pill; another 95 such volunteers are given a placebo. The study lasts for three months; all participants are also required to perform a regimen of aerobic exercises of thirty minutes per day.

The researchers find that of the 100 subjects who took the medicine, the average weight loss was 29 pounds (with a standard deviation of 15 pounds), but of the 95 subjects who took the placebo, the average weight loss was only 12 pounds (with a standard deviation of 22 pounds). Is there sufficient evidence that the weight loss pill is effective?

Solution. Our two hypotheses, replete with more descriptive subscripts than "1" and "2," are

$H_0 : \mu_m - \mu_p = 0$ (The mean weight loss among the medicine takers is equal to the mean weight loss among the placebo takers)

$H_0 : \mu_m - \mu_p > 0$ (The mean weight loss among the medicine takers is greater than the mean weight loss among the placebo takers)

Before finding a test statistic, p-value, and confidence interval, we have to check the conditions for inference.

This study had randomization in the assignment of subjects to the treatment groups, though volunteers were used. The sample sizes of both groups are well above thirty individuals. And the samples are independent of each other, since a single subject did not receive both the weight loss pill and the placebo.

Feel free to calculate the test statistic and p-value by hand, but technology makes that unnecessary. (By the way, if referring to the t distribution table, use 94 degrees of freedom—one less than the smaller sample size—which you'll then subsequently round down to the nearest available row.) The calculator function for the **two-sample t test for the difference between two means** is 2-SampTTest. A sample calculator screenshot is shown below.

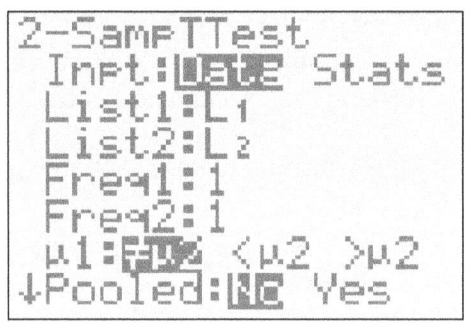

The "Pooled" option is referencing the variances of the populations. If "Yes" is selected, we're giving the calculator consent to proceed as if both population variances are equal (population standard deviations would be utilized in the test statistic formula). But it is unrealistic that we would either know the values of the population variances or know if those variances are equal in measure. Although this "pooled variance assumption" increases the test's statistical power, as well as altering the value of the test statistic, we won't be making use of it in these pages.

Using 2-SampTTest, we find a test statistic of 6.27 and a p-value equal to 0^+, significant at the 5% level. A 95% confidence interval for the data can be found using the 2-SampTInt calculator function: (11.649, 22.351). Since zero, the conjectured value, is not in the interval at the 95% level, we reject the null hypothesis at the 5% level. Also note how far from zero both the lower and upper bounds of the interval sit: we can be 95% confident that the difference in the population average weight loss between the two treatments is between 11.649 and 22.351 pounds. Thus, there is enough evidence to claim that the mean weight loss among the medicine takers is greater than the mean weight loss among the placebo takers.

§9f. *Chi-Square*

The Italian engineer and scientist Vilfredo Pareto is perhaps best known for his Pareto Principle. Real-world examples abound of his eponymous principle, as Vincent Deary explains in *How We Are*:

> It turns out that we didn't do a lot of killing in the first two world wars. A lot of people died, but not many killed. We just couldn't do it, not when it came down to it, not when we came face-to-face and hand-to-hand with the other life that we were meant to extinguish. We just couldn't do it, most of us. The statistic is the usual one, illustrating the economist Vilfredo Pareto's principle of the vital few, also known as the 80-20 law. Twenty per cent of people own 80 per cent of the wealth; 20 per cent of the population take 80 per cent of the sick time; 20 per cent of the criminals are responsible for 80 per cent of the crime. There's always a determined minority hogging the lion's share of any common property.

Pareto was one of the earliest scientists to study the distribution of wealth—something very much under the economic microscope today, though it has a different name: *wealth inequality*. According to the Federal Reserve Board Survey of Consumer Finances (2014), the "Ownership of U.S. Financial Assets, 2013, by Family Wealth Class" is distributed as such:

	Bottom Half	Next 25%	Next 15%	Top 10%
Wealth	0.8%	3.0%	11.5%	84.5%

The inequality gap is astounding, and is more extreme in the U.S. now than ever before—which can be seen simply by comparing distributions from the past with those of today.

When looking at these *one-way tables*—"one-way," since they house categorical data collected from a single variable—we see how closely what we *expect* matches with the data we've *collected*. The match doesn't have to be perfect; after all, the foundational statistical concept of variability must always be a consideration. (For instance, a distribution of the wealth statistics from residents in the city you live in probably won't perfectly match the one-way table of wealth distribution above.) But, within limits, it has to be close, otherwise we have evidence to claim that the data gathered does not sync up with the expected distribution. Let's revisit a familiar example, albeit slightly modified.

Example 9f.-1

Suppose a factory claims that 25% of all vegetables packaged into their bags are frozen peas.

Carrots make up 25%, corn constitutes 40%, and pieces of broccoli account for the remaining 10%. A single bag of 8 oz. frozen vegetables has 100 vegetables contained inside.

(a) How many vegetables would you expect of each type in a single bag of 100? Are you guaranteed to obtain these counts? Justify your answer.

Twenty-five peas, 25 carrots, 40 pieces of corn, and 10 pieces of broccoli. You are not guaranteed to obtain these counts because there is variability from bag to bag.

(b) Suppose a bag of vegetables contained 10 peas, 30 carrots, 50 pieces of corn, and 10 pieces of broccoli. Would you have reason to suspect that the factory's stated distribution of vegetables in each bag is incorrect?

Perhaps. More analysis is required.

We need to have a systematic way to quantify (and, ultimately, classify) how divergent from a distribution is *too* divergent for a given data set. Because there are a number of proportions to compare—after all, we're not just examining peas this time—we can't use multiple tests of two proportions, comparing two vegetables at a time; not only would this be terribly wasteful and cumbersome, but it would also result in what's termed the *problem of multiple comparisons.*[*]

Instead, we will lean on a new distribution, called the chi-square distribution.[†] A table of chi-distribution values is presented at the back of this primer; this table's use is very similar to the *t* distribution table's: degrees of freedom clue you in to the correct row to use, and the columns relay the appropriate *p*-value. But the chi-square distribution,[‡] unlike the *t* distribution or normal distribution, skewed to the right; like the *t* distribution, though, a different chi-square distribution exists for each degree of freedom—the higher the df, the less right skewed the shape, with the mean of any specific distribution located at the df value.

With a one-way table, the degrees of freedom is equal to one fewer than the number of categories. The *expected counts*—the frequency of observations we would anticipate seeing once we collect data (based off of an a priori theoretical relative frequency distribution)[§]—must, in every category, be equal to at least five observations, otherwise one of the technical conditions for in-

---

[*] As strange as this may sound, the multiple comparisons problem led to *dead* salmon showing statistically significant evidence of responding to images of people in various social situations (due to statistical error in fMRI data). As comedian Patricia Marx, in her book *Let's Be Less Stupid*, writes, "The researchers did not say what the dead fish was thinking, but my guess is, 'I'd rather be lox.'" For more about the dead fish study, please see the following article: http://blogs.scientificamerican.com/scicurious-brain/ignobel-prize-in-neuroscience-the-dead-salmon-study/

[†] "Chi," referencing the Greek letter, is pronounced "kye," not "chee."

[‡] The chi-square distribution is a special case of the *gamma distribution*, which is itself a special case of the exponential distribution (recall the connection between the exponential and Poisson distributions discussed earlier).

[§] If any expected counts end up having decimal components, round as little as possible.

ference is not met. (The others are randomization, independence, and the 10% Condition if sampling without replacement.) The expected and *observed counts* (i.e., the data collected and summarized in the one-way table) are formally compared using the *chi-square test statistic*:

$$\chi^2 = \sum \frac{(\text{Observed} - \text{Expected})^2}{\text{Expected}}$$

Once the test statistic is calculated, the *p*-value is found by traveling to the correct row in the chi-square distribution table (based on the df), locating, within the given critical values, the test statistic as closely as possible, and scrolling upward to the associated tail probability.

The closer that the chi-square test statistic is to zero, the less chance there is of rejecting the null hypothesis—and vice versa. Here's why: if the observed and the expected counts are equal—meaning that the data gathered matches the distribution anticipated perfectly, without any variability whatsoever—the numerators of the terms of the test statistic will all equal zero, resulting in a test statistic of zero, and non-significant data. Contrariwise, if the observed and expected counts are light years apart, this divergence will result in the test statistic blowing up—and statistical significance the only possible conclusion to be drawn.

Chi-square hypothesis testing was jointly developed by Fisher and Pearson but unfortunately led to many public *ad hominem* attacks between them.

Example 9f.-2

Reconsider the frozen vegetables example.

(a) Create a one-way table of the expected counts for a bag containing 100 vegetables.

	Peas	Carrots	Corn	Broccoli
Expected Counts	25	25	40	10

(b) Are all of the expected counts at least 5, satisfying the technical condition?

Yes, easily.[*]

(c) A fresh bag of frozen vegetables is opened. In it, there are 10 peas, 30 carrots, 50 pieces of corn, and 10 pieces of broccoli. Revise your one-way table to account for both the observed and expected counts.

---

[*] Suppose that one category contained fewer than 5 observations. Statisticians, to keep the procedure of inference valid, might combine this small category's results with another category, resulting in the technical condition being met.

	Peas	Carrots	Corn	Broccoli
Observed Counts	10	30	50	10
Expected Counts	25	25	40	10

(d) What would the null hypothesis be?

The factory's distribution of types of vegetables in a frozen vegetable bag is correct.

(e) What would the alternative hypothesis be?

The factory's distribution of types of vegetables in a frozen vegetable bag is incorrect.[*]

(f) Check the remaining technical conditions.

We can assume that the vegetables were packaged randomly, subject to the factory's distribution of vegetable type. In addition, since there are more than ten times as many vegetables in the entire population as there are in a single bag,[†] the 10% Condition holds.

(g) Calculate the chi-square test statistic.

$$\chi^2 = \frac{(10-25)^2}{25} + \frac{(30-25)^2}{25} + \frac{(50-40)^2}{40} + \frac{(10-10)^2}{10} = 12.5$$

(h) Which category contributes the most to the test statistic?

The category contributing the most has the largest differential between the observed and expected counts: here, the peas.

(i) Find the p-value.

The degrees of freedom is one fewer than the number of categories; since there are four categories, the df is 3.
    Next, go to the appropriate row, and search for the numbers sandwiching 12.5:

---

[*] Writing out null and alternative hypotheses in symbols is possible, but too tedious and time-consuming and is ultimately unnecessary.

[†] In other words, there are at least 10 x 100 = 1,000 vegetables in the host population.

				Right Tail Probability					
df	0.25	0.20	0.15	0.10	0.05	0.03	0.02	0.01	0.005
3	4.11	4.64	5.32	6.25	7.81	9.35	9.84	11.34	12.84
4	5.39	5.99	6.74	7.78	9.49	11.14	11.67	13.28	14.86

The *p*-value is between 0.005 and 0.01.

(j) What is your test decision?

We reject the null hypothesis at the 5% significance level.

(k) Relay any conclusions in context.

We have enough evidence to claim that the factory's distribution of types of vegetables in a frozen vegetable bag is incorrect.

Some graphing calculators have the capability to find the *p*-value from a **chi-square goodness-of-fit test**, but others don't—specifically, newer TI-84s can do it, but 83s and some 84s can't.

Check to see if you have the function: hit the STAT button, scroll to the TESTS menu, and see if the GOF test is present.

If the GOF test is there, you'll need to enter in both the observed counts as a list (tip: use L1) and the expected counts as a list (tip: use L2). Then you can utilize the GOF test to rapidly find the test statistic and *p*-value.

Suppose you have collected numerical data of some sort and want to investigate the probabilities of certain leading digits. Does 5 come up more than 6, or 7? How often does 1 come up? Organizing the data into a one-way table, with all nine leading digits as categories (zero is not present), you may anticipate an expected relative frequency distribution, assuming a uniformity of frequencies, to look like this:

Leading Digit	1	2	3	4	5	6	7	8	9
Probability	11.1%	11.1%	11.1%	11.1%	11.1%	11.1%	11.1%	11.1%	11.1%

Benford's law would say otherwise, however. Benford's law, named after physicist Frank Benford but not first discovered by him (Stigler's law of eponymy rears its head again), is a "law" describing the expected relative distribution of leading digits obtained from real-world data. Although you might expect the first digit of data to have an equal probability of being 1 through 9, it turns out that 1 is more than five times as frequently occurring as 9; in fact, in ascending order, starting from 1, each digit becomes less and less likely to be observed.

Benford's law predicts data gathered from income distributions, stock prices, and population figures, among other data sets, reasonably well. Built off of a logarithmic scale, the law has the following distribution:

Leading Digit	1	2	3	4	5	6	7	8	9
Probability	30.1%	17.6%	12.5%	9.7%	7.9%	6.7%	5.8%	5.1%	4.6%

Comparing the leading digits of collected data to their expected counts—courtesy of Benford's law—and then using a chi-square goodness-of-fit test to determine statistical significance should, at this juncture, strike you as straightforward.

What if the data come in the form of a two-way table, however? Can the chi-square distribution still be utilized to obtain a *p*-value? The answer is yes—but there are a number of qualifications.

For one, the expected counts are calculated differently; rather than relying on some outside distribution, the expected counts are found by considering only the marginal totals. Also, the degrees of freedom are calculated by taking the product of one fewer than the number of rows and one fewer than the number of columns. And the technical conditions are modified a bit. Let's work through an example.

Example 9f.-3

Consider this two-way table of political affiliation and the type of car one drives:

	Political Affiliation			
	Democrat	Republican	Other	**Total**
American	32	21	15	**68**
Foreign	55	40	8	**103**
Total	**87**	**61**	**23**	**171**

(a) Would you say that there's an association between political affiliation and the type of car one drives? If so, describe the strength and direction of the association and explain why this could be.

Perhaps. It seems like Democrats, by a slightly higher margin than Republicans, prefer foreign cars.

In a two-way table, oftentimes the data in the cells don't match up with our expectations. If the observed counts match the expected counts exactly, then the data is completely independent, meaning that one variable has absolutely no effect on the other.

We have to figure out a way to calculate what is expected. To determine the expected counts, calculate the following for each cell:

$$\text{Expected Count of the Cell} = \frac{(\text{Row Marginal Total}) \cdot (\text{Column Marginal Total})}{\text{Grand Total}}$$

(b) Find the number of Democrats expected to buy American cars.

$$\text{Democrats, American Cars} = \frac{68 \cdot 87}{171} = 34.5$$

(c) Fill in the following two-way table, this time finding the expected counts for each of the cells.

	Political Affiliation			
	Democrat	Republican	Other	**Total**
American	*34.5*	*24.3*	*9.1*	**68**
Foreign	*52.4*	*36.7*	*13.9*	**103**
Total	**87**	**61**	**23**	**171**

(d) State the null and alternative hypotheses.

Null: Political affiliation and type of car one drives are independent (or not associated).
Alternative: Political affiliation and type of car one drives are dependent (or associated).

(e) Check the technical conditions.

Although not stated, assume the data are from an SRS. All expected counts are at least 5. And the 10% Condition is met.

(f) Calculate the chi-square test statistic.

$$\chi^2 = \frac{(32-34.5)^2}{34.5} + \frac{(21-24.3)^2}{24.3} + \cdots + \frac{(8-13.9)^2}{13.9} = 7.27$$

(g) Which category contributes the most to the test statistic?

The "Other" "American" category.

(h) Calculate the degrees of freedom.

$$(2-1)\cdot(3-1) = 2$$

(i) Find the *p*-value.

From the chi-square distribution table, the *p*-value is between 0.025 and 0.05.

(j) Report the test decision.

Reject the null hypothesis at the 5% significance level.

(k) Relays any conclusions in context.

We have enough evidence to claim that political affiliation and type of car one drives are dependent (or associated).

In the previous example, data were collected from a single random sample, and a **chi-square test for independence** was conducted to determine if the results were statistically significant. However, that's not our only option with chi-square: we might instead gather data from *multiple* simple random samples, all of which need to be taken independently of one another, and conduct a significance test. This inference procedure is called a **chi-square test for homogeneity**, and differs little from the chi-square test for independence. The key distinction is that the test centers on whether or not the proportions between the groups are homogenous.

Your graphing calculator has the built-in capability to conduct any sort of **chi-square significance test** of bivariate data. The calculator is set to relay the chi-square statistic and the *p*-value.

First, we need to enter in two matrices into our calculator: one for the observed counts and another for the expected counts of our two-way table. To enter a matrix into the calculator, find the MATRIX button/option and scroll over to the EDIT menu. Let's use matrix [A] to represent our observed counts and matrix [B] to represent our expected counts. Here's what's nice, though: you don't actually need to enter in the expected counts—the calculator computes them automatically! But, when entering in the values for any matrices, make sure the dimensions of the matrices are correct: the format is *rows x columns*, and marginal total rows and columns aren't included.

When the observed and expected counts' matrices are stored, hit the STAT button, then scroll over to the TESTS menu. We wish to use the χ^2 – Test option, as shown in the screenshot below.

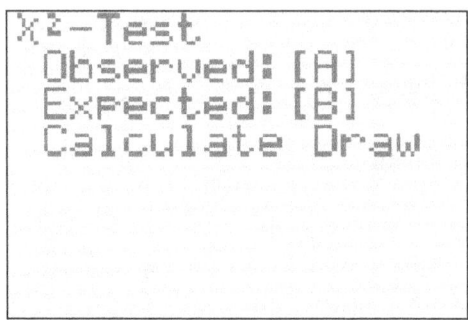

Example 9f.-4

Reconsider this two-way table of political affiliation and the type of car one drives:

	Political Affiliation			
	Democrat	Republican	Other	**Total**
American	32	21	15	**68**
Foreign	55	40	8	**103**
Total	**87**	**61**	**23**	**171**

Use the calculator to find the *p*-value.

Solution. Since the calculator is not tied to looking up values from a table, it can obtain an exact *p*-value by using the chi-square distribution directly. The (rounded) *p*-value is 0.026.

Finally, note that performing a chi-square test with two categories per variable (in a 2*x*2, or fourfold, table) is effectively equivalent to two-sided *z* test of two proportions. The value of χ^2 ends up equaling z^2, and the *p*-values are identical.

§9g. *Correlation Coefficient Redux*

The chi-square distribution permits significance testing of categorical bivariate data. Significance testing can also be applied to quantitative bivariate data. For instance, sample correlation coefficients can be utilized in hypothesis tests of population correlation coefficients.

For example, consider radon, an odorless and colorless gas. According to the Environmental Protection Agency (EPA), radon is the second-leading cause of lung cancer deaths in the U.S., behind smoking. The EPA states that a "safe" level of radon indoors in 4 picoCuries[*] per liter (pCi/L). Radon emanates from many sources: water, soil, and other materials. How can we test a geographic area for significant correlations between radon emissions and geologic or atmosphere characteristics?

Much ink has been spilled in research articles examining the effects of radon. We need to understand how real-world, thorough statistical analysis of a problem is actually employed. So we'll briefly look at one article called "Airborne Radon in Homes in Summit County, Ohio: A Geographic Analysis" (1997):

---

[*] These radioactivity measurement units are named after Marie Curie, who conducted some of the first groundbreaking (literally) studies on radioactive elements. Curie, one of the greatest scientific researchers of all time, won two Nobel prizes (in physics and in chemistry) but also suffered debility and death as a result of toxic levels of radiation exposure.

Statistical correlations were examined between many variables. Positive correlations were found between radon activity and air temperature, soil permeability, surface uranium concentration, and proximity to underground mines. A negative correlation was found between radon reading and barometric pressure.

Even though correlations were found, were *statistically significant* correlations found? In other words, were the correlations between the variables unlikely to be due to chance alone? A dot-map, along with a table summarizing the results, is presented in the article. Statistically significant correlations were found between radon levels and (1) air pressure; (2) depth to bedrock; (3) rainfall; (4) soil permeability; and (5) air temperature, among several other factors. However, statistically significant correlations were not found between radon levels and (1) depth to groundwater; and (2) season.

But how is statistical significance determined when working with quantitative bivariate data? The best line of attack is to work through an example.

Example 9g.-1

A random sample of 10 high school students produced the following results for the number of hours of television watched per week and GPA.

TV hours:	12	21	8	20	16	16	24	0	11	18
GPA:	3.2	2.3	3.7	2.5	3.0	2.1	2.7	3.8	2.9	2.6

(a) Which is the explanatory variable and which is the response variable?

TV hours is explanatory; GPA is response.

Input the data into your calculator (call the lists TV and GPA), and view a scatterplot of the data. To do this, press 2nd and then Y=, and select PLOT1. Make sure the plot is ON, the scatterplot icon is selected, and the correct "Xlist" and "Ylist" are inputted. Then press the ZOOM key, and select ZoomStat.

(b) Does the scatterplot reveal an association between time spent watching television and GPA? If so, is the association positive or negative? Weak, moderate, or strong? Linear or nonlinear?

There is a linear association, which is moderately negative.

Make sure DiagnosticOn is selected (it should already be) using the catalog feature. Next, press the STAT button, scroll over to the CALC menu, and use the LinReg(a+bx) option to obtain the correlation coefficient and the coefficient of variation.

(c) Report the values of the correlation coefficient and coefficient of variation.

$r = -0.814$ and $r^2 = 0.662$.

In order to perform a test of significance on the population correlation coefficient (denoted by ρ), we need to find the probability of gathering the sample data *if* there is no linear correlation between the hours spent watching television and the GPA among the *population* of high school students.

(d) Write out the null and alternative hypotheses.

Null: There is no linear association between TV hours watched and GPA.
Alternative: There is a negative linear association between TV hours watched and GPA.

(e) State the null and alternative hypotheses symbolically.

$H_0 : \rho = 0$
$H_a : \rho < 0$

The technical conditions (or assumptions or requirements) are (1) the data are from a random sample; (2) the data points look relatively linear; and (3) there are no outliers.

(f) Are all the technical conditions met?

Yes.

The test statistic for this inference procedure is $t = \dfrac{r\sqrt{n-2}}{\sqrt{1-r^2}}$. We will make use of the t distribution with $n-2$ degrees of freedom, since we are examining two variables rather than one.

(g) Calculate the test statistic.

$$t = \frac{-0.814\sqrt{10-2}}{\sqrt{1-0.662}} = -3.96$$

(h) Find the p-value.

The degrees of freedom is 8. On the t distribution table, looking up the absolute value of -3.96 is equivalent to looking up 3.96. Therefore, the p-value is between 0.001 and 0.0025.

(i) Report your test decision and context-specific conclusion.

We reject the null hypothesis at 5%. There is enough evidence to claim that there is a negative linear association between TV hours watched and GPA.

Let's hold off on using the graphing calculator to directly find the test statistic and the *p*-value of a correlation test of significance until after reviewing linear regression.

§9h. *Linear Regression Redux*

L et's review linear regression and then integrate into it the ideas of statistical inference.

Example 9h.-1

Again, a random sample of 10 high school students produced the following results for the number of hours of television watched per week and GPA.

TV hours:	12	21	8	20	16	16	24	0	11	18
GPA:	3.2	2.3	3.7	2.5	3.0	2.1	2.7	3.8	2.9	2.6

If you haven't already done so, please input this data into your calculator, calling the first list TV and the second list GPA.

(a) Recall that $r = -0.814$ and $r^2 = 0.662$. Interpret both measures in context.

Correlation coefficient: there is a negative, strong association between TV hours watched and GPA.

Coefficient of determination: 66.2% of the variation in GPA can be accounted for (or explained) by TV hours watched.

Instead of performing a hypothesis test on the population correlation coefficient, we'll do something equivalent: a hypothesis test on the population slope since, after all, a slope of zero indicates no linear relationship.

(b) Find the equation of the line of best fit using the calculator's LinReg(a+bx) option. Report descriptive variable names when writing the regression equation.

$\hat{GPA} = 3.82 - 0.065 hours$

(c) Interpret the slope.

For every additional hour of television watched, a student's GPA goes down by 0.065 of a point.

(d) If another random sample of ten high school students was obtained from the same population, and the data from this new sample was examined, would the exact same regression equation you found above be reproduced? Briefly explain.

No—sampling variability.

(e) If many, many random samples of ten high school students were drawn from the same population, and the sample slope from each sample of data was calculated, where would the mean of those slopes be centered?

At the population parameter—in this case, the population slope. (Note that the question effectively describes taking a sampling distribution.)

Next, we will perform a test of significance on the population slope. The sample slope is denoted by the letter b, while the population slope is represented by β—yes, the same β signifying the probability of a type II error.

(f) What is the best estimate we have for β?

The sample slope b, which is equal to -0.065; b is an unbiased point estimate of β.

(j) Write out the null and alternative hypotheses.

Null: There is no linear association between TV hours watched and GPA.
Alternative: There is a negative linear association between TV hours watched and GPA.

(k) State the null and alternative hypotheses symbolically.

$$H_0 : \beta = 0$$
$$H_a : \beta < 0$$

Like any other significance test, recall that there are technical conditions we must check before performing the test.

The technical conditions (or assumptions or requirements) here are (1) the data are from a random sample; (2) the data points look relatively linear; (3) the residual plot looks random; (4) the data set is independent; and (5) the distribution of the residuals is normal.

(g) Check the technical conditions.

The data are from an SRS. A scatterplot of the data appears linear. The residual plot can be examined by using the residual list $_L$RESID, automatically created by the calculator after LinReg(a+bx) is run—it indeed looks random. We'll assume that no student's television viewing habits (or GPA) affects the others in the random sample, so the condition of independence is met. And, finally, constructing an NPP of $_L$RESID gives us confidence of the normality of the distribution.

Now that we've stated the null and alternate hypotheses and checked the technical conditions, here's where it gets a bit tricky. Although H_0 and H_a are relatively easy to come up with, the test statistic, which is $t = \dfrac{b-0}{SE_b}$, is not. The numerator, as always with any test statistic, accounts for the sample data: in this case, the slope coefficient b. But the denominator is labeled SE_b, or the standard error of the sample slope b.

We will use the calculator to obtain SE_b, the standard error of the sample slope. Unfortunately, the calculator cannot produce this value directly; instead, we first note that SE_b is equal to:

$$SE_b = \frac{\text{The sum of the squared residuals}}{\left(\text{The sample standard deviation}\right) \cdot \sqrt{n-1}}$$

We need to find the numerator and the denominator of the formula separately in order to calculate SE_b.

(h) The sample standard deviation (in the denominator) is easy to find; simply run the 1-Var Stats function on the explanatory variable. Record the value below.

$s_x = 7.07$

(i) To obtain the numerator, hit the $\boxed{\text{STAT}}$ button, scroll over to the TESTS menu, and scroll down to find LinRegTTest. Input your x and y list names, select your alternate hypothesis, and then hit the Calculate feature. Scroll down to find the s value. That's the sum of the squared residuals. Record this value below.

$s = 0.346$

(j) Calculate SE_b, based on your answers to parts (h) and (i).

$$SE_b = \frac{0.346}{7.07\sqrt{10-1}} = 0.0163$$

(k) Now calculate the test statistic.

$$t = \frac{-0.065 - 0}{0.0163} = -3.99$$

(l) Find the *p*-value.

The degrees of freedom is 8 (two less than the sample size). Looking up the absolute value of the test statistic on the *t* distribution table gives a *p*-value between 0.001 and 0.0025.

(m) Report your test decision and context-specific conclusion.

We reject the null hypothesis at 5%. There is enough evidence to claim a negative linear association between TV hours watched and GPA.

(n) Because the conjected value of interest—in this hypothesis test, the population slope—is set to zero, there is an easier way of obtaining SE_b using LinRegTTest.[*] You can back into the value by taking the slope *b* and dividing it by the test statistic. Record that number below, comparing it to your answer to part (j).

$b = -0.065$ and $t = -3.96$, so $SE_b = \dfrac{b}{t} = \dfrac{-0.065}{-3.96} = 0.0163$, which is the same value as was calculated above.

Look again at your calculator's output screens after running LinRegTTest with the TV-GPA data:

```
LinRegTTest
  y=a+bx
  ß<0 and ρ<0
  t=-3.958716557
  p=.0020922775
  df=8
↓a=3.822646536
```

```
LinRegTTest
  y=a+bx
  ß<0 and ρ<0
↑b=-.0645648313
  s=.3461312444
  r²=.6620399482
  r=-.8136583731
```

Notice how not only are *s*, *r*, r^2, *a*, *b*, and df (degrees of freedom—two less than the sample size) reported, but the test statistic and the *p*-value are also displayed (although the standard error of *b* is not shown, you can back into it by taking *b* divided by *t* if the conjected population slope is zero).

---

[*] The conjectured value of the population slope might not always be equal to zero. For example, the conjectured slope might be equal to one. However, LinRegTTest always assumes a value of zero.

We cannot rely on LinRegTTest to carry us through *any* linear regression inference problem, though, because there are two things your calculator will not do: construct confidence intervals and interpret computer printouts. (In all fairness, some later TI-84s *can* find a confidence interval—the function is unsurprisingly called LinRegTInt—but not all of them.)

It's unsurprising that we can obtain a confidence interval for the population slope β. This can give us a range of values of where we expect β to reside. The formula is $b \pm t * SE_b$. This shouldn't be surprising, given that the format for *any* confidence interval is

$$\text{statistic} \pm \text{critical value} * (\text{standard deviation of statistic})$$

Example 9h.-2

Reconsider the random sample of 10 high school students TV hours watched and GPA.

(a) Find a 95% confidence interval for the population slope.

$$-0.065 \pm 2.306 \cdot 0.0163 = \left(-0.103, -0.027\right)$$

(b) Interpret the confidence interval.

We are 95% confident that for every additional hour of television, a student's GPA decreases by between 0.027 and 0.103 points.

Example 9h.-3

Here's what a Minitab computer printout would like for the TV hours and GPA data.

Predictor	Coef	StDev	T	P
Constant	3.823	0.132	2.81	0.342
TVhours	-0.065	0.016	-3.96	0.002
S = 0.346		R-sq = 66.2%		

Instead of "StDev," the column heading may read "SE Coef," but *both* measures refer to SE_b.

(a) Calculate the test statistic using two numbers from the computer printout; then, compare it to the test statistic displayed in the computer printout. Why are the values a bit different?

$$t = \frac{-0.065}{0.016} = -4.0625; \text{ the value in the printout is } -3.96, \text{ which is different because of}$$

rounding.

(b) Using only the computer printout, report the *p*-value of the study.

The *p*-value is 0.002.[*]

§9i. *Analysis of Variance*

Suppose we wish to compare the means of not two, but three independent samples of quantitative data sets. If we simply pair off sample means—testing mean one with mean two, mean one with mean three, and mean two with mean three—we put ourselves in jeopardy of rejecting a true null—i.e., we increase our chances of a type I error. Thus, we need to stick with a single test, one which takes into account all of the means simultaneously.

A *One-Way Analysis of Variance* test, or ANOVA, is a formal statistical inference procedure that tests the equality of three or more population means and was originally developed by Ronald Fisher. Each ANOVA test requires a *factor*, or a way to distinguish the different populations from each other. Probability values of ANOVA are found using an *F*-test statistic (read off of the *F* distribution, which is skewed to the right).

Let's wade our way through the procedure by hand first, and then shortcut most of the tedium using the calculator.

The ANOVA procedure relies on calculations of samples means and sample variances. Suppose there are n elements in each sample, and there are m samples of data. We find the mean for each sample, and calculate the average of these means—and then represent this average as $\bar{x}$.

Next, we calculate the *sample variance of the means of the samples*, represented by S^2. This functions as a measure of spread *between* the m samples of data.

The sample variance of *each* of the m samples of data also needs to be found. From these m sample variances, a new measure is calculated: the mean of the m sample variances, denoted by $\bar{x}_s$.

The *F*-test statistic is the ratio of the variation among the sample means (S^2) to the variation among the sample variances ($\bar{x}_s$); you can also think of the *F*-test statistic as the variance between the samples to the variance within the samples. More precisely, the test statistic is given as

$$F = \frac{n \cdot S^2}{\bar{x}_s}$$

If the sample means are close together, then there is little variance between the samples—hence, the test statistic is close to zero and the *p*-value is large, indicating no significant difference. But if, instead, the sample means are far apart, then there is much variability between the

---

[*] If we had performed a hypothesis test of the population *y*-intercept of the TV-GPA data, our *p*-value would have been 0.342 (as shown in the printout).

samples—leading to a large test statistic, a small p-value, and a declaration of statistical significance.

If an F distribution table was included at the back of this primer, then you'd find the p-value by looking up the F-test statistic with $m-1$ and $m \cdot (n-1)$ degrees of freedom. But an F distribution table is not included. If you really wish to practice the process by hand, F distribution tables are widely available online.[*]

The technical conditions for the inference procedure are (1) the samples all come from normally distributed populations; (2) all sample sizes are approximately the same size; (3) the samples were all obtained randomly; (4) the samples are independent of one another; and (5) the samples are all categorized in only "one way"—meaning by one variable, such as by time or distance or weight.

Now we'll make use of the graphing calculator's ANOVA feature.

Example 9i.-1

A researcher wants to see if there is a difference in the mean weight lost among at least one of three current diet pills out on the market. Volunteers suffering from obesity are instructed to randomly take one of three diet pills. The weight lost over a six-month time period among each of the volunteers is recorded. Assume each population is normally distributed. The results are shown below.

Pill 1	12	31	23	8	
Pill 2	13	5	17	21	10
Pill 3	15	19	22	9	29

At the 1% level of significance, is there enough evidence to claim that at least one of the pills is different from the rest?

Solution. The null hypothesis is that the population means of all three pills are the same (i.e., that $H_0 : \mu_1 = \mu_2 = \mu_3$. The alternative hypothesis is that at least one population mean is different from the rest. (The ANOVA test, as it stands here, it not sufficient to tell *which* of the means, if any, are different; in addition, the notion of "different" could imply that the particular pill is significantly more *or* less effective than the others.)

The data sets are categorized in one way—by the weight of the participants in the study. Although the participants were not obtained randomly, but were rather volunteers, there is randomization of subjects to treatments. Furthermore, it is stated that the populations are normally distributed, the samples are roughly equal in size, and independence is satisfied—since an individual taking one pill is also not taking another one.

Using the graphing calculator, we can find the F-test statistic and the p-value. To do this, all three lists must be entered into the calculator; call the lists PILL1, PILL2, and PILL3. Then,

---

[*] Such as at http://www.socr.ucla.edu/applets.dir/f_table.html

press the STAT key, scroll over to the TESTS column, and find the ANOVA(option. Enter in the three lists, separated by commas, and press ENTER. Your screen should look as follows:

```
One-way ANOVA
 F=.7492418422
 p=.4953868418
 Factor
  df=2
  SS=96.2571429
↓ MS=48.1285714
```

With that large of a p-value, we fail to reject the null hypothesis at 1%. We do not have enough evidence to claim that at least one of the pills is different from the rest.

Even though no pill was shown to be different above, by examining confidence intervals or visuals displays of the data we might get some idea of how the different samples compare. But these are strictly informal means of comparison; a statistically rigorous approach would involve performing *multiple comparison tests*, like Tukey's test (if all samples sizes are the same) or Scheffe's test (if the samples sizes are not the same), to ascertain which—if any—of the means are different from the rest. These *pairwise comparisons* add another layer of Byzantine calculations to ANOVA and are best left for computer software, such as SPSS, to perform. Beyond that, ANOVA doesn't have to be restricted to just one factor—there are complex inference procedures for two-way ANOVA, in which two independent variables are explored, as well.

§9j. *Nonparametric Methods*

All of the *parametric* procedures for inference we've explored so far rely on set distributions, whether they be the normal, t, chi-square, or F. A class of *nonparametric*, or distribution-free, methods also exist, although they are ipso facto less powerful than parametric procedures since there are fewer built-in assumptions—and less power means less chance of rejecting a false null hypothesis. But nonparametric methods allow us to venture into a statistical analysis tabula rasa, in effect, and that is indeed powerful—in the colloquial sense of the term, at least. Let us briefly outline some of these nonparametric procedures here.

- The *sign test* is oftentimes used for before-and-after studies. Perhaps you need to test the effectiveness of an SAT prep course. A pretest is given to all students at the beginning of the course; after taking the course, a posttest is also required of all students. The differences in the scores for each student is recorded and categorized as either *positive*, *negative*, or *no difference*. The no difference scores are discarded from the analysis. The proportion of positive scores as compared to the total of positive and negative scores is the sample statistic—and is assumed to be an unbiased estimator of the parameter (the population proportion of positive differences). From there, the normal distribution reappears so that a test statistic and corresponding *p*-value can be calculated. Notice that the population distribution of positive and negative differences, however, is not relevant to the analysis.

- Likewise, the *rank-sum test* (also called the *Wilcoxon Rank Sum test*) can handle any population distribution. The rank-sum test is useful when comparing two small (i.e., less than thirty observations) sets of independent quantitative data where the host populations are not known to be normally distributed. First, the samples must be "ranked"—from smallest to greatest in total, ignoring distinctions between the samples (data points from tied ranks result in associated mean values). Next, distinct ranks are summed, and a mean, standard deviation, and test statistic are calculated, just in time to lean on the normal distribution yet again to produce a *p*-value.

- Ranking is also necessary when utilizing the *Spearman test of rank correlation*. A monotone relationship—that is, as *x* increases, *y* also increases (or vice versa)—can be detected; but monotonicity, in and of itself, tells us nothing about the linearity of the variables being compared.

- Using a *resampling* method is also distribution-free, since subsets of the sampled data are repeatedly sampled with replacement and then ordered or organized in some way for analysis. (The genesis of these repeated sampling experiments is Buffon's needle problem.)[*] For example, consider *bootstrapping* with a set of quantitative bivariate data: Effectively, each of the (*x*, *y*) data pairs will be randomly selected a set number times with replacement until a new distribution is formed, and the correlation coefficient of that new distribution will be plotted; then, the process will repeat, thousands of times. From the distribution of correlation coefficients, we will produce a 95% confidence interval by culling the 2.5th and 95.5th percentiles of the distribution of samples.[†] Courtesy of duality, the confidence interval resulting from the bootstrapping will allow us to make a test decision of whether to reject or fail to reject the null hypothesis. Computer software such as SPSS and R allow for bootstrapping procedures as well as other varieties of resampling methods.[‡]

---

[*] Count Buffon, the French mathematician and naturalist, posed this probability question: "Suppose we have a floor made of parallel strips of wood, each the same width, and we drop a needle onto the floor. What is the probability that the needle will lie across a line between two strips?" Although the solution, which involves π, can be found using algebraic methods, the problem spurred the development of Monte Carlo methods of simulation involving repeated sampling.

[†] For instance, if we generate 10,000 samples, the correlation coefficients of the 250th and the 9,750th samples will function as the lower and upper bounds, respectively, of the resulting 95% confidence interval.

[‡] Which include *jackknifing* (fleshed out and named by Tukey) and *permutation tests* (also called *exact* or *randomization tests*), among many others.

§9k. *Logic vs. Probability*

Not all statisticians and scientists agree that hypothesis testing should be used in scientific research, at least in the form it takes now. To start with, Fisher himself didn't particularly like the arbitrariness of the significance level as a stringent dividing line between so-called significant and non-significant data. (In *Scientific American*, journalist Regina Nuzzo memorably calls significance testing a "scientific spam filter.") In addition, many people misunderstand the definition of the *p*-value, missing that *conditional probability* plays a part in the calculation: What are the chances of seeing the sample result, *given* that the null hypothesis is true? Even that, though, isn't the whole story, since the a priori probability that the null hypothesis is true isn't obvious—the null hypothesis is often just an arbitrary construct by the researcher, especially in the social sciences and even more especially in non-experimental studies.

Publication bias also factors into this discussion. Recall how eager Fisher was to dismiss the results of the observational studies about smoking, claiming that studies which demonstrated a smoking effect tended to be published, while those that didn't were likely rejected. Even today, scientific publications have a selection bias, since those studies that show statistical significance are much more likely to be published while statistically "insignificant" studies are either rejected for publication or never even submitted for consideration (this is called the *file-drawer problem*).

But even in studies that ostensibly demonstrate statistically significant results, data can be manipulated or simply ignored to support a priori beliefs—whether consciously or not. Gould devotes a portion of *The Mismeasure of Man* to detailing the strange case of Paul Broca, an anatomist with a special interest in craniology; even when Broca's measurements of brains didn't conform to his prior assumptions (bigger = smarter, European > African), he would manufacture reasons for the deviations. The problem is one of what Gould terms "creative interpretation": a reminder that statistics, and science in general, isn't performed by dispassionate computers, but human beings. The tools of hypothesis testing, and the interpretation of measurements, fall under human caprice and are subject to human prejudice. Gould again:

> Theories are built upon the interpretation of numbers, and interpreters are often trapped by their own rhetoric. They believe in their own objectivity, and fail to discern the prejudice that leads them to one interpretation among many consistent with their numbers.... Shall we believe that science is different today [than in Broca's time] simply because we share the cultural context of most practicing scientists and mistake its influence for objective truth?

Scientists today often treat the most basic results of hypothesis testing as sacrosanct, figuring it unnecessary to delve any deeper into what has the imprimatur of science and thus the gloss of objective truth. But ignoring *effect sizes*—like correlation, regression, and other parameters—as well as confidence intervals, in favor of simply reporting *p*-values and sample sizes in published articles, have become the bane of opponents (and even agnostics) of hypothesis testing everywhere. The *American Psychological Manual* implores researchers not to ignore these statistical ideas when assembling papers. The *APA* also urges its readers to "….take seriously the statistical power considerations associated with the tests of hypotheses" and "[w]herever possible, base discussion and interpretation of results on point and interval estimates." The concept of statistical power should not be ignored, and statistical detail from studies should be reported.

Authors of the article "High Impact = High Statistical Standards? Not Necessarily So" (February 13, 2013), in the journal *PLoSOne*, note a number of limitations with null hypothesis significance testing (NHST), such as:

- "NHST does not provide any information about precision, meaning the likely error in an estimate of a parameter, such as a mean, proportion, or correlation. Any estimate based on a physical, biological or behavioral measure will contain error, and it is fundamental to know how large this error is likely to be."
- "NHST does not give an estimate of the difference from H_0, which is a measure of effect size, even when the answer is 'Yes, there is a difference from zero.'"
- "[T]he *p*-value is very likely to be quite different if an experiment is repeated. For example if a two-tailed result gives $p = 0.05$, there is an 80% chance the one-tailed *p*-value from a replication will fall in the interval (.00008, .44), a 10% chance that $p < .00008$, and fully a 10% chance that $p > .44$. In other words, a *p*-value provides only extremely vague information about a result's repeatability."

Picking up from that last point, journalist Faye Flam, author the *Philadelphia Inquirer* article "Researchers Show Ease of Finding Dubious Results" (June 6, 2012), said,

> The problem isn't just that samples are too small, [Wharton researcher Uri] Simonsohn said. Researchers are also picking through their data in ways that make it too easy for them to find statistical flukes that look like real patterns. Current reporting practice allows this to go undetected....
>
> Unfortunately, [Simonsohn] said, there's no ethic of replicating studies in psychology, so other equally flawed studies are incorporated into established wisdom....
>
> Scientists often decide when they have enough samples to stop taking data. Instead of going through 100 trials, say, they stop as soon as they get a statistically significant result, [Simonsohn] said. (This is akin to leaving a poker game just after a lucky hand won you a big pot and put you ahead. There's a reason this is considered bad practice.)

Lest you think that such dubious statistical practices don't have real-world consequences, consider what Melody Petersen, in *Our Daily Meds*, says about the head-scratching implications of repeated clinical trials:

> How do we interpret positive results in the context of several more studies that fail to demonstrate [an] effect [of a medication]? I am not sure I have an answer to that.... That would mean, in a sense, that the sponsor could just do studies until the cows come home until he gets two of them that are statistically significant by chance alone.

Point taken—at least according to John K. Kruschke, who penned a pro-Bayesian, anti-NHST essay (called simply "Bayesian Data Analysis," from 2010). To wit:

> ...NHST is based on the intentions of the researcher and analyst: the *p*-value depends entirely on the assumed intention. If data are collected for a certain duration instead of for a certain sample

size, the *p*-value changes. If some data are lost by accident or attrition or declaration of outliers, the *p*-value changes. If the analyst wants to be thorough and investigate multiple comparisons, the *p*-value changes. If the researcher might possibly collect more data in the future that could be compared with the present data, then the *p*-value of the present data changes.

Unlike NHST, though, Bayesian data analysis—which, recall, is related to an iterative process of probabilities of belief—doesn't suffer the same limitations and thus might be an effective antidote to the rampant misuse and abuse of NHST.

Perhaps the issues with NHST cut even deeper, though. There is much systemic criticism that centers on the foundations of hypothesis testing (and confidence intervals): namely, the frequentist approach to probability.

Imagine a lottery with 1,000 tickets, one of which is a winner.[*] The probability of the first ticket being a winner is 0.001; because the probability is so small, we reject that the ticket is a winner. Same with the second ticket, and the third, and the fourth.... Logically, since every ticket has been rejected as a winner, none will win—contradicting the setup of the problem.

In February of 2015, the *Basic and Applied Social Psychology* (BASP), a blind peer review academic journal, issued an editorial that caused quite a stir amongst academicians: they banned hypothesis testing and confidence intervals, including "all vestiges of the NHTSP [null hypothesis significance testing procedure]," such as *p*-values and confidence intervals, entirely from their pages.

But why? Here's how they explain the problem:

> In the NHSTP, the problem is in traversing the distance from the probability of the finding, given the null hypothesis, to the probability of the null hypothesis, given the finding.... Regarding confidence intervals, the problem is that, for example, a 95% confidence interval does not indicate that the parameter of interest has a 95% probability of being within the interval.... Analogous to how the NHSTP fails to provide the probability of the null hypothesis, which is needed to provide a strong case for rejecting it, confidence intervals do not provide a strong case for concluding that the population parameter of interest is likely to be within the stated interval.

The BASP editors go on to note their wait-and-see attitude with respect to alternative Bayesian procedures.

In an online post lamenting the ban, noted statistician George Cobb writes, "I have yet to encounter a statistical method that can't be used stupidly, but I can't imagine that banning the use of any statistical tools will eliminate stupidity about data.... Statistical methods—whether Fisherian *p*-values, Neyman-Pearson operating characteristics, or Bayesian posterior intervals, are at best an aid to thinking about data, never a substitute."

David Salsburg explains that the ambiguity arising from hypothesis testing is tied to its reliance on probability, rather than on logic: "probability introduces the idea that some propositions are probably or almost true," contrary to the strict dichotomy in logic of true or false. Resulting NHST paradoxes, such as the lottery paradox, arise from this "little bit of resulting unsureness." Logic and probability, like oil and water, just don't mix.

---

[*] This example comes from *The Lady Tasting Tea*, but is attributed to L. Jonathan Cohen.

And, finally, even if results from a hypothesis test are shown to be extremely improbable, this improbability doesn't necessarily translate to real-world, *practical* significance. For example, consider Michael Drosnin's popular book *The Bible Code*, which summarized statistical research into supposedly "predictive" patterns, called equidistant letter sequences (ELSs), found in the Old and New Testaments. Though the *p*-values were indeed very small, the end of the world luckily didn't arrive in 2006—as predicted.

With time and much effort, though, these many issues with hypothesis testing will be sorted out—but there will be much kicking and screaming along the way.[*] Opponents of hypothesis testing have one inarguable point, however: using statistical procedures and the results they produce mindlessly is not only bad statistics—it is bad science.[†] Good scientific theories are subjected to the most rigorous experimental and empirical standards, and only the most robust of these theories survive. Statistical tools should be no different. The mathematical model should conform to fit the phenomena, not the other way around. Remember the sage words of John Tukey: "For if its [statistics'] methods fail the test of experience—not the test of logic—they are discarded."

---

[*] There are numerous problems with multiple regression analysis (MRA) as well. Studies that use MRA "control" for factors—like age, socioeconomic status, prior health, and a host of other independent (or predictor) variables—before determining the "net effect" of one independent variable on the dependent (or output) variable. Epidemiology, the study of disease in populations, makes heavy use of MRA. But disentangling the effects of independent variables isn't an easy task, let alone enumerating all possible independent variables in a study; in addition, MRA is most often used with samples that have been self-selected, not with studies in which cases are randomly assigned to treatment groups. Results from well-designed controlled randomized experiments will always be superior to conclusions drawn from MRA. As psychologist Richard Nisbett laments, "What nature has joined together, multiple regression analysis cannot put asunder."

[†] This is especially true among psychologists and other social scientists, who often shy away from gaining a more rigorous understanding of the mathematical methods they nevertheless frequently employ.

The Antidote to Apophenia

⚭

Mathematics in general, and especially mathematical logic in particular, is replete with paradoxes, statements that when subjected to the scrutiny of careful examination collapse under their own labyrinthine complexity: *This statement is a lie* and *I'm lying right now* are Kafkaesque in their opacity but overt in their meaninglessness. And yet under such scrutiny so much of language—being pointers to referents rather than the referents themselves—reveals itself as an ongoing shell game. As David Hume wrote, "[T]o believe that whatever received a name must be an entity or being, having an independent existence of its own," is a cognitive fallacy. Such is the symbol-grounding problem brought to life.

No doubt image, like language, suffers as well when subjected to careful examination. Recall René Magritte's famous painting of a pipe, in which, underneath and with a mocking flourish, he scribbles: *Ceci n'est pas une pipe*, This is not a pipe. Of course it's not a pipe—it's a *painting* of a pipe. But look even closer at the image, with a magnifying glass or a microscope, or electronically zoom in on a digital copy until only a handful of its pixels fills your computer screen, and even the notion of the representation of the "pipe" must be called into question.

Now concatenate the dizzying referent complexities of language with the mushy image-representation problems—our senses' interpretation of experiences must immediately be called into question, into doubt. The ancient Greeks knew this, and tried to counteract their puzzlement with a "Platonic heritage, with its emphasis in clear distinctions and separated immutable entities," as Stephen Jay Gould observed; they unsuccessfully attempted to gerrymander the complexity of the world. Plato conveniently explained away the representation problem with his Parable of the Cave. Aristotle placed a primacy on visual perception in his *Metaphysics*, writing that

> All men by nature desire to know. An indication of this is the delight we take in our senses; for even apart from their usefulness they are loved for themselves; and above all others the sense of sight. For not only with a view to action, but even when we are not going to do anything, we prefer seeing (one might say) to everything else. The reason is that this, most of all the senses, makes us know and brings to light many differences between things.

The great thinkers of the Enlightenment disagreed with Aristotle, and tried to work wonders with mathematics, or mathematical-type philosophy, to anesthetize their creeping doubts about the presence of the divine despite the apparent predictability of the natural world (à la Galileo and Newton and Pascal). Natural law, rather than divine law, was ready to assume its place at the head of the table. For example, Descartes was radically skeptical about anything outside of the self (he espoused what author Matthew Crawford calls "epistemic individualism"), and his famous formulation *cogito ergo sum* serves as a rejoinder to those who would argue that true knowledge flows as through a sieve from the external world to one's sensory perception. In a fit of contorted mental gymnastics, Descartes managed to produce a proof of the existence of God by analogy with deductive logic:

> I found that the existence of the [Perfect] Being was comprised in the idea in the same way that the equality of its three angles to two right angles is comprised in the idea of a triangle, or as in the idea of a sphere, the equidistance of all points on its surface from the centre, or even still more clearly; and that consequently it is at least as certain that God, who is this Perfect Being, is, or exists, as any demonstration of Geometry can be.

But by the last century, the God argument's bottom fell out, triggering a full-on assault of the nature of language, meaning, symbology, and mathematics itself (which was detailed in the earlier "Foundations" §). Keynes, in his *Treatise on Probability*, wrestled with the ideas of perception when he noted, "Part of our knowledge we obtain direct; and part by argument," setting the stage for the still-unsettled inductive-versus-deductive acquisition-of-knowledge debate, as Gavin Kennedy explains in *Invitation to Statistics*.

With appeals to the divine no longer conceivable end causes, would exhaustive enumeration of symbols and referents help us to move forward? Nobel Prize-winning physicist Ernest Rutherford famously quipped, "All science is either physics or stamp collecting." Thanks to the taxonomists who made themselves at home in modern academic psychology departments, a full flowering of (perhaps arbitrary) pathological diagnoses took hold—in the form of the early neurotic imaginings of Freud to the later iterations of the *DSM* and elsewhere—ready to pounce on any deviancy of interpretation between symbol and possible referent.

All of us are unique, and each of us possesses a freedom to think and act that cannot, ultimately, be anticipated by numbers or reduced to equations. As Viktor Frankl, in *Man's Search for Meaning*, explains,

> [E]very human being has the freedom to change at any instant. Therefore, we can predict his future only within the large framework of a statistical survey referring to a whole group; the individual personality, however, remains essentially unpredictable…. How can we dare to predict the behavior of man? We may predict the movements of a machine, of an automaton; more than this, we may even try to predict the mechanisms or "dynamisms" of the human *psyche* as well. But man is more than *psyche*.

Which brings us back to where we started: *apophenia*, of perceiving patterns of meaning where there are none. To claim a cure implies the presence of an affliction, or a deviation, of some kind. While the statistical methods of the three axes (variation, collection, and interpretation) explored in this primer, the result of the work of mathematicians and statisticians over many generations, will indeed help separate the signal from the noise, there is no surefire "antidote" to apophenia, short of engaging in a persistent habit of thinking skeptically. By that force of habit, we become "authorities in the practical and scientific spheres, by so many acts and hours of work," as William James explained. Only a continued vigilance, a rapt attention to detail, a ready willingness to learn, and an openness to surprise can counteract a stubbornness of mind and body—allowing us to be right just a little more often than we are wrong.

Statistical Tables

∞

Table A1

φ

The Normal Distribution − Negative Z-Scores

Standard Normal Probabilities Table										
z	0	0.01	0.02	0.03	0.04	0.05	0.06	0.07	0.08	0.09
-3.4	0.0003	0.0003	0.0003	0.0003	0.0003	0.0003	0.0003	0.0003	0.0003	0.0002
-3.3	0.0005	0.0005	0.0005	0.0004	0.0004	0.0004	0.0004	0.0004	0.0004	0.0003
-3.2	0.0007	0.0007	0.0006	0.0006	0.0006	0.0006	0.0006	0.0005	0.0005	0.0005
-3.1	0.0010	0.0009	0.0009	0.0009	0.0008	0.0008	0.0008	0.0008	0.0007	0.0007
-3.0	0.0013	0.0013	0.0013	0.0012	0.0012	0.0011	0.0011	0.0011	0.0010	0.0010
-2.9	0.0019	0.0018	0.0018	0.0017	0.0016	0.0016	0.0015	0.0015	0.0014	0.0014
-2.8	0.0026	0.0025	0.0024	0.0023	0.0023	0.0022	0.0021	0.0021	0.0020	0.0019
-2.7	0.0035	0.0034	0.0033	0.0032	0.0031	0.0030	0.0029	0.0028	0.0027	0.0026
-2.6	0.0047	0.0045	0.0044	0.0043	0.0041	0.0040	0.0039	0.0038	0.0037	0.0036
-2.5	0.0062	0.0060	0.0059	0.0057	0.0055	0.0054	0.0052	0.0051	0.0049	0.0048
-2.4	0.0082	0.0080	0.0078	0.0075	0.0073	0.0071	0.0069	0.0068	0.0066	0.0064
-2.3	0.0107	0.0104	0.0102	0.0099	0.0096	0.0094	0.0091	0.0089	0.0087	0.0084
-2.2	0.0139	0.0136	0.0132	0.0129	0.0125	0.0122	0.0119	0.0116	0.0113	0.0110
-2.1	0.0179	0.0174	0.0170	0.0166	0.0162	0.0158	0.0154	0.0150	0.0146	0.0143
-2.0	0.0228	0.0222	0.0217	0.0212	0.0207	0.0202	0.0197	0.0192	0.0188	0.0183
-1.9	0.0287	0.0281	0.0274	0.0268	0.0262	0.0256	0.0250	0.0244	0.0239	0.0233
-1.8	0.0359	0.0351	0.0344	0.0336	0.0329	0.0322	0.0314	0.0307	0.0301	0.0294
-1.7	0.0446	0.0436	0.0427	0.0418	0.0409	0.0401	0.0392	0.0384	0.0375	0.0367
-1.6	0.0548	0.0537	0.0526	0.0516	0.0505	0.0495	0.0485	0.0475	0.0465	0.0455
-1.5	0.0668	0.0655	0.0643	0.0630	0.0618	0.0606	0.0594	0.0582	0.0571	0.0559
-1.4	0.0808	0.0793	0.0778	0.0764	0.0749	0.0735	0.0721	0.0708	0.0694	0.0681
-1.3	0.0968	0.0951	0.0934	0.0918	0.0901	0.0885	0.0869	0.0853	0.0838	0.0823
-1.2	0.1151	0.1131	0.1112	0.1093	0.1075	0.1056	0.1038	0.1020	0.1003	0.0985
-1.1	0.1357	0.1335	0.1314	0.1292	0.1271	0.1251	0.1230	0.1210	0.1190	0.1170
-1.0	0.1587	0.1562	0.1539	0.1515	0.1492	0.1469	0.1446	0.1423	0.1401	0.1379
-0.9	0.1841	0.1814	0.1788	0.1762	0.1736	0.1711	0.1685	0.1660	0.1635	0.1611
-0.8	0.2119	0.2090	0.2061	0.2033	0.2005	0.1977	0.1949	0.1922	0.1894	0.1867
-0.7	0.2420	0.2389	0.2358	0.2327	0.2296	0.2266	0.2236	0.2206	0.2177	0.2148
-0.6	0.2743	0.2709	0.2676	0.2643	0.2611	0.2578	0.2546	0.2514	0.2483	0.2451
-0.5	0.3085	0.3050	0.3015	0.2981	0.2946	0.2912	0.2877	0.2843	0.2810	0.2776
-0.4	0.3446	0.3409	0.3372	0.3336	0.3300	0.3264	0.3228	0.3192	0.3156	0.3121
-0.3	0.3821	0.3783	0.3745	0.3707	0.3669	0.3632	0.3594	0.3557	0.3520	0.3483
-0.2	0.4207	0.4168	0.4129	0.4090	0.4052	0.4013	0.3974	0.3936	0.3897	0.3859
-0.1	0.4602	0.4562	0.4522	0.4483	0.4443	0.4404	0.4364	0.4325	0.4286	0.4247
0.0	0.5000	0.4960	0.4920	0.4880	0.4840	0.4801	0.4761	0.4721	0.4681	0.4641

The next page of *Table A* displays the area/proportion/probability for *positive* z-scores.

Table A2

ɸ

The Normal Distribution – Positive Z-Scores

z	0	0.01	0.02	0.03	0.04	0.05	0.06	0.07	0.08	0.09
					Standard Normal Probabilities Table					
0.0	0.5000	0.5040	0.5080	0.5120	0.5160	0.5199	0.5239	0.5279	0.5319	0.5359
0.1	0.5398	0.5438	0.5478	0.5517	0.5557	0.5596	0.5636	0.5675	0.5714	0.5753
0.2	0.5793	0.5832	0.5871	0.5910	0.5948	0.5987	0.6026	0.6064	0.6103	0.6141
0.3	0.6179	0.6217	0.6255	0.6293	0.6331	0.6368	0.6406	0.6443	0.6480	0.6517
0.4	0.6554	0.6591	0.6628	0.6664	0.6700	0.6736	0.6772	0.6808	0.6844	0.6879
0.5	0.6915	0.6950	0.6985	0.7019	0.7054	0.7088	0.7123	0.7157	0.7190	0.7224
0.6	0.7257	0.7291	0.7324	0.7357	0.7389	0.7422	0.7454	0.7486	0.7517	0.7549
0.7	0.7580	0.7611	0.7642	0.7673	0.7704	0.7734	0.7764	0.7794	0.7823	0.7852
0.8	0.7881	0.7910	0.7939	0.7967	0.7995	0.8023	0.8051	0.8078	0.8106	0.8133
0.9	0.8159	0.8186	0.8212	0.8238	0.8264	0.8289	0.8315	0.8340	0.8365	0.8389
1.0	0.8413	0.8438	0.8461	0.8485	0.8508	0.8531	0.8554	0.8577	0.8599	0.8621
1.1	0.8643	0.8665	0.8686	0.8708	0.8729	0.8749	0.8770	0.8790	0.8810	0.8830
1.2	0.8849	0.8869	0.8888	0.8907	0.8925	0.8944	0.8962	0.8980	0.8997	0.9015
1.3	0.9032	0.9049	0.9066	0.9082	0.9099	0.9115	0.9131	0.9147	0.9162	0.9177
1.4	0.9192	0.9207	0.9222	0.9236	0.9251	0.9265	0.9279	0.9292	0.9306	0.9319
1.5	0.9332	0.9345	0.9357	0.9370	0.9382	0.9394	0.9406	0.9418	0.9429	0.9441
1.6	0.9452	0.9463	0.9474	0.9484	0.9495	0.9505	0.9515	0.9525	0.9535	0.9545
1.7	0.9554	0.9564	0.9573	0.9582	0.9591	0.9599	0.9608	0.9616	0.9625	0.9633
1.8	0.9641	0.9649	0.9656	0.9664	0.9671	0.9678	0.9686	0.9693	0.9699	0.9706
1.9	0.9713	0.9719	0.9726	0.9732	0.9738	0.9744	0.9750	0.9756	0.9761	0.9767
2.0	0.9772	0.9778	0.9783	0.9788	0.9793	0.9798	0.9803	0.9808	0.9812	0.9817
2.1	0.9821	0.9826	0.9830	0.9834	0.9838	0.9842	0.9846	0.9850	0.9854	0.9857
2.2	0.9861	0.9864	0.9868	0.9871	0.9875	0.9878	0.9881	0.9884	0.9887	0.9890
2.3	0.9893	0.9896	0.9898	0.9901	0.9904	0.9906	0.9909	0.9911	0.9913	0.9916
2.4	0.9918	0.9920	0.9922	0.9925	0.9927	0.9929	0.9931	0.9932	0.9934	0.9936
2.5	0.9938	0.9940	0.9941	0.9943	0.9945	0.9946	0.9948	0.9949	0.9951	0.9952
2.6	0.9953	0.9955	0.9956	0.9957	0.9959	0.9960	0.9961	0.9962	0.9963	0.9964
2.7	0.9965	0.9966	0.9967	0.9968	0.9969	0.9970	0.9971	0.9972	0.9973	0.9974
2.8	0.9974	0.9975	0.9976	0.9977	0.9977	0.9978	0.9979	0.9979	0.9980	0.9981
2.9	0.9981	0.9982	0.9982	0.9983	0.9984	0.9984	0.9985	0.9985	0.9986	0.9986
3.0	0.9987	0.9987	0.9987	0.9988	0.9988	0.9989	0.9989	0.9989	0.9990	0.9990
3.1	0.9990	0.9991	0.9991	0.9991	0.9992	0.9992	0.9992	0.9992	0.9993	0.9993
3.2	0.9993	0.9993	0.9994	0.9994	0.9994	0.9994	0.9994	0.9995	0.9995	0.9995
3.3	0.9995	0.9995	0.9995	0.9996	0.9996	0.9996	0.9996	0.9996	0.9996	0.9997
3.4	0.9997	0.9997	0.9997	0.9997	0.9997	0.9997	0.9997	0.9997	0.9997	0.9998

The previous page of *Table A* displays the area/proportion/probability for *negative z*-scores.

Table B

φ

The Random Number Table

	1	2	3	4	5	6	7	8	9	10
Row 1	91679	10156	19469	04819	83693	47287	72991	48875	85179	64178
Row 2	53571	66494	36631	72310	37386	54900	74615	58006	11604	52556
Row 3	66867	72461	99058	41276	65754	66117	58492	99923	35823	39134
Row 4	17891	65465	91123	21920	89059	18734	01313	63844	36358	40871
Row 5	83824	18407	33528	82055	26810	43173	38326	81527	92973	80162
Row 6	77574	32161	45551	20132	12384	65398	78668	00802	83720	94555
Row 7	43276	27990	11061	83674	34594	28056	85196	95301	97035	26771
Row 8	32778	98492	53227	74146	56972	04005	22822	48002	39371	05152
Row 9	68393	63250	85060	34943	76362	31044	27831	91674	39440	82217
Row 10	53375	47146	86954	16210	18296	78800	20897	46211	02689	93785
Row 11	99734	87699	34515	31497	40836	76702	76436	45908	47605	91077
Row 12	87663	92456	38086	97826	80567	80827	66734	66418	97823	02961
Row 13	79604	82725	62678	08405	26293	35910	06764	60608	47733	08383
Row 14	73073	04683	08419	22293	71427	32760	30975	29314	22015	73880
Row 15	77171	87439	04555	06100	79493	25819	27007	72895	71670	07736
Row 16	68010	00253	96296	01155	65926	69696	18764	43420	83876	67494
Row 17	16755	94613	02277	09653	73579	83112	17795	35906	10365	83055
Row 18	53913	88335	40501	09824	62494	52766	20624	40211	30078	69402
Row 19	70983	00065	83819	37477	39167	23430	32803	18921	89648	08832
Row 20	92422	10742	12589	23807	94806	23157	99109	85384	89741	20085
Row 21	71413	75798	34817	20571	35196	32154	80016	64964	42606	35929
Row 22	88861	56842	57442	40841	80247	45291	03655	54639	91122	52990
Row 23	05879	12844	36411	77685	86248	37114	72139	25142	64126	30273
Row 24	93902	22767	83715	41747	46802	00211	24461	11145	97605	27014
Row 25	32337	82151	61545	19312	50223	78265	00319	58904	70432	24217

Table C

φ

The Student's t Distribution Table

df	\multicolumn{12}{c}{Right Tail Probability}
	0.25	0.2	0.15	0.1	0.05	0.025	0.02	0.01	0.005	0.0025	0.001	0.0005
1	1.000	1.376	1.963	3.078	6.314	12.710	15.890	31.820	63.660	127.300	318.300	636.600
2	0.816	1.061	1.386	1.886	2.920	4.303	4.849	6.965	9.925	14.090	22.330	31.600
3	0.765	0.978	1.250	1.638	2.353	3.182	3.482	4.541	5.841	7.453	10.210	12.920
4	0.741	0.941	1.190	1.533	2.132	2.776	2.999	3.747	4.604	5.598	7.173	8.610
5	0.727	0.920	1.156	1.476	2.015	2.571	2.757	3.365	4.032	4.773	5.893	6.869
6	0.718	0.906	1.134	1.440	1.943	2.447	2.612	3.143	3.707	4.317	5.208	5.959
7	0.711	0.896	1.119	1.415	1.895	2.365	2.517	2.998	3.499	4.029	4.785	5.408
8	0.706	0.889	1.108	1.397	1.860	2.306	2.449	2.896	3.355	3.833	4.501	5.041
9	0.703	0.883	1.100	1.383	1.833	2.262	2.398	2.821	3.250	3.690	4.297	4.781
10	0.700	0.879	1.093	1.372	1.812	2.228	2.359	2.764	3.169	3.581	4.144	4.587
11	0.697	0.876	1.088	1.363	1.796	2.201	2.328	2.718	3.106	3.497	4.025	4.437
12	0.695	0.873	1.083	1.356	1.782	2.179	2.303	2.681	3.055	3.428	3.930	4.318
13	0.694	0.870	1.079	1.350	1.771	2.160	2.282	2.650	3.012	3.372	3.852	4.221
14	0.692	0.868	1.076	1.345	1.761	2.145	2.264	2.624	2.977	3.326	3.787	4.140
15	0.691	0.866	1.074	1.341	1.753	2.131	2.249	2.602	2.947	3.286	3.733	4.073
16	0.690	0.865	1.071	1.337	1.746	2.120	2.235	2.583	2.921	3.252	3.686	4.015
17	0.689	0.863	1.069	1.333	1.740	2.110	2.224	2.567	2.898	3.222	3.646	3.965
18	0.688	0.862	1.067	1.330	1.734	2.101	2.214	2.552	2.878	3.197	3.611	3.922
19	0.688	0.861	1.066	1.328	1.729	2.093	2.205	2.539	2.861	3.174	3.579	3.883
20	0.687	0.860	1.064	1.325	1.725	2.086	2.197	2.528	2.845	3.153	3.552	3.850
21	0.686	0.859	1.063	1.323	1.721	2.080	2.189	2.518	2.831	3.135	3.527	3.819
22	0.686	0.858	1.061	1.321	1.717	2.074	2.183	2.508	2.819	3.119	3.505	3.792
23	0.685	0.858	1.060	1.319	1.714	2.069	2.177	2.500	2.807	3.104	3.485	3.768
24	0.685	0.857	1.059	1.318	1.711	2.064	2.172	2.492	2.797	3.091	3.467	3.745
25	0.684	0.856	1.058	1.316	1.708	2.060	2.167	2.485	2.787	3.078	3.450	3.725
26	0.684	0.856	1.058	1.315	1.706	2.056	2.162	2.479	2.779	3.067	3.435	3.707
27	0.684	0.855	1.057	1.314	1.703	2.052	2.158	2.473	2.771	3.057	3.421	3.690
28	0.683	0.855	1.056	1.313	1.701	2.048	2.154	2.467	2.763	3.047	3.408	3.674
29	0.683	0.854	1.055	1.311	1.699	2.045	2.150	2.462	2.756	3.038	3.396	3.659
30	0.683	0.854	1.055	1.310	1.697	2.042	2.147	2.457	2.750	3.030	3.385	3.646
40	0.681	0.851	1.050	1.303	1.684	2.021	2.123	2.423	2.704	2.971	3.307	3.551
60	0.679	0.848	1.045	1.296	1.671	2.000	2.099	2.390	2.660	2.915	3.232	3.460
80	0.678	0.846	1.043	1.292	1.664	1.990	2.088	2.374	2.639	2.887	3.195	3.416
100	0.677	0.845	1.042	1.290	1.660	1.984	2.081	2.364	2.626	2.871	3.174	3.390
1000	0.675	0.842	1.037	1.282	1.646	1.962	2.056	2.330	2.581	2.813	3.098	3.300
∞	0.674	0.841	1.036	1.282	1.645	1.960	2.054	2.326	2.576	2.807	3.091	3.291

Table D

⌘

The Chi-Square Distribution Table

df	Right Tail Probability								
	0.25	0.20	0.15	0.10	0.05	0.025	0.02	0.01	0.005
1	1.32	1.64	2.07	2.71	3.84	5.02	5.41	6.63	7.88
2	2.77	3.22	3.79	4.61	5.99	7.38	7.82	9.21	10.60
3	4.11	4.64	5.32	6.25	7.81	9.35	9.84	11.34	12.84
4	5.39	5.99	6.74	7.78	9.49	11.14	11.67	13.28	14.86
5	6.63	7.29	8.12	9.24	11.07	12.83	13.39	15.09	16.75
6	7.84	8.56	9.45	10.64	12.59	14.45	15.03	16.81	18.55
7	9.04	9.80	10.75	12.02	14.07	16.01	16.62	18.48	20.28
8	10.22	11.03	12.03	13.36	15.51	17.53	18.17	20.09	21.95
9	11.39	12.24	13.29	14.68	16.92	19.02	19.68	21.67	23.59
10	12.55	13.44	14.53	15.99	18.31	20.48	21.16	23.21	25.19
11	13.70	14.63	15.77	17.28	19.68	21.92	22.62	24.72	26.76
12	14.85	15.81	16.99	18.55	21.03	23.34	24.05	26.22	28.30
13	15.98	16.98	18.20	19.81	22.36	24.74	25.47	27.69	29.82
14	17.12	18.15	19.41	21.06	23.68	26.12	26.87	29.14	31.32
15	18.25	19.31	20.60	22.31	25.00	27.49	28.26	30.58	32.80
16	19.37	20.47	21.79	23.54	26.30	28.85	29.63	32.00	34.27
17	20.49	21.61	22.98	24.77	27.59	30.19	31.00	33.41	35.72
18	21.60	22.76	24.16	25.99	28.87	31.53	32.35	34.81	37.16
19	22.72	23.90	25.33	27.20	30.14	32.85	33.69	36.19	38.58
20	23.83	25.04	26.50	28.41	31.41	34.17	35.02	37.57	40.00
21	24.93	26.17	27.66	29.62	32.67	35.48	36.34	38.93	41.40
22	26.04	27.30	28.82	30.81	33.92	36.78	37.66	40.29	42.80
23	27.14	28.43	29.98	32.01	35.17	38.08	38.97	41.64	44.18
24	28.24	29.55	31.13	33.20	36.42	39.36	40.27	42.98	45.56
25	29.34	30.68	32.28	34.38	37.65	40.65	41.57	44.31	46.93
26	30.43	31.79	33.43	35.56	38.89	41.92	42.86	45.64	48.29
27	31.53	32.91	34.57	36.74	40.11	43.19	44.14	46.96	49.64
28	32.62	34.03	35.71	37.92	41.34	44.46	45.42	48.28	50.99
29	33.71	35.14	36.85	39.09	42.56	45.72	46.69	49.59	52.34
30	34.80	36.25	37.99	40.26	43.77	46.98	47.96	50.89	53.67
40	45.62	47.27	49.24	51.81	55.76	59.34	60.44	63.69	66.77
60	66.98	68.97	71.34	74.40	79.08	83.30	84.58	88.38	91.95
80	88.13	90.41	93.11	96.58	101.90	106.60	108.10	112.30	116.30
100	109.10	111.70	114.70	118.50	124.30	129.60	131.10	135.80	140.20

Selected Bibliography

Adam, D. (2015). *The man who couldn't stop: OCD and the true story of a life lost in thought*. New York: Sarah Crichton Books.

Albert, J. (2003). *Teaching statistics using baseball*. Washington, DC: Mathematical Association of America.

Bergamini, D. (1963). *Mathematics*. New York: Time.

Bernstein, P. (1996). *Against the gods: The remarkable story of risk*. New York: John Wiley & Sons.

Borowski, E., & Borwein, J. (2002). *Dictionary of mathematics* (2nd ed.). London: Collins.

Buchholz, T. (1989). *New ideas from dead economists: An introduction to modern economic thought*. New York: New American Library.

Cobb, G. (2007). The Introductory Statistics Course: A Ptolemaic Curriculum? *EScholarship, University of California*. Retrieved from https://escholarship.org/uc/item/6hb3k0nz

Crawford, M. (2015) *The world beyond your head: On becoming an individual in an age of distraction*. New York: Farrar, Straus and Giroux.

Creswell, J. (2003). *Research design: Qualitative, quantitative, and mixed method approaches* (2nd ed.). Thousand Oaks, Calif.: Sage Publications.

Dawkins, R. (1998). *Unweaving the rainbow: Science, delusion, and the appetite for wonder*. Boston: Houghton Mifflin.

Deary, V. (2014). *How we are: Book one of the how to live trilogy*. New York: Farrar, Straus and Giroux.

Devlin, K. (1998). *The language of mathematics: Making the invisible visible*. New York: W.H. Freeman.

Ellenberg, J. (2015). *How not to be wrong: The power of mathematical thinking*. New York: Penguin.

Feynman, R., & Leighton, R. (1985). *"Surely you're joking, Mr. Feynman!": Adventures of a curious character*. New York: W.W. Norton.

Ford, M. (2015). *Rise of the robots: Technology and the threat of a jobless future*. New York: Basic Books.

Frankl, V. (1959). *Man's search for meaning*. Boston: Beacon.

Gerovitch, S. (2013). The man who invented modern probability. *Nautilus*.

Gertner, J. (2012). *The idea factory: Bell Labs and the great age of American innovation*. New York: Penguin.

Gigerenzer, G. (2014). *Risk savvy: How to make good decisions*. New York: Penguin.

Goldstein, D. (2014). *The teacher wars: A history of America's most embattled profession*. New York: Doubleday.

Goldstein, R. (2006). *Betraying Spinoza: The renegade Jew who gave us modernity*. New York: Next book.

Gore, A. (2007). *The assault on reason*. New York: Penguin Press.

Gorroochurn, P. (2012). Some laws and problems of classical probability and how Cardano anticipated them. *Chance*, 13-20.

——— (2014). Thirteen correct solutions to the "problem of points" and their histories. *The Mathematical Intelligencer*, 56-64.

Gould, S. (1981). *The mismeasure of man*. New York: Norton.

——— & McGarr, P. (2007). *The richness of life: The essential Stephen Jay Gould*. New York: W.W. Norton.

Groisman, B. (2008). The End of Sleeping Beauty's Nightmare. *The British Journal for the Philosophy of Science*, 409-416.

Hájek, A. (2008). A philosopher's guide to probability. In G. Bammer & M. Smithson (eds.), *Uncertainty and Risk: Multidisciplinary Perspectives*. Routledge.

Hock, R. (2009). *Forty studies that changed psychology: Explorations into the history of psychological research* (6th ed.). Upper Saddle River, N.J.: Pearson/Prentice Hall.

Hogg, R., & Tanis, E. (2009). *Probability and statistical inference*. New York: Macmillan.

Huff, D., & Geis, I. (1954). *How to lie with statistics*. New York: Norton.

Igo, S. (2007). *The averaged American: Surveys, citizens, and the making of a mass public*. Cambridge, Mass.: Harvard University Press.

Isaacson, W. (2014). *The innovators: How a group of hackers, geniuses, and geeks created the digital revolution*. New York: Simon & Shuster.

Jauhar, S. (2015). *Doctored: The disillusionment of an American physician*. New York: Farrar, Straus and Giroux.

Johnson, S. (2007). *The ghost map: The story of London's most terrifying epidemic—and how it changed science, cities, and the modern world*. New York: Riverhead.

Kahneman, D. (2011). *Thinking, fast and slow*. New York: Farrar, Straus and Giroux.

Katz, V. (1998). *A history of mathematics: An introduction* (2nd ed.). New York: Addison Wesley.

Kennedy, G. (1983). *Invitation to statistics*. Oxford: M. Robertson.

Keynes, J. (1921). *A treatise on probability*. London: Macmillan and, Limited.

Kline, M. (1980). *Mathematics, the loss of certainty*. New York: Oxford University Press.

Lehrer, J. (2009). *How we decide*. Boston: Houghton Mifflin Harcourt.

Levinovitz, A. (2015). *The gluten lie: And other myths about what you eat*. New York: Regan Arts.

Levitt, S., & Dubner, S. (2005). *Freakonomics: A rogue economist explores the hidden side of everything*. New York: William Morrow.

Levitt, S., & Dubner, S. (2009). *Superfreakonomics: Global cooling, patriotic prostitutes, and why suicide bombers should buy life insurance*. New York: William Morrow.

Ma, J. (2010). *The house advantage: Playing the odds to win big in business*. New York: Palgrave Macmillan.

Madden, E. (1957). Aristotle's treatment of probability and signs. *Philosophy of Science* 24 (2): 167-172

Malkiel, B. (2015). *A random walk down Wall Street* (11th ed.). New York: Norton.

McLuhan, M. (1964). *Understanding media: The extensions of man*. Canada: McGraw-Hill.

Mischel, W. (2014). *The marshmallow test: Mastering self-control*. New York: Little, Brown.

Mukherjee, S. (2010). *The emperor of all maladies: A biography of cancer*. New York: Scribner.

Nisbett, R. (2015) *Mindware: Tools for smart thinking*. New York: Farrar, Straus and Giroux.

Parcells, B., & Demasio, N. (2014). *Parcells: A football life*. New York: Crown Archetype.

Park, R. (2008). *Superstition: Belief in the age of science*. Princeton, N.J.: Princeton University Press.

Paulos, J. (1988). *Innumeracy: Mathematical illiteracy and its consequences*. New York: Hill and Wang.

Peck, R., & Olsen, C. (2005). *Introduction to statistics and data analysis* (2nd ed.). Belmont, CA: Thomson Brooks/Cole.

Petersen, M. (2008). *Our daily meds: How the pharmaceutical companies transformed themselves into slick marketing machines and hooked the nation on prescription drugs*. New York: Farrar, Straus and Giroux.

Pinker, S. (2014). *The sense of style: The thinking person's guide to writing in the 21st century*. New York: Viking.

Rabinovitch, N. (1969). Studies in the history of probability and statistics. XXH. Probability in the Talmud. *Biometrika*, 437-441.

Roose, K. (2015). *Young money: Inside the hidden world of Wall Street's post-crash recruits*. New York: Grand Central.

Rossman, A., Oehsen, J., & Chance, B. (2002). *Workshop statistics: Discovery with data and the graphing calculator* (2nd ed.). New York: Springer.

Sacks, O. (2015). *On the move: A life*. New York: Alfred A. Knopf.

Salsburg, D. (2001). *The lady tasting tea: How statistics revolutionized science in the twentieth century*. New York: W.H. Freeman.

Sandel, M. (2009). *Justice: What's the right thing to do?* New York: Farrar, Straus and Giroux.

Seife, C. (2011). *Proofiness: How you're being fooled by the numbers*. New York: Penguin.

Silver, N. (2012). *The signal and the noise: Why so many predictions fail—but some don't*. New York: Penguin.

Starnes, D., Yates, D., & Moore, D. (2012). *The practice of statistics for AP* (4th ed.). New York: W. H. Freeman.

Taleb, N. (2007). *The black swan: The impact of the highly improbable*. New York: Random House.

Todhunter, I. (1903). *The Elements of Euclid for the use of schools and colleges*. London: Macmillan.

Triola, M. (2011). *Elementary statistics* (11th ed.). Reading, Mass.: Addison-Wesley.

Wallace, D. (2010). *Everything and more: A compact history of infinity*. New York: W. W. Norton.

Webster, I. (2004). *Merriam-Webster's school dictionary*. Springfield, Mass.: Merriam-Webster.

Wood, M. (2015). *Alfred Hitchcock: The man who knew too much*. Boston, Mass.: New Harvest.

Zalta, E. (2015). *Stanford encyclopedia of philosophy*. Stanford, Calif.: Metaphysics Research Lab, Center for the Study of Language and Information, Stanford University.

Acknowledgments

ϕ

The encouragement and support of a number of individuals helped to bring this ever-evolving manuscript to life—including my erstwhile teachers Ned, Louis, Carlos, Tom, and Lisa, as well as my friends Yan, Scott, and Mariel—but four people in particular deserve special recognition. Over the years, Ashlie, Ali, Aspen, and especially Aashika, all former university students of mine, urged me to repurpose my statistics lecture presentations and class materials into a thorough, refined, stand-alone text. Their sage advice, along with innumerable hours of toil (and joy), has resulted in this primer—which, I can only hope, makes fully manifest what they originally had in mind.

Any mistakes in the text are solely my responsibility.

About the Author

❖

MARK JONES LORENZO, a teacher, is the author of two previous books: *Affront to Meritocracy*, a collection of short works, and *Not Ok*, a guide to computer programming in the BASIC language. He lives in Pennsylvania with his dogs.

www.ingramcontent.com/pod-product-compliance
Lightning Source LLC
Chambersburg PA
CBHW080633180526
45168CB00008B/3153